Lecture Notes in Mathematics 1613

Subseries: Institut de Mathématiques, Université de Strasbourg

Advisor: J.-L. Loday

Springer
Berlin
Heidelberg
New York
Barcelona
Budapest
Hong Kong
London
Milan
Paris
Santa Clara
Singapore
Tokyo

J. Azéma M. Emery
P. A. Meyer M. Yor (Eds.)

Séminaire de Probabilités XXIX

Springer

Editors

Jacques Azéma
Marc Yor
Laboratoire de Probabilités
Université Pierre et Marie Curie
Tour 56, 3 ème étage
4, Place Jussieu, F-75252 Paris Cedex, France

Michel Emery
Paul André Meyer
Institut de Recherche Mathématique Avancée
Université Louis Pasteur
7, rue René Descartes, F-67084 Strasbourg, France

Cataloging-in-Publication Data.

Die Deutsche Bibliothek - CIP-Einheitsaufnahme

Séminaire de probabilités ... - Berlin ; Heidelberg ; New York ; London ; Paris ; Tokyo ; Hong Kong ; Barcelona ; Budapest : Springer.
ISSN 0720-8766

29 (1995)
(Lecture notes in mathematics ; Vol. 1613)
ISBN 3-540-60219-4 (Berlin ...)
NE: GT

Mathematics Subject Classification (1991): 60GXX, 60HXX, 60JXX

ISBN 3-540-60219-4 Springer-Verlag Berlin Heidelberg New York

Printed in Germany

Typesetting: Camera-ready TEX output by the authors
SPIN: 10479528 46/3142-543210 - Printed on acid-free paper

SÉMINAIRE DE PROBABILITÉS XXIX

TABLE DES MATIÈRES

On quantum extensions of the Azéma martingale semigroup

by A.M. Chebotarev and F. Fagnola

1. Introduction

In this note we study some quantum extensions of classical Markovian semigroups related to the Azéma martingales with parameter β ($\beta \neq 0$, $\beta \neq -1$) (see [1], [5], [6]). The formal infinitesimal generator given by

$$(\mathcal{L}_0 f)(x) = (\beta x)^{-2}(f(cx) - f(x) - \beta x f'(x))$$

on bounded smooth functions f can be written formally as follows (see [9])

$$\mathcal{L}(m_f) = Gm_f + L^* m_f L + m_f G^*$$

where m_f denotes the multiplication operator by f, the operator G is the infinitesimal generator of a strongly continuous contraction semigroup on $L^2(\mathbb{R};\mathbb{C})$ (see Section 2) and L is related to G by the formal condition $G + G^* + L^*L = 0$. The associated minimal quantum dynamical semigroup, can be easily constructed, for example as in [2], [3], [4], [8]. We show that this semigroup is conservative if $\beta < \beta_*$ and it is not if $\beta > \beta_*$ where β_* is the unique solution of the equation

$$\exp(\beta) + \beta + 1 = 0.$$

Therefore it is a natural conjecture that the minimal quantum dynamical semigroup is a ultraweakly continuous extension to $\mathcal{B}(h)$ of the Azéma martingale semigroup when $\beta < \beta_*$. However we can not prove this fact because the characterisation of the classical infinitesimal generator is not known. The above quantum dynamical semigroup is not such an extension when $\beta > \beta_*$ because the corresponding classical Markovian semigroup is identity preserving.

We were not able to study the critical case $\beta = \beta_*$ although it seems reasonable that conservativity holds also in this case. In fact, as shown by Emery in [5], the Azéma martingale with parameter β starting from $x \neq 0$ can hit 0 in finite time only if $\beta > \beta_*$. The operators G and L we consider are singular at the point 0, hence, in this case, boundary conditions on G at 0 should appear to describe the behaviour of the process.

The cases when $\beta < \beta_*$ and $\beta > \beta_*$ are studied in Section 3 by checking a necessary and sufficient condition obtained in [2]. In Section 5 we apply a sufficient condition for conservativity obtained in [3]. This condition yields the previous result when $\beta \leq -1.5$; since $\beta_* = -1.278...$, it is quite "close" to the necessary and sufficient one.

2. Notation and preliminary results

Let β be a fixed real number with $\beta \neq -1, \beta \neq 0$ and let $c = \beta+1$. Let $h = L^2(\mathbb{R};\mathbb{C})$ and denote by $\mathcal{B}(h)$ the *-algebra of all bounded operators on h. Let us consider the strongly continuous contraction semigroup P on h defined by

$$(P(t)u)(x) = \begin{cases} \left(1 - \frac{2t}{\beta x^2}\right)^{(1-\beta)/4\beta} u\left(x\left(1-\frac{2t}{\beta x^2}\right)^{1/2}\right) & \text{if } 1 - \frac{2t}{\beta x^2} > 0 \\ 0 & \text{if } 1 - \frac{2t}{\beta x^2} \leq 0 \end{cases}$$

The dual semigroup P^* can be easily computed

$$(P^*(t)u)(x) = \begin{cases} \left(1 + \frac{2t}{\beta x^2}\right)^{-(1+\beta)/4\beta} u\left(x\left(1+\frac{2t}{\beta x^2}\right)^{1/2}\right) & \text{if } 1 + \frac{2t}{\beta x^2} > 0 \\ 0 & \text{if } 1 + \frac{2t}{\beta x^2} \leq 0 \end{cases}$$

Let D_0 be the linear manifold of infinitely differentiable functions with compact support vanishing in a neighbourhood of 0. The infinitesimal generators G and G^* of the semigroups P and P^* satisfy

$$Gu(x) = -\frac{1}{\beta x}u'(x) + \frac{\beta-1}{2\beta^2 x^2}u(x), \qquad G^*u(x) = \frac{1}{\beta x}u'(x) - \frac{\beta+1}{2\beta^2 x^2}u(x)$$

for all $u \in D_0$. In fact P has been obtained integrating the first order partial differential equation

$$\frac{\partial w(t,x)}{\partial t} = -\frac{1}{\beta x}\frac{\partial w(t,x)}{\partial x} + \frac{\beta-1}{2\beta^2 x^2}w(t,x)$$

by the characteristics method. Consider the operator M on h defined by

$$D(M) = \{ u \in h \mid x^{-1}u(x) \in h \}, \qquad Mu(x) = (\beta x)^{-1}u(x).$$

and let S be the unitary operator on h

$$Su(x) = |c|^{-1/2}u(c^{-1}x).$$

The form $\mathcal{L}$ on h given by

$$\langle v, \mathcal{L}(f)u\rangle = \langle G^*v, fu\rangle + \langle SMv, fSMu\rangle + \langle v, fG^*u\rangle$$

for all $u, v \in D_0$, transforms $f \in D_0$ into the multiplication operator on h by

$$(\mathcal{L}f)(x) = (\beta x)^{-2}\left(f(cx) - f(x) - \beta x f'(x)\right).$$

Thus the restriction of $\mathcal{L}$ to D_0 coincides with the restriction to D_0 of the infinitesimal generator of an Azéma-Emery martingale with parameter β (see [5]).

The domain D_0, however, may be too small for the operators G and G^*. In fact it can be shown that D_0 is a core for G if and only if $\beta \geq -1/2$ and is a core for G^* if and only if $\beta \leq 1/2$. The domains of G and G^* however can be described as the range of the resolvent operators $R(1;G) = (1-G)^{-1}$ and $R(1;G^*) = (1-G^*)^{-1}$.

Proposition 2.1. *For all $u \in h$ define a function u_0 by*

$$u_0(x) = \begin{cases} \int_x^{\mathrm{sgn}(x)\infty} \exp(-\beta y^2/2)|y|^{-(\beta+1)/2\beta} y u(y) dy & \text{if } \beta > 0 \\ -\int_0^x \exp(-\beta y^2/2)|y|^{-(\beta+1)/2\beta} y u(y) dy & \text{if } \beta < 0 \end{cases}$$

Then the operator $R(1; G^*)$ is given by

$$\left(R(1; G^*)u\right)(x) = \beta|x|^{(\beta+1)/2\beta} \exp(\beta x^2/2) u_0(x). \tag{2.1}$$

Proof. For all $u, v \in h$ we have

$$\langle v, R(1; G^*)u\rangle = \int_{\mathbb{R}} v(x)dx \int_0^\infty e^{-t}\left(1 + \frac{2t}{\beta x^2}\right)^{-(\beta+1)/4\beta} u\left(x\sqrt{1 + \frac{2t}{\beta x^2}}\right) dt.$$

By the change of variables $y = x\sqrt{1 + 2t/\beta x^2}$ in the integral with respect to t we obtain the representation formula (2.1). □

Proposition 2.2. *The domain of the operator M contains the domain of the operator G^*. Moreover the necessary condition for T to be conservative,*

$$\langle G^* v, u\rangle + \langle SMv, SMu\rangle + \langle v, G^* u\rangle = 0 \tag{2.2}$$

for all vectors u, v in the domain of G^*, is fulfilled.

Proof. Clearly, to establish the inclusion $D(G^*) \subset D(M)$, it suffices to show that, for all $u \in h$, the integral

$$\int_{\mathbb{R}} x^{-2} \left|\left(R(1; G^*)u\right)(x)\right|^2 dx \tag{2.3}$$

is convergent. To this end, let us first fix $r \in (0, 1)$ and denote by $I(r)$ the set $(-r^{-1}, -r) \cup (r, r^{-1})$. Integrating by parts we have

$$\begin{aligned} \int_{I(r)} (\beta x)^{-2} \left|\left(R(1; G^*)u\right)(x)\right|^2 dx &= \int_{I(r)} (\beta x)^{-1} |x|^{1/\beta} \cdot \beta \mathrm{sgn}(x) \exp(\beta x^2) |u_0(x)|^2 dx \\ &= \beta r^{-1/\beta} \exp(\beta r^{-2}) \left(|u_0(r^{-1})|^2 - |u_0(-r^{-1})|^2\right) \\ &+ \beta r^{1/\beta} \exp(\beta r^2) \left(|u_0(-r)|^2 - |u_0(r)|^2\right) \\ &- 2\int_{I(r)} \beta^2 |x|^{1+1/\beta} \exp(\beta x^2) |u_0(x)|^2 dx \\ &+ 2\Re e \int_{I(r)} \beta |x|^{(1+\beta)/2\beta} \exp(\beta x^2/2) \bar{u}(x) u_0(x) dx \end{aligned}$$

The first two terms vanish as r tends to zero. In fact consider, for example, the case $\beta > 0$, then, using the Schwartz inequality, we can write the estimate

$$r^{1/\beta} \exp(\beta r^2) |u_0(r)|^2 \le \|u\|^2 \exp(\beta r^2) r^{1/\beta} \int_r^\infty \exp(-\beta y^2) y^{1-1/\beta} dy. \tag{2.4}$$

Clearly, when $\beta > 1/2$, the right-hand side integral is bounded, therefore (2.4) vanishes as r tends to 0. On the other hand, when $\beta \in (0, 1/2]$, by the De L'Hôpital rule, we have

$$\lim_{r\to 0+} r^{1/\beta} \int_r^\infty \exp(-\beta y^2) y^{1-1/\beta} dy = \lim_{r\to 0+} \beta r^{1+1/\beta} \exp(\beta r^2) r^{1-1/\beta} = 0.$$

In a similar way one can compute the other limits and show that the first two terms vanish. The third and fourth term clearly converge to

$$-2\left\|R(1;G^*)u\right\|^2, \qquad 2\Re e\,\langle u, R(1;G^*)u\rangle$$

respectively. Therefore the integral (2.3) is convergent. Moreover, by the identity $R(1;G^*) - I = G^* R(1;G^*)$, we have

$$\begin{aligned}\left\|MR(1;G^*)u\right\|^2 &= -2\left\|R(1;G^*)u\right\|^2 + 2\Re e\,\langle u, R(1;G^*)u\rangle \\ &= -2\Re e\,\langle R(1;G^*)u, G^* R(1;G^*)u\rangle\end{aligned}$$

Therefore we proved also the identity (2.2), with $v = u$, because the operator S is unitary. The proof for arbitrary v, u is the same. □

Having found a Lindblad form for the infinitesimal generator of the classical process we investigate whether the corresponding minimal quantum dynamical semigroup (abbreviated to *m.q.d.s.* in the sequel) on $\mathcal{B}(h)$ is identity preserving i.e. conservative. Recall that the m.q.d.s. $\mathcal{T}$ is the ultraweakly continuous semigroup on $\mathcal{B}(h)$ defined as follows (see [2], [3], [4], [8]). For all positive element X of $\mathcal{B}(h)$, let us consider the increasing sequence

$$\left\langle u, \mathcal{T}_t^{(0)}(X)u\right\rangle = \langle P^*(t)u, XP^*(t)u\rangle$$

$$\begin{aligned}\left\langle u, \mathcal{T}_t^{(n+1)}(X)u\right\rangle &= \langle P^*(t)u, XP^*(t)u\rangle \\ &\quad + \int_0^t \left\langle SMP^*(t-s)u, \mathcal{T}_s^{(n)}(X)SMP^*(t-s)u\right\rangle ds.\end{aligned}$$

The bounded operator $\mathcal{T}_t(X)$ is given by

$$\mathcal{T}_t(X) = \sup_{n\geq 0} \mathcal{T}_t^{(n)}(X).$$

Proposition 2.3. *The abelian subalgebra $L^\infty(\mathbb{R};\mathbb{C})$ of $\mathcal{B}(h)$ is invariant for the m.q.d.s. $\mathcal{T}$.*

Proof. In fact, for every $X \in L^\infty(\mathbb{R};\mathbb{C})$, a straightforward computation shows that

$$\mathcal{T}_t^{(n)}(X) \in L^\infty(\mathbb{R};\mathbb{R})$$

for all $t \geq 0$ and all integer $n \geq 0$. □

Let β_* be the unique solution of the equation

$$\exp(\beta) + \beta + 1 = 0.$$

It is easy to check the inequality

$$-1.2785 < \beta_* < -1.2784.$$

In the following sections we shall prove the

Theorem 2.4. *The m.q.d.s. is conservative if $\beta < \beta_*$ and is not conservative if $\beta > \beta_*$.*

We recall the necessary and sufficient condition for the m.q.d.s. to be conservative obtained in [2]. Let $\mathcal{Q} : \mathcal{B}(h) \to \mathcal{B}(h)$ be the normal and monotone map defined by

$$\langle v, \mathcal{Q}(X)u\rangle = \int_0^\infty e^{-t} \langle SMP^*(t)v, XSMP^*(t)u\rangle \, dt.$$

Theorem 2.5. *Let G^*, M and S be the above defined operators. The following conditions are equivalent:*
i) the semigroup $\mathcal{T}$ is conservative,
ii) the sequence $(\mathcal{Q}^n(I))_{n\geq 1}$ converges strongly to 0,
iii) the equation $\mathcal{L}(X) = X$ has no nonzero positive solution in $\mathcal{B}(h)$.

We refer to [7] Th. 3.3, Prop. 3.5, 3.6 for the proof. The technical condition (B) used there can always be assumed without loss of generality as shown in [3] Lemma 2.4.

3. The case $\beta < \beta_*$

We shall check the condition *ii)* of Theorem 2.5. As a first step we establish a useful formula.

Lemma 3.1. *Let f be a positive bounded measurable function on $\mathbb{R}$. Identify f with the corresponding multiplication operator. The operator $\mathcal{Q}(f)$ agrees with the multiplication operator by the positive measurable function*

$$(\mathcal{Q}(f))(y) = (-2\beta)^{-1} \int_0^\infty ds \exp\left(\beta y^2 s/2\right) (1+s)^{-1+1/(2\beta)} f\left(cy(1+s)^{1/2}\right) \quad (3.1)$$

Proof. Let u be a smooth function with compact support contained in $\mathbb{R} - \{0\}$. Denote by $q(\cdot,\cdot)$ and $p(\cdot,\cdot)$ the functions

$$q(t,y) = \left(1 - 2t/(\beta y^2)\right)^{1/2}, \quad p(t,x) = \begin{cases} (1 + 2t/(\beta x^2))^{1/2} & \text{if } x^2 > -2t/\beta \\ 0 & \text{if } x^2 \leq -2t/\beta. \end{cases}$$

A straightforward computation yields

$$\begin{aligned}\langle u, \mathcal{Q}(f)u\rangle &= \int_0^\infty e^{-t} \langle MP^*(t)u, (S^* fS)MP^*(t)u\rangle \, dt \\ &= \int_0^\infty e^{-t}dt \int_{\mathbb{R}} dx \frac{1}{\beta^2 x^2} p(t,x)^{-1-1/\beta} f(cx) \left|u(xp(t,x))\right|^2.\end{aligned}$$

By the change of variables $xp(t,x) = y$, the right-hand side can be written in the form

$$\begin{aligned}&\int_0^\infty e^{-t}dt \int_{\mathbb{R}} dy|u(y)|^2 (\beta y)^{-2} \left(q(t,y)\right)^{-2+1/\beta} f(cyq(t,y)) \\ &= \int_{\mathbb{R}} dy|u(y)|^2 (\beta y)^{-2} \int_0^\infty dt\, e^{-t} \left(q(t,y)\right)^{-2+1/\beta} f(cyq(t,y)).\end{aligned}$$

Changing the variable t to $s = -2t/(\beta y^2)$ we obtain the formula (3.1). □

Let $\mathcal{F}$ be the cone of positive measurable function on $I\!R$ bounded on open subsets of $I\!R$ not containing 0. In view of Lemma 3.1 we can extend the map $\mathcal{Q}$ to $\mathcal{F}$ defining $\mathcal{Q}(f)$ as the unique positive selfadjoint operator such that

$$\langle u, \mathcal{Q}(f)u\rangle = \sup_{n\geq 1} \langle u, \mathcal{Q}(f \wedge n)u\rangle$$

for all $u \in D_0$. Clearly the operator $\mathcal{Q}(f)$ agrees with the moltiplication operator by the function $(\mathcal{Q}(f))$ given by (3.1). Moreover we have the following

Lemma 3.2. *Let f, g be two elements of $\mathcal{F}$ satisfying the inequality $f \leq g$. Then the operators $\mathcal{Q}(f)$, $\mathcal{Q}(g)$ satisfy the inequality*

$$\mathcal{Q}(f) \leq \mathcal{Q}(g).$$

Let q_η denote the function

$$q_\eta : I\!R - \{0\} \to I\!R, \qquad q_\eta(x) = |x|^{-\eta}$$

for all $\eta > 0$. This function will be often identified with the corresponding positive self-adjoint multiplication operator.

Lemma 3.3. *The operator $\mathcal{Q}(I)$ satisfies the following inequalities*

$$\mathcal{Q}(I) \leq 1, \qquad \mathcal{Q}(I) \leq (1+\beta)^{-2}(1-2\beta)^{-1}q_2.$$

Proof. Use the formula (3.1), f being the constant function 1. To establish the second inequality it suffices to estimate the first factor in the integrand by the constant 1 and compute the integral. The first inequality is well known; however it can be checked here in the same way. □

Lemma 3.4. *For all $\eta > 0$ the following inequality holds*

$$\mathcal{Q}(q_\eta) \leq [(1-\eta\beta)|1+\beta|^{\eta}]^{-1} q_\eta.$$

Proof. In fact, by Lemma 3.1, we have

$$\begin{aligned}
(\mathcal{Q}(q_\eta))(y) &= |y|^{-\eta} \cdot |c|^{-\eta}(-2\beta)^{-1} \int_0^\infty ds \exp(\beta y^2 s/2)\,(1+s)^{-1+1/(2\beta)-\eta/2} \\
&\leq |y|^{-\eta} \cdot |c|^{-\eta}(-2\beta)^{-1} \int_0^\infty ds\,(1+s)^{-1+1/(2\beta)-\eta/2} \\
&= ((1-\eta\beta)|1+\beta|^{\eta})^{-1} q_\eta(y).
\end{aligned}$$

This proves the Lemma. □

We can now prove the first part of Theorem 2.4.

Proposition 3.5. *Suppose $\beta < \beta_*$. Then the m.q.d.s. $\mathcal{T}$ is conservative.*

Proof. Since β is smaller than β_* we have the inequality

$$\lim_{\eta\to 0+} (1-\eta\beta)^{1/\eta} |\beta+1| = \exp(-\beta)|\beta+1| > 1.$$

Therefore we can choose a real number $\eta \in (0,2]$ such that

$$(1-\eta\beta)|\beta+1|^{\eta} > 1.$$

From Lemma 3.3 it follows that there exists a positive constant ξ, depending only on β, such that

$$\mathcal{Q}(I) \leq \xi q_\eta.$$

Thus, using Lemma 3.4, we can easily establish the following estimate by induction

$$\mathcal{Q}^n(I) \leq \xi \left[(1-\eta\beta)|\beta+1|^\eta\right]^{-n} q_\eta.$$

Letting n tend to infinity we check the condition ii) of Theorem 2.5 is fulfilled. □

4. The case $\beta > \beta_*$

We shall show that the condition *iii)* of Theorem 2.5 is not satisfied. Fix $\lambda = 1$. We consider first $\mathcal{L}$ as the differential operator $\mathcal{L}_d$ on some function space given by

$$(\mathcal{L}_d(f))(x) = (\beta x)^{-2}\left(f(cx) - f(x) - \beta x f'(x)\right) \tag{4.1}$$

and construct a nonzero positive bounded continuous function f on $\mathbb{R}$ solving the differential equation $f = \mathcal{L}_d(f)$. Then we show that the function f satisfies the condition

$$\langle v, fu\rangle = \langle G^*v, fu\rangle + \langle SMv, fSMu\rangle + \langle v, fG^*u\rangle = \langle v, \mathcal{L}(f)u\rangle \tag{4.2}$$

for all vectors v, u in the domain of G^*.

For every open subset E of $\mathbb{R}$ we denote by $\mathcal{C}^k(E;\mathbb{R})$ (resp. $\mathcal{C}^k_b(E;\mathbb{R})$) the vector space of real-valued continuous (resp. bounded continuous) functions on E with continuous (resp. bounded continuous) derivatives of the first k orders.

Lemma 4.1. *Let g be an element of $\mathcal{C}^0_b(\mathbb{R}-\{0\};\mathbb{R})$ and let f be an element of $\mathcal{C}^0_b(\mathbb{R}-\{0\};\mathbb{R})$. If $\beta > 0$ the following conditions are equivalent:*

a) *$f \in \mathcal{C}^1(\mathbb{R}-\{0\};\mathbb{R})$ and, for all $x \in \mathbb{R}-\{0\}$, we have*

$$f(x) - (\beta x)^{-2}\left(f(cx) - f(x) - \beta x f'(x)\right) = g(x), \tag{4.3}$$

b) *for all $x \in \mathbb{R}-\{0\}$ we have*

$$f(x) = |x|^{-1/\beta}\exp(-\beta x^2/2)\int_0^x |t|^{1/\beta}\exp(\beta t^2/2)\left[\beta t g(t) + (\beta t)^{-1} f(ct)\right] dt. \tag{4.4}$$

If $\beta < 0$ the condition a) is equivalent to the following condition:
c) for all $x \in I\!R - \{0\}$ we have

$$f(x) = -|x|^{-1/\beta} \exp(-\beta x^2/2) \int_x^{\mathrm{sgn}(x)\infty} |t|^{1/\beta} \exp(\beta t^2/2) \left[\beta t g(t) + (\beta t)^{-1} f(ct)\right] dt \tag{4.5}$$

Proof. The differential equation (4.3) can be written in the normal form

$$f'(x) = -\left(\beta x + (\beta x)^{-1}\right) f(x) + \left[\beta x g(x) + (\beta x)^{-1} f(cx)\right].$$

Therefore, integrating this first order differential equation, for all $x, x_0 \in I\!R$ with $0 < x_0 < x$ or $x < x_0 < 0$ we have

$$\begin{aligned} f(x) &= f(x_0)|x_0|^{1/\beta} \exp(\beta x_0^2/2)|x|^{-1/\beta} \exp(-\beta x^2/2) \\ &\quad + |x|^{-1/\beta} \exp(-\beta x^2/2) \int_{x_0}^{x} |t|^{1/\beta} \exp(\beta t^2/2) \left[\beta t g(t) + (\beta t)^{-1} f(ct)\right] dt. \end{aligned}$$

The function f being bounded, if $\beta > 0$, we can let x_0 go to 0 and obtain (4.4). If $\beta < 0$, we can exchange x and x_0, let x_0 go to $\mathrm{sgn}(x)\infty$ and obtain (4.5). Conversely, differetiating (4.4) and (4.5) we obtain (4.3). $\square$

Let $C_b^0(I\!R - \{0\}; I\!R_+)$ denote the cone of nonnegative functions in $C_b^0(I\!R - \{0\}; I\!R)$. The following proposition gives essentially the construction of the so-called minimal solution to the Feller-Kolmogorov equation of a classical stochastic process.

Proposition 4.2. *There exists a map*

$$\mathcal{R} : C_b^0(I\!R - \{0\}; I\!R_+) \to C_b^0(I\!R - \{0\}; I\!R_+) \cap C^1(I\!R - \{0\}; I\!R)$$

with the following properties:
a) for every $g \in C_b^0(I\!R - \{0\}; I\!R_+)$, the function $\mathcal{R}(g)$ satisfies the equation (4.3),
b) for every $g \in C_b^0(I\!R - \{0\}; I\!R_+)$, we have the inequality

$$\|\mathcal{R}(g)\|_\infty \le \|g\|_\infty,$$

c) for every $g, \tilde{g} \in C_b^0(I\!R - \{0\}; I\!R_+)$, such that $g \le \tilde{g}$ and every function $\tilde{f} \in C_b^0(I\!R - \{0\}; I\!R_+) \cap C^1(I\!R - \{0\}; I\!R)$ satisfying the equation (4.3) with $g = \tilde{g}$ we have the inequality

$$\mathcal{R}(g) \le \tilde{f}.$$

Proof. Consider, for example, the case $\beta > 0$. Let $(f_n)_{n \ge 0}$ be the sequence of elements of $C_b^0(I\!R - \{0\}; I\!R_+)$ defined by

$$\begin{aligned} f_0(x) &= 0, \\ f_{n+1}(x) &= |x|^{-1/\beta} \exp(-\beta x^2/2) \int_0^x |t|^{1/\beta} \exp(\beta t^2/2) \left[\beta t g(t) + (\beta t)^{-1} f_n(ct)\right] dt. \end{aligned}$$

We can easily show by induction the inequality

$$\|f_n\|_\infty \le \|g\|_\infty . \tag{4.6}$$

In fact, (4.6) holds when $n = 0$. Suppose it has been established for an integer n. Then, for all $x \in I\!R - \{0\}$, we have the inequalities

$$\begin{aligned} |f_{n+1}(x)| &\le \|g\|_\infty \, |x|^{-1/\beta} \exp(-\beta x^2/2) \int_0^x |t|^{1/\beta} \exp(\beta t^2/2) \left[\beta t + (\beta t)^{-1}\right] dt \\ &= \|g\|_\infty \, |x|^{-1/\beta} \exp(-\beta x^2/2) \int_0^x \frac{d}{dt}\left(|t|^{1/\beta} \exp(\beta t^2/2)\right) dt \\ &\le \|g\|_\infty , \end{aligned}$$

which prove the inequality (4.6) for the integer $n+1$. Since g is nonnegative-valued, we can also show by induction that the sequence $(f_n)_{n\ge 0}$ is increasing. Let us consider the function $\mathcal{R}(g)$ defined by

$$(\mathcal{R}(g))(x) = \sup_{n\ge 0} f_n(x).$$

Clearly $\mathcal{R}(g)$ is nonnegative and satisfies the condition b). Moreover $\mathcal{R}(g)$ satisfies the equation (4.4). Hence it belongs to $C_b^0(I\!R - \{0\}; I\!R) \cap C^1(I\!R - \{0\}; I\!R)$ and satisfies the equation (4.3). Finally let $g, \tilde{g}$ and f be as in c) and let $(f_n)_{n\ge 0}$ be the above defined sequence. Notice that, for all integer n and all $x \in I\!R$ with $x \ne 0$ the difference $\tilde{f}(x) - f_n(x)$ an be written in the from

$$|x|^{-1/\beta} \exp(-\beta x^2/2) \int_0^x |t|^{1/\beta} \exp(\beta t^2/2) \left[(\beta t)^{-1} (\tilde{f} - f_n)(ct) + \beta t\, (\tilde{g}(t) - g(t))\right] dt.$$

Therefore, since $\tilde{f} \ge 0 = f_0$, we can prove by induction the inequality $\tilde{f} \ge f_n$ for every n. By the definition of $\mathcal{R}_\lambda$ the property c) follows. The case when $\beta < 0$ can be dealt with in the same way. □

Lemma 4.3. *For all* $\beta > \beta_*$ *there exists* $\eta \in (0,1]$ *such that*

$$|\beta + 1|^\eta - (1 + \eta\beta) = 0. \tag{4.7}$$

Proof. The function

$$\varphi : [0,1] \to I\!R, \qquad \varphi(x) = |\beta + 1|^x - 1 - \beta x$$

has the following properties

$$\varphi(0) = 0, \qquad \varphi(1) \le 0, \qquad \varphi'(0) = \log|\beta + 1| - \beta < 0,$$

because $\beta > \beta_*$. This proves the Lemma. Notice that, in the case $\beta \ge -1$, we can choose $\eta = 1$ since $\varphi(1) = |\beta + 1| - \beta - 1 = 0$. □

Proposition 4.4. *Suppose $\beta > \beta_*$ and consider a number $\eta \in (0,1]$ fulfilling (4.7). Let g be an element of $C_b^0(\mathbb{R};\mathbb{R}_+)$ satisfying the inequality*

$$0 \leq g(x) \leq k|x|^\eta/(1+|x|^\eta)$$

for all $x \in \mathbb{R}$, k being a positive constant. Then the function $\mathcal{R}(g)$ satisfies the inequality

$$0 \leq \mathcal{R}(g) \leq k|x|^\eta/(1+|x|^\eta).$$

Proof. Consider the function

$$\tilde{f} : \mathbb{R} \to \mathbb{R}, \qquad \tilde{f}(x) = k|x|^\eta/(1+|x|^\eta).$$

Clearly (4.7) implies the inequality $|\beta+1|^\eta(1-\eta\beta) < 1$. Then a straightforward computation yields

$$\left(\mathcal{L}_d\left(\tilde{f}\right)\right)(x) = k|x|^\eta \frac{(|\beta+1|^\eta - (1+\eta\beta)) + (|\beta+1|^\eta(1-\eta\beta)-1)|x|^\eta}{(\beta x)^2(1+|c|^\eta|x|^\eta)(1+|x|^\eta)^2} \leq 0 \quad (4.8)$$

for all $x \in \mathbb{R} - \{0\}$. Let us consider the function $g \in C_b^0(\mathbb{R}-\{0\},\mathbb{R}_+)$ defined by

$$\tilde{g} = \tilde{f} - \mathcal{L}_d\left(\tilde{f}\right).$$

Because of (4.8) and the definition of $\tilde{f}$ we have the inequalities

$$\tilde{g} \geq \tilde{f} \geq g.$$

Applying the Proposition 4.2 c), we complete the proof. □

Proposition 4.5. *Suppose $\beta > \beta_*$. Let $g \in C_b^0(\mathbb{R};\mathbb{R})$ be a Lipschitz continuous function. Then there exists a function $f \in C_b^0(\mathbb{R};\mathbb{R}) \cap C^1(\mathbb{R}-\{0\};\mathbb{R})$ such that $f(0) = g(0)$ and*

$$f(x) - (\mathcal{L}_d(f))(x) = g(x) \tag{4.9}$$

for all $x \neq 0$.

Proof. Fix $\eta \in (0,1]$ satisfying (4.7) and write g in the form

$$g(0) + (g-g(0))^+ - (g-g(0))^-.$$

Since g is Lipschitz continuous there exists a positive constant k such that

$$0 \leq (g-g(0))^+(x) \leq \frac{k|x|^\eta}{1+|x|^\eta}, \qquad 0 \leq (g-g(0))^-(x) \leq \frac{k|x|^\eta}{1+|x|^\eta}.$$

Consider the function f defined by $f(0) = g(0)$ and

$$f(x) = g(0) + \left(\mathcal{R}\left((g-g(0))^+\right)\right)(x) - \left(\mathcal{R}\left((g-g(0))^-\right)\right)(x)$$

for all $x \neq 0$. This function satisfies the equation (4.9) because of Proposition 4.2 a) and the fact that (4.9) is linear. Moreover, from Proposition 4.4, we have the inequality

$$|f(x) - g(0)| \leq k|x|^\eta/(1+|x|^\eta)$$

which shows that f is continuous at the point 0. □

Proposition 4.6. *Suppose $\beta > \beta_*$. Then there exists a function $f \in C_b^0(\mathbb{R};\mathbb{R}) \cap C^0(\mathbb{R}-\{0\};\mathbb{R})$ satisfying the differential equation*

$$f(x) = (\mathcal{L}_d(f))(x) \tag{4.10}$$

for all $x \neq 0$ and the condition $f(0) > 0$.

Proof. Let us consider the function

$$f_1 : \mathbb{R} \to \mathbb{R}, \qquad f_1(x) = (1+x^2)^{-1},$$

and the function $g \in C_b^0(\mathbb{R};\mathbb{R})$ defined as the only continuous extension of the function

$$g(x) = f_1(x) - (\mathcal{L}_d(f_1))(x)$$

for all $x \neq 0$. A straightforward computation yields

$$g(x) = \frac{2 + (c^2 - 2c)x^2 + c^2x^4}{(1+c^2x^2)(1+x^2)^2}.$$

By Proposition 4.5 there exists a function $f_2 \in C_b^0(\mathbb{R};\mathbb{R}) \cap C^1(\mathbb{R}-\{0\};\mathbb{R})$ satisfying the equation (4.9) and the condition $f_2(0) = g(0) = 2$. The function

$$f = f_2 - f_1$$

satisfies the differential equation (4.10) and the condition $f(0) = 1$. □

The following propositions show that a function f satisfying the conditions of Proposition 4.6 is non-negative everywhere.

Proposition 4.7. *Let f be an element of $C_b^0(\mathbb{R}-\{0\};\mathbb{R}) \cap C^1(\mathbb{R}-\{0\};\mathbb{R})$ satisfying the differential equation (4.10). Then*

$$\lim_{|x|\to\infty} f(x) = 0. \tag{4.11}$$

Proof. In the case $\beta > 0$, for all $x \neq 0$, we have the inequality

$$|f(x)| \leq \|f\|_\infty \frac{\left|\int_0^x |t|^{1/\beta} \exp(\beta t^2/2)\, dt\right|}{\beta |x|^{1/\beta} \exp(\beta x^2/2)}.$$

Therefore (4.11) follows computing the limits, for example by the De L'Hôpital rule. The proof in the case $\beta < 0$ is similar. □

Proposition 4.8. *Let f be an element of $C_b^0(\mathbb{R};\mathbb{R}) \cap C^1(\mathbb{R}-\{0\};\mathbb{R})$ satisfying the differential equation (4.9) and the condition $f(0) > 0$. Then f is non-negative everywhere.*

Proof. We use the well-known minimum principle. Suppose that there exists $a \in \mathbb{R}$ such that $f(a) < 0$. Then, by Proposition 4.7, there exists $b \in \mathbb{R} - \{0\}$ such that

$$\min_{x\in\mathbb{R}} f(x) = f(b) < 0.$$

The point b does not coincide with 0, hence f is differentiable at that point and $f'(b)$ vanishes. Then we have the contradiction

$$f(b) = (\beta b)^{-2} (f(cb) - f(b)) \geq 0.$$

This shows that f must be non-negative everywhere. □

Proposition 4.9. *Let f be an element of $C_b^0(\mathbb{R};(0,+\infty)) \cap C^1(\mathbb{R}-\{0\};(0,+\infty))$ satisfying the differential equation (4.10). Then f satisfies also the identity (4.2) for all $v, u \in D(G^*)$.*

Proof. Suppose, for example, $\beta > 0$. Let us consider two vectors in the domain of G^* represented in the form $R(1;G^*)v, R(1;G^*)u$ with $v, u \in h$. For all $r \in (0,1)$ denote by I_r the set $(-r^{-1},-r)\cup(r,r^{-1})$. The scalar product $\langle R(1;G^*)v, fR(1;G^*)u\rangle$ can be written as

$$\lim_{r\to 0}\int_{I_r} (R(1;G^*)\bar{v})(x)\,(\beta x)^{-2}\, f(cx)\,(R(1;G^*)u)(x)\,dx$$

$$-\lim_{r\to 0}\int_{I_r} (R(1;G^*)\bar{v})(x)(\beta x)^{-2} f(x)\,(R(1;G^*)u)(x)\,dx$$

$$-\lim_{r\to 0}\int_{I_r} (R(1;G^*)\bar{v})(x)(\beta x)^{-1} f'(x)\,(R(1;G^*)u)(x)\,dx.$$

Integrating by parts the third integral, we can write the above sum of the second and third term in the form

$$\begin{aligned}
&\langle SMR(1;G^*)v, fSMR(1;G^*)u\rangle \\
&+ 2\langle R(1;G^*)v, fR(1;G^*)\rangle - \langle v, fR(1;G^*)u\rangle - \langle R(1;G^*)v, fu\rangle \\
&+ \beta \lim_{r\to 0} |r|^{1/\beta} \exp(\beta r^2)\bar{v}_0(r)u_0(r)f(r) \\
&- \beta \lim_{r\to 0} |r|^{-1/\beta} \exp(\beta r^{-2})\bar{v}_0(r^{-1})u_0(r^{-1})f(r^{-1}) \\
&- \beta \lim_{r\to 0} |r|^{1/\beta} \exp(\beta r^2)\bar{v}_0(-r)u_0(-r)f(-r) \\
&+ \beta \lim_{r\to 0} |r|^{-1/\beta} \exp(\beta r^{-2})\bar{v}_0(-r^{-1})u_0(-r^{-1})f(-r^{-1})
\end{aligned}$$

where v_0 and u_0 are defined as in Proposition 2.1. It is easy to see as in the proof of Proposition 2.2 that all the above limits vanish. Then the proof can be completed using the identity $R(1;G^*) - I = G^*R(1;G^*)$. □

Thus we have proved the condition *iii)* of Theorem 2.5 is not fulfilled.

Proposition 4.10. *The m.q.d.s. $\mathcal{T}$ is not conservative if $\beta > \beta_*$.*

5. Applying the sufficient condition of [3]

In this section we will show that the m.q.d.s. $\mathcal{T}$ is conservative if $\beta \leq -3/2$ by checking a sufficient condition obtained in [3]. We transform first the form $\mathcal{L}$ defined in Section 2 shifting the spectrum operator G^* by $-1/2$ and considering the operators

$$L_1 = I, \qquad L_2 = SM.$$

The "shifted" m.q.d.s. generated by the form

$$(v, Xu) \to \langle G^*v, Xu\rangle + \langle L_1v, XL_1u\rangle + \langle L_2v, XL_2u\rangle + \langle v, XG^*u\rangle$$

is conservative if and only if the "unshifted" one is also. Let us recall that, if $\beta < 0$, $R(1; G^*)$ is modified as follows

$$(R(1; G^*)u)(x) = \beta|x|^{(\beta+1)/2\beta} \exp(3\beta x^2/4)u_0(x)$$

where

$$u_0(x) = -\int_0^x \exp(-3\beta y^2/4)|y|^{-(\beta+1)/2\beta} yu(y)dy$$

Consider the selfadjoint operator defined by

$$D(C) = \{ u \in h \mid x^{-2}u(x) \in h \}, \qquad Cu(x) = \left(1 + (\beta x)^{-2}\right) u(x).$$

The domain of its positive square root $C^{1/2}$ coincides with the domain of the operator M. Therefore, by Proposition 2.2, it is contained in the domain of G^* and, for all $v, u \in D(G^*)$, we have

$$-\langle G^*v, u\rangle - \langle v, G^*u\rangle = \left\langle C^{1/2}v, C^{1/2}u\right\rangle.$$

Moreover C has a bounded inverse which is bounded from below by the identity operator. We shall denote by C_ε (for $\varepsilon > 0$) the bounded operator $(\varepsilon I + C^{-1})^{-1}$.

Let $\mathcal{Q} : \mathcal{B}(h) \to \mathcal{B}(h)$ be the normal and monotone map defined by

$$\langle v, \mathcal{Q}(X)u\rangle = \int_0^\infty e^{-t}\left(\langle P^*(t)v, XP^*(t)u\rangle + \langle SMP^*(t)v, XSMP^*(t)u\rangle\right)dt.$$

Theorem 5.1. *Suppose that there exists a positive constant b such that:*
i) for all $u \in D(G^*)$ the following inequality holds

$$-2\Re\left\langle G^*u, C^{-1}u\right\rangle \le b\|u\|^2, \tag{5.1}$$

ii) for all $\varepsilon > 0$ and all u in a dense subset of h contained in the domain of $C^{1/2}$ we have

$$\langle u, \mathcal{Q}(C_\varepsilon)u\rangle \le \left\|C^{1/2}u\right\|^2 + b\|u\|^2. \tag{5.2}$$

Then the m.q.d.s. $\mathcal{T}$ is conservative.

We refer to [3] Th. 4.2 for the proof and check the conditions *i)*, *ii)*.

Lemma 5.2. *Suppose $\beta < 0$. Then the inequality (5.1) holds with $b = 3$.*

Proof. We use vectors in the domain G^* represented in the form $R(1; G^*)$ with $u \in h$. The identity $G^*R(1; G^*) = R(1; G^*) - I$ yields

$$\begin{aligned} &-2\Re\left\langle G^*R(1; G^*)u, C^{-1}R(1; G^*)u\right\rangle \\ &\quad = -2\Re\langle R(1; G^*)u, C^{-1}R(1; G^*)u\rangle + 2\Re\left\langle u, C^{-1}R(1; G^*)u\right\rangle. \end{aligned}$$

The operator C^{-1} is positive, hence it suffices to show that the second term is bounded from above. Let r be an element of $(0,1)$ and denote by I_r the set $(-r^{-1},-r)\cup(r,r^{-1})$. Integrating by parts we have

$$\begin{aligned}
&\langle u, C^{-1}R(1;G^*)u\rangle \\
&= \beta \lim_{r\to 0}\int_{I_r}\left[x|x|^{-\frac{\beta+1}{2\beta}}\mathrm{e}^{-3\beta x^2/4}\bar{u}(x)\right]\frac{\beta^2x^2}{1+\beta^2x^2}\operatorname{sgn}(x)|x|^{\frac{1}{\beta}}\mathrm{e}^{3\beta x^2/2}u_0(x)dx \\
&= \lim_{r\to 0}\Big\{\frac{\beta^3 r^2}{1+\beta^2r^2}r^{\frac{1}{\beta}}\exp(3\beta r^2/2)\left(|u_0(r)|^2-|u_0(-r)|^2\right) \\
&\quad+\frac{\beta^3 r^{-2}}{1+\beta^2 r^{-2}}r^{-\frac{1}{\beta}}\exp(3\beta r^2/2)\left(|u_0(-r^{-1})|^2-|u_0(r^{-1})|^2\right) \\
&\quad+\beta\int_{I_r}\frac{2\beta^2}{(1+\beta^2x^2)^2}|x|^{1+\frac{1}{\beta}}\exp(3\beta r^2/2)|u_0(x)|^2dx \\
&\quad+\beta^2\int_{I_r}\frac{1}{1+\beta^2x^2}|x|^{1+\frac{1}{\beta}}\exp(3\beta r^2/2)|u_0(x)|^2dx \\
&\quad+3\beta^2\int_{I_r}\frac{\beta^2x^2}{1+\beta^2x^2}|x|^{1+\frac{1}{\beta}}\exp(3\beta r^2/2)|u_0(x)|^2dx \\
&\quad-\beta\int_{I_r}\frac{\beta^2x^2}{1+\beta^2x^2}|x|^{\frac{1+\beta}{2\beta}}\exp(3\beta r^2/4)\bar{u}_0(x)u(x)dx\Big\}.
\end{aligned}$$

As in the proof of Proposition 2.2 we can prove that
- the first and second term vanish,
- the third term is negative because $\beta<0$,
- the sum of the fourth and fifth term is bounded from above by $3\,\|R(1;G^*)u\|^2$,
- the sixth term converges to $-\langle R(1;G^*)u, C^{-1}u\rangle$.

This proves the lemma. □

Remark. A similar proof shows that the inequality (5.1) holds also when $\beta>0$ with $b=3(1+\beta)$.

Lemma 5.3. *The condition ii) of Theorem 5.1 holds when $\beta\le -3/2$.*

Proof. Let $\varepsilon>0$ and let u be a smooth function with compact support contained in $\mathbb{R}-\{0\}$. Remark that C_ε coincides with the multiplication operator by the function m_ε given by

$$m_\varepsilon(x)=\left(\varepsilon+\left(1+\beta^{-2}x^{-2}\right)^{-1}\right)^{-1}.$$

Let $q(\cdot,\cdot)$ and $p(\cdot,\cdot)$ be the functions defined in the proof of Lemma 3.1. A straightforward computation yields

$$\begin{aligned}
&\int_0^\infty \mathrm{e}^{-t}\left(\langle SMP^*(t)u, C_\varepsilon SMP^*(t)u\rangle+\langle P^*(t)u, C_\varepsilon P^*(t)u\rangle\right)dt \\
&=\int_0^\infty \mathrm{e}^{-2t}dt\int_{\mathbb{R}}dx\left(\frac{m_\varepsilon(cx)}{\beta^2x^2}+m_\varepsilon(x)\right)p(t,x)^{-1-\frac{1}{\beta}}\left|u\left(xp(t,x)\right)\right|^2.
\end{aligned}$$

By the change of variables $xp(t,x) = y$, the right-hand side can be written in the form

$$\int_0^\infty e^{-2t}dt \int_{\mathbb{R}} dy|u(y)|^2 (q(t,y))^{\frac{1}{\beta}} \left(\frac{m_\varepsilon(cyq(t,y))}{\beta^2 y^2 (q(t,y))^2} + m_\varepsilon(yq(t,y))\right)$$
$$= \int_{\mathbb{R}} dy|u(y)|^2 \int_0^\infty dt\, e^{-2t} (q(t,y))^{\frac{1}{\beta}} \left(\frac{m_\varepsilon(cyq(t,y))}{\beta^2 y^2 (q(t,y))^2} + m_\varepsilon(yq(t,y))\right)$$

Changing the variable t to $s = -2t/(\beta y^2)$ and letting ε tend to 0 we can estimate the integral with respect to t by

$$\frac{1}{-2\beta^3 c^2 y^2} \int_0^\infty \exp(\beta y^2 s)\,(1+s)^{-2+1/(2\beta)}\, ds$$
$$+ \frac{2}{-2\beta} \int_0^\infty \exp(\beta y^2 s)\,(1+s)^{-1+1/(2\beta)}\, ds$$
$$- \frac{1}{2} \int_0^\infty \beta y^2 \exp(\beta y^2 s)\,(1+s)^{-1+1/(2\beta)}\, ds.$$

Since β is negative, this sum can be estimated by

$$\frac{1}{-2\beta^3 c^2 y^2} \int_0^\infty (1+s)^{-2+1/(2\beta)}\, ds + \frac{2}{-2\beta} \int_0^\infty (1+s)^{-1+1/(2\beta)}\, ds$$
$$- \frac{1}{2} \int_0^\infty \beta y^2 \exp(\beta y^2 s)\, ds$$
$$= \left((1-2\beta)(1+\beta)^2 \beta^2 y^2\right)^{-1} + 5/2.$$

Therefore, recalling the definition of the operator C given in this section, the condition *ii)* of Theorem 5.1 holds, in the case $\beta < 0$ whenever

$$(1-2\beta)(1+\beta)^2 \geq 1$$

i.e. $\beta \leq -3/2$. □

Theorem 5.4. *The m.q.d.s. $\mathcal{T}$ is conservative if $\beta \leq -3/2$.*

Acknowlegdement. This work was done whilst the second author was visiting the Moscow Institute for Electronics and Mathematics in June 1993. Financial supports from the host institution and MURST-Fondi 40% are gratefully acknowledged.

References

[1] J. Azéma, M. Yor: Etude d'une martingale remarquable. *Sém. Prob.* **XXIII** (1989), 88–130.

[2] A.M. Chebotarev: The theory of conservative dynamical semigroups and its applications. Preprint MIEM n.1 March 1990.

[3] A.M. Chebotarev, F. Fagnola: Sufficient conditions for conservativity of quantum dynamical semigroups. *J. Funct. Anal.* **118** (1993), 131–153.

[4] E.B. Davies: Quantum dynamical semigroups and the neutron diffusion equation. *Rep. Math. Phys.* **11** (1977), 169–188 .

[5] M. Emery: On the Azéma martingales. *Sém. Prob.* **XXIII** (1989), 67–87.

[6] M. Emery: Sur les martingales d'Azéma (Suite). *Sém. Prob.* **XXIV** (1990), 442–447.

[7] F. Fagnola: Unitarity of solutions to quantum stochastic differential equations and conservativity of the associated semigroups. *Quantum Probability and Related Topics* **VII** pp. 139–148.

[8] A. Mohari, K.B. Sinha: The minimal quantum dynamical semigroup and its dilations. *Proc. Indian Ac. Sc.* **102** n.3 (1992), 159–173.

[9] K.R. Parthasarathy: Remarks on the quantum stochastic differential equation $dX = (c-1)Xd\Lambda + dQ$. I.S.I. Technical Report n. 8809. (1988).

Alexandr M. Chebotarev
Moscow Institute for
Electronics and Mathematics
Applied Mathematics Department
Bolshoi Vusovski per. 3/12
RUSSIA - 109028 Moscow

Franco Fagnola
Università di Pisa
Dipartimento di Matematica
Via F. Buonarroti, 2
I - 56127 PISA
ITALY

An Inequality for the Predictable Projection of an Adapted Process

F. Delbaen and W. Schachermayer

Department of Mathematics, Vrije Universiteit Brussel

Institut für Statistik der Universität Wien

Abstract. Let $(f_n)_{n=1}^N$ be a stochastic process adapted to the filtration $(\mathcal{F}_n)_{n=0}^N$. Denoting by $(g_n)_{n=1}^N$ the predictable projection of this process, i.e., $g_n = E_{n-1}(f_n)$ we show that the inequality

$$\left[E(\sum_{n=1}^N | g_n |^q)^{p/q}\right]^{1/p} \leq 2 \left[E(\sum_{n=1}^N | f_n |^q)^{p/q}\right]^{1/p}$$

or, in more abstract terms

$$\|(g_n)_{n=1}^N\|_{L^p(l^q_N)} \leq 2\|(f_n)_{n=1}^N\|_{L^p(l^q_N)}$$

holds true for $1 \leq p \leq q \leq \infty$ (with the obvious interpretation in the case of $p = \infty$ or $q = \infty$).

Several similar results, pertaining also to the case $p > q$, are known in the literature. The present result may have some interest in view of the following reasons: (1) the case $p = 1$ and $2 < q \leq \infty$ seems to be new; (2) we obtain 2 as a uniform constant which is sharp in the case $p = 1$, $q = \infty$ and (3) the proof is very easy.

We denote by $(\mathcal{F}_n)_{n=0}^N$ a filtration on a probability space $(\Omega, \mathcal{F}, \mathbb{P})$ and let E_n be the conditional expectation with respect to $\mathcal{F}_n$. Let $(f_n)_{n=1}^N$ be a stochastic process which will be assumed in most of this note to be adapted to $(\mathcal{F}_n)_{n=1}^N$ and to satisfy the appropriate integrability conditions so that the subsequent statements make sense; we denote by $(g_n)_{n=1}^N$ the predictable projection of $(f_n)_{n=1}^N$, i.e., $g_n = E_{n-1}(f_n)$.

For $1 \leq p, q \leq \infty$ we define

$$\|(f_n)_{n=1}^N\|_{p,q} = \|(f_n)_{n=1}^N\|_{L^p(l^q_N)} = \left[E(\sum_{n=1}^N | f_n |^q)^{p/q}\right]^{1/p}.$$

We shall prove the following inequality.

Acknowledgment: We thank S. Kwapień, P.A. Meyer, P. Müller, G. Schechtman and M. Yor for helpful discussions and relevant information on the existing literature.

1. Lemma. *For $1 \le p \le q \le \infty$ and an adapted process $(f_n)_{n=1}^N$ we have*

$$\|(g_n)_{n=1}^N\|_{L^p(l_N^q)} \le 2\|(f_n)_{n=1}^N\|_{L^p(l_n^q)}.$$

The constant 2 is sharp for the case $p = 1, q = \infty$.

Let us first give some motivation for this inequality and relate it to known results.

Our interest in this inequality stems from an application in the field of Mathematical Finance [D-S], where the present authors needed the case $p = 2, q = \infty$. Note that the case $q = \infty$ may be rephrased in terms of the maximal functions f^* and g^* of the processes $(f_n)_{n=1}^N, (g_n)_{n=1}^N$, while the case $q = 2$ is related to the square function.

M. Yor kindly pointed out a possible connection to inequality 3 below ([L],[Y]) and S. Kwapień, P. Müller and G. Schechtman pointed out other known variants of the above inequality. Special thanks go to S. Kwapień who suggested to us — among other valuable remarks — how to adapt the proof from the case $p = 1, q = \infty$ to the general situation $1 \le p \le q \le \infty$.

For the convenience of the reader we summarize the existing results. The starting point is the subsequent Stein's inequality ([S], th. 3.8).

2. The Inequality of E. Stein. *Let $1 < p < \infty$ and $1 \le q \le \infty$ and $(f_n)_{n=1}^N$, a (not necessarily adapted) process. There is a constant $C_p > 0$ such that*

$$\|(g_n)_{n=1}^N\|_{p,q} \le C_p\|(f_n)_{n=1}^N\|_{p,q}.$$

In fact, Stein formulated the above result for $q = 2$ only, but his proof shows the result for all $1 \le q \le \infty$ (see [D] for an exposition of this more general setting).

The setting of Stein's inequality is more general than Lemma 1 in the sense that it does not require that $(f_n)_{n=1}^N$ is adapted to the filtration $(\mathcal{F}_n)_{n=1}^N$ and it also pertains to the case $p > q$. On the other hand, as shown by easy examples [S], Stein's inequality breaks down as $p \to 1$ (and $q \ne 1$) and $p \to \infty$ (and $q \ne \infty$).

Stein's inequality was also considered by W. Johnson, B. Maurey, G. Schechtman and L. Tzafriri [J-M-S-T], who extended it by replacing L^p by a rearrangement invariant space X on [0,1], for which the Boyd indices verify $0 < \beta_X \le \alpha_X < 1$.

D. Lépingle and M. Yor noted that there is an interesting difference in the problem of the constant C_p in Stein's inequality between $p \to 1$ and $p \to \infty$. In the case $p \to \infty$ (and $q \ne \infty$) it is easy to give examples $(f_n)_{n=1}^N$ of *adapted* processes for which C_p becomes big as $p \to \infty$ (see [S] and [L]). For the convenience of the reader we sketch the typical example: Let $f_n = 2\chi_{A_n}$, where $(A_n)_{n=1}^N$, are disjoint sets with $P(A_n) = 2^{-n}$. Letting $\mathcal{F}_n = \sigma(f_1, \dots, f_n)$ we obtain that $\|(f_n)_{n=1}^N\|_{\infty,q} = 2$, while $\|(g_n)_{n=1}^N\|_{\infty,q} = N^{\frac{1}{q}}$, which tends to infinity if $q < \infty$.

On the other hand, the situation is different for $p \to 1$ and here more can be said for *adapted* processes (see e.g.,[J], th. 1.6):

3. Inequality of D. Lépingle and M. Yor. *Let $p = 1, q = 2$ and $(f_n)_{n=1}^N$ be adapted to $(\mathcal{F}_n)_{n=1}^N$. Then*

$$\|(g_n)_{n=1}^N\|_{p,2} \le 2\|(f_n)_{n=1}^N\|_{p,2}.$$

In a remarkable paper by J. Bourgain ([B], prop. 5) the same result was obtained with a constant $K_1 = 3$ as an auxiliary step (preceded by the phrase: "The next inequality is probably known, but we include its proof here for the sake of completeness").

It is interesting to compare the proofs: E. Stein uses Doob's maximal inequality and interpolation, Lépingle and Yor adapted the argument of C. Herz for the proof of the $H^1 - BMO$ duality (see Garsia [G], p.9) while J. Bourgain directly uses the atomic decomposition of H^1. In the two last proofs it seems crucial to restrict to the case $q = 2$. Let us also point out that in the case $1 < p < \infty$ the above inequality may also be deduced from a convexity lemma of D. Burkholder ([Bu], Lemma 16.1); it also can be derived in full generality (i.e., also in the case $p = 1$, but with a constant which is possibly worse) from results on decoupled conditionally independent tangent sequences as presented in the book of S. Kwapień and W. Woyczyński ([K-W], th. 5.2.1 and 5.6.2). See also S. Dilworth [D] for a clear exposition of the case $1 < p < \infty$.

We now pass to the proof of Lemma 1.

PROOF OF LEMMA 1. We first prove the case $p = 1$. For $1 \leq q < \infty$ the function

$$\varphi_q(x, h) = (x^q + h^q)^{\frac{1}{q}} - h \qquad x \geq 0, h \geq 0$$

and, in the case $q = \infty$, the function

$$\varphi_\infty(x, h) = (x \vee h) - h = (x - h)_+ \qquad x \geq 0, h \geq 0$$

is, for fixed $h \geq 0$, convex in $x \in \mathbb{R}$.

If $\mathcal{G}$ is a σ-algebra contained in the σ-algebra $\mathcal{F}$, $f \in L^1(\mathcal{F})$, $h \in L^1(\mathcal{G})$, then by applying the conditional version of Jensen's inequality we have for each $1 \leq q \leq \infty$

$$\varphi_q(\mathbb{E}(f \mid \mathcal{G}), h) \leq \mathbb{E}(\varphi_q(f, h) \mid \mathcal{G}) \qquad \text{P.a.s.}$$

Hence, denoting by g the conditioned expectation of f with respect $\mathcal{G}$,

$$g = \mathbb{E}(f \mid \mathcal{G}),$$

we obtain, for $1 \leq q < \infty$,

$$\mathbb{E}(\varphi_q(g, h)) \leq \mathbb{E}(\varphi_q(f, h))$$

and, for $q = \infty$,

$$\mathbb{E}((g - h)_+) \leq \mathbb{E}((f - h)_+).$$

Now fix an adapted sequence $(f_n)_{n=1}^N$ and its predictable projection $(g_n)_{n=1}^N$ as in the statement of Lemma 1. We have to show that, for $1 \leq q \leq \infty$

$$\|(g_1, \ldots, g_N)\|_{L^1(l^q_N)} \leq 2\|(f_1, \ldots, f_N)\|_{L^1(l^q_N)}$$

which will readily finish the proof.

We may assume that $f_n \geq 0$. Denote, for $1 \leq n \leq N, 1 \leq q < \infty$,

$$\hat{f}_n = (f_1^q + \cdots + f_n^q)^{\frac{1}{q}},$$
$$\hat{g}_n = (g_1^q + \cdots + g_n^q)^{\frac{1}{q}},$$
$$\hat{h}_n = (f_1^q + g_1^q + \cdots + f_n^q + g_n^q)^{\frac{1}{q}},$$

and, in the case $q = \infty$,

$$f_n^* = f_1 \vee \cdots \vee f_n,$$
$$g_n^* = g_1 \vee \cdots \vee g_n,$$
$$h_n^* = f_n^* \vee g_n^*.$$

In order to show that

$$\mathbb{E}(\hat{g}_N) \leq 2\mathbb{E}(\hat{f}_N) \quad \text{and} \quad E(g_N^*) \leq 2E(f_N^*)$$

we shall proceed inductively showing that, in fact, for each $n = 1, \ldots, N$

$$\mathbb{E}(\hat{h}_n) \leq 2\mathbb{E}(\hat{f}_n) \quad \text{and} \quad E(h_n^*) \leq 2E(f_n^*), \qquad (*)$$

For $n = 1$ this holds true as $h_1^* \leq \hat{h}_1 \leq f_1 + g_1$ and $E(f_1) = E(g_1)$.

So suppose that $(*)$ holds true for $n-1$. We give the proof for the cases $1 \leq q < \infty$ and $q = \infty$ separately. In the case $1 \leq q < \infty$ we obtain from the argument at the beginning of the proof that

$$\mathbb{E}(\varphi_q(g_n, \hat{h}_{n-1}) \leq \mathbb{E}\varphi_q(f_n, \hat{h}_{n-1}).$$

Note that

$$\begin{aligned} \hat{f}_n &= \hat{f}_{n-1} + (f_n^q + \hat{f}_{n-1}^q)^{\frac{1}{q}} - \hat{f}_{n-1} \\ &= \hat{f}_{n-1} + \varphi_q(f_n, \hat{f}_{n-1}). \end{aligned}$$

As, for fixed $x \geq 0$, the function $\varphi_q(x, h)$ is decreasing in h, we may estimate

$$\hat{h}_n \leq \hat{h}_{n-1} + \varphi_q(f_n, \hat{h}_{n-1}) + \varphi_q(g_n, \hat{h}_{n-1})$$

so that

$$\begin{aligned} \mathbb{E}(\hat{h}_n) &\leq \mathbb{E}(\hat{h}_{n-1}) + 2\mathbb{E}\varphi_q(f_n, \hat{h}_{n-1}) \\ &\leq 2[\mathbb{E}(\hat{f}_{n-1}) + \mathbb{E}\varphi_q(f_n, \hat{f}_{n-1})] \\ &= 2\mathbb{E}(\hat{f}_n). \end{aligned}$$

In the case $q = \infty$ we proceed similarly. From the argument of the beginning of the proof, we obtain that

$$E\left[(g_n - h_{n-1}^*)_+\right] \leq E\left[(f_n - h_{n-1}^*)_+\right].$$

Noting that

$$f_n^* = f_{n-1}^* + (f_n - f_{n-1}^*)_+$$

and

$$h_n^* \leq h_{n-1}^* + (f_n - h_{n-1}^*)_+ + (g_n - h_{n-1}^*)_+$$

we obtain that

$$\begin{aligned}E(h_n^*) &\le E(h_{n-1}^*) + 2E\left[(f_n - h_{n-1}^*)_+\right] \\ &\le 2\left[E(f_{n-1}^*) + E(f_n - f_{n-1}^*)_+)\right] \\ &= 2E(f_n^*),\end{aligned}$$

which finishes the proof in the case $p = 1$.

For the case $1 < p \le q \le \infty$ fix an adapted sequence $(f_n)_{n=1}^N, f_n \in L^p(\mathcal{F}_n)$ and $(g_n)_{n=1}^N$ its predictable projection. Letting $\bar{f}_n = f_n^p \in L^1(\mathcal{F}_n), \bar{g}_n = g_n^p \in L^1(\mathcal{F}_{n-1})$, note that by Jensen's inequality $\bar{g}_n \le \mathbb{E}(\bar{f}_n \mid \mathcal{F}_{n-1})$. Apply the first part of the proof to $r = \frac{q}{p}$ to obtain

$$\mathbb{E}(\bar{g}_1^r + \cdots + \bar{g}_N^r)^{\frac{1}{r}} \le 2\mathbb{E}(\bar{f}_1^r + \cdots + \bar{f}_N^r)^{\frac{1}{r}}$$

or

$$\mathbb{E}(g_1^q + \cdots + g_N^q)^{\frac{1}{q}} \le 2^{\frac{1}{p}}\mathbb{E}(f_1^q + \cdots + f_N^q)^{\frac{1}{q}}$$

which shows Lemma 1 also for the case $1 < p \le q \le \infty$, with the constant 2 replaced by $2^{\frac{1}{p}}$. □

REMARK. (1) Let us show that the constant 2 is sharp for $p = 1, q = \infty$. Choose $\Omega = [0,1], \mathcal{F}$ the Lebesgue-measurable subsets and P to be Lebesgue measure. Let $\mathcal{F}_0 = \{\emptyset, \Omega\}$ and $\mathcal{F}_1 = \mathcal{F}_2 = \mathcal{F}$. For $\varepsilon > 0$, let $f_1 = f_2 = \varepsilon^{-1}\chi_{[0,\varepsilon]}$, for which we get $\|(f_n)_{n=1}^2\|_{1,\infty} = 1$. The predictable projection is given by $g_1 = 1$ and $g_2 = f_2$ for which we obtain $\|(g_n)_{n=1}^2\|_{1,\infty} = 2 - \varepsilon$.

We did not succeed in determining the sharp constant in the other cases $1 \le p \le q \le \infty$ (except for the trivial case $p = q$ where the sharp constant clearly is 1).

(2) One might try to prove the inequality

$$E(\sum_{n=1}^N \mid g_n \mid^q)^{p/q} \le C_p E(\sum_{n=1}^N \mid f_n \mid^q)^{p/q}$$

by applying Jensen's inequality in an even more direct and brutal way:

$$\mid g_n \mid^q = \mid E_{n-1}(f_n) \mid^q \le E_{n-1}(\mid f_n \mid^q)$$

whence

$$\sum_{n=1}^N \mid g_n \mid^q \le \sum_{n=1}^N E_{n-1}(\mid f_n \mid^q)$$

and

$$E(\sum_{n=1}^N \mid g_n \mid^q) \le E(\sum_{n=1}^N \mid f_n \mid^q).$$

But at this point this reasoning comes to an end as there is no way to derive from this last inequality the original one. Indeed, this argument already breaks down in the case $N = 1$, for $p < q$, as in this case one would need that conditional expectation is a bounded operator on L^r, where $r = p/q < 1$, which is not the case.

(3) One might want to apply interpolation to derive the inequality from the trivial cases $1 \le p = q \le \infty$ (where it holds with a constant 1) and the extreme case $p = 1, q = \infty$ to obtain it for the general case $1 \le p \le q \le \infty$ (thus avoiding the distinction of cases in the above proof).

Indeed the interpolation theorem of Benedek-Panzone [B-P] implies that an operator T defined on $L^p(\Omega, \mathcal{F}, l_N^q)$ which is bounded for the extreme cases ($p = 1$, $q = 1$), ($p = \infty, q = \infty$), ($p = 1, q = \infty$) ist bounded for all $1 \le p \le q \le \infty$. But there is a difficulty: The present operator T which assigns to every sequence $(f_n)_{n=1}^N$ the sequence $g_n = \mathbb{E}(f_n \mid \mathcal{G}_n)$ is only bounded (by a constant 2) on the suspace of $L^1(\Omega, \mathcal{F}, l_N^q)$ formed by the sequences $(f_n)_{n=1}^N$ adapted to $(\mathcal{F}_n)_{n=1}^N$. It is not obvious (at least not to the authors) how to modify the proof of Benedek-Panzone to adapt it to this subspace of $L^p(\Omega, \mathcal{F}, l_N^q)$.

(4) Finally we want to point out that we formulated Lemma 1 only in the case of finite discrete time, but as noted by D. Lépingle [L], "le passage au temps continu ne présente pas de difficulté" (compare also [D-S], th. 2.3).

Let us sketch this for the case $q = \infty$ (in the case of $q < \infty$ the proper setting for the continuous time case is the concept of "processus mince", see, e.g. [J], th. 1.6.). Let $X = (X_t)_{t \in \mathbb{R}_+}$ be a positive optionel process defined on $(\Omega, \mathcal{F}, (\mathcal{F}_t)_{t \in \mathbb{R}_+}, \mathbb{P})$ and $X = (X_t)_{t \in \mathbb{R}_+}$ a positive predictable process such that Y is less then or equal to the predictable projection ${}^pX$ of X. The assertion of the lemma is

$$\mathbb{E}(Y^*) \le 2\mathbb{E}(X^*).$$

We may write

$$Y^* = \sup(Y_{T_1} \vee Y_{T_2} \vee \cdots \vee Y_{T_n})$$

where the sup is taken over all increasing finite sequences $T_1 \le T_2 \le \cdots \le T_n$ of predictable stopping limes. As it suffices to take a countable number of such sequences, there arises no problem in which sense the above supremum has to be interpreted.

From Lemma 1 we obtain

$$\begin{aligned} \mathbb{E}(Y_{T_1} \vee \cdots \vee Y_{T_n}) &\le 2\mathbb{E}(X_{T_1} \vee \cdots \vee X_{T_n}) \\ &\le 2\mathbb{E}(X^*), \end{aligned}$$

which readily reduces the continuous time case to the discrete time case.

P.A. Meyer kindly pointed out to us a very elegant proof of the inequality

$$\mathbb{E}(Y^*) \le 2\mathbb{E}(X^*),$$

which directly works in the continuous time setting and which we now reproduce. We may write

$$\mathbb{E}(Y^*) = \sup \mathbb{E}(Y_S)$$

where S runs through the nonnegative random variables (again it suffices to consider a sequence $(S_n)_{n=1}^\infty$ in the definition of the above sup).

Fix S and denote by $A = (A_t)_{t \in \mathbb{R}_+}$ the dual predictable projection of the process $\chi_{[S,\infty[}$. This is an increasing process whose potential

$$\mathbb{E}(A_\infty - A_t \mid \mathcal{F}_t)$$

is bounded by one. As the jumps of A are also bounded by one we have that its left potential

$$Z_t = \mathbb{E}(A_\infty - A_{t-} \mid \mathcal{F}_t)$$

is bounded by 2.

Considering cadlag versions of the processes X^* und Y^* we then may estimate

$$\begin{aligned}
\mathbb{E}(Y_S) &= \mathbb{E}\Big(\int_0^\infty Y_t dA_t\Big) \\
&\leq \mathbb{E}\Big(\int_0^\infty X_t dA_t\Big) \\
&\leq \mathbb{E}\Big(\int_0^\infty X_t^* dA_t\Big) \\
&= \mathbb{E}\Big(\int_0^\infty (A_\infty - A_{t-}) dX_t^*\Big) \\
&= \mathbb{E}\Big(\int_0^\infty Z_t dX_t^*\Big) \\
&\leq 2\mathbb{E}(X_\infty^*)
\end{aligned}$$

which finishes the proof in the case $p = 1$. For the case $1 \leq p \leq \infty$ one applies Jensen's inequality as in the proof of Lemma 1 above to obtain

$$\|Y^*\|_p \leq 2^{\frac{1}{p}} \|X^*\|_p.$$

References

[B]. J. Bourgain, *Embedding L^1 into L^1/H^1*, TAMS (1983), p.689-702.

[B-P]. A. Benedek, R. Panzone, *The spaces L^p with Mixed Norms*, Duke Math. J. **28** (1961), 301-324.

[Bu]. D. Burkholder, *Distribution Function Inequalities for Martingales*, Annals of Probability 1 (1973), 19 – 42.

[D-S]. F. Delbaen, W. Schachermayer, *A General Version of the Fundamental Theorem of Asset Pricing, submitted.*

[D]. S. J. Dilworth, *Some probabilistic inequalities with applications to functional analysis*, Banach Spaces (Bor-Luh Lin, W. B. Johnson ed.), Contemp. Math., AMS (1992).

[G]. A. M. Garsia, *Martingale Inequalities*, W. A. Benjamin, Reading, Mass., 1973.

[J]. T. Jeulin, *Semimartingales et Grossissement d'une Filtration*, Springer LNM **833** (1980).

[J-M-S-T]. W.B. Johnson, B. Maurey, G. Schechtman, L. Tzafriri, *Symmetric structures in Banach spaces*, Mem. AMS no. **217** (1979), vol. 19.

[K-W]. S. Kwapień, W. Woyczyński, *Random Series and Stochastic Integrals: Single and Multiple*, Birkhäuser, Boston-Basel-Berlin, 1992.

[L]. D. Lépingle, *Une inégalité de martingales*, Sém. de Proba XII, Springer LNM **649** (1978), 134-137.

[S]. E.M. Stein, *Topics in Harmonic Analysis*, Ann. of Math. Studies **63**, **Princeton Univ. Press** (1970).

[Y]. M. Yor, *Inégalités entre processus minces et applications*, CRAS Paris **286**, **Serie A** (1978), 799-801.

Department of Mathematics — Institute of Actuarial Sciences Vrije Universiteit Brussel, Pleinlaan 2, B-1050 Brussel

Institut für Statistik der Universität Wien, Brünnerstrasse 72, A-1210 Wien, Austria.

E-mail: FDELBAEN @ TENA2.VUB.AC.BE, wschach@stat1.bwl.univie.ac.at

A MARTINGALE PROOF OF THE KHINCHIN ITERATED LOGARITHM LAW FOR WIENER PROCESSES

N.V. KRYLOV

Let w_t be a d-dimensional Wiener process. Our goal here is to give a martingale proof of the following celebrated Khinchin log log law:

Theorem 1. *With probability one*

$$\limsup_{t\to\infty} \frac{|w_t|}{\sqrt{2t\log\log t}} = 1.$$

The "standard" proof of this result can be found in many places (cf. for instance [1]). Our martingale proof is based on formula (4), which might be interesting by itself. In particular, this formula allows us to give a rather short proof of the Kolmogorov–Petrovskii criterion allowing one to recognize when a given function $\alpha(t)$ is an "upper" or "lower" function.

Let $\alpha(t)$ be a strictly positive continuously differentiable function on $[0,\infty)$. For $t>0$, $x\in\mathbb{R}^d$ define

$$h(t,x) = \frac{1}{t^{d/2}} e^{|x|^2/(2t)}, \quad \beta(t) = \frac{1}{t^{d/2}} e^{\alpha^2(t)/(2t)}, \quad \gamma(t) = (\frac{1}{\beta(t)})',$$

$$\tau(t) = \inf\{s\ge t : |w_s| \ge \alpha(s)\}.$$

Lemma 2. *For any $\varepsilon>0$ the process*

$$m_\varepsilon(t) := h(t+\varepsilon, w_t)\beta^{-1}(t+\varepsilon) - \int_0^t h(r+\varepsilon, w_r)\gamma(r+\varepsilon)\,dr$$

is a martingale.

To prove the lemma it suffices to observe that for $0\le s\le t$ we have

$$\beta^{-1}(t+\varepsilon) - \beta^{-1}(s+\varepsilon) = \int_s^t \gamma(r+\varepsilon)\,dr,$$

then take a normal $(0,(t+\varepsilon)I)$ variable ξ independent of w and notice that

$$E\{h(t+\varepsilon,w_t)|\mathcal{F}_s^w\} = Eh(t+\varepsilon, x+\xi\left(\frac{t-s}{t+\varepsilon}\right)^{1/2})|_{x=w_s} =$$

$$\frac{1}{(2\pi)^{d/2}(t+\varepsilon)^d}\int_{\mathbb{R}^d} \exp\frac{1}{2(t+\varepsilon)}[|w_s + y\left(\frac{t-s}{t+\varepsilon}\right)^{1/2}|^2 - |y|^2]\,dy,$$

$$\frac{1}{2(t+\varepsilon)}[|w_s + y\left(\frac{t-s}{t+\varepsilon}\right)^{1/2}|^2 - |y|^2] =$$

$$\frac{1}{2(s+\varepsilon)}|w|_s^2 - \frac{1}{2(t+\varepsilon)}|y\left(\frac{s+\varepsilon}{t+\varepsilon}\right)^{1/2} + w_t\left(\frac{t-s}{s+\varepsilon}\right)^{1/2}|^2,$$

$$E\{h(t+\varepsilon,w_t)|\mathcal{F}_s^w\} =$$

The work was partially supported by NSF Grant DMS-9302516

$$\frac{1}{(2\pi)^{d/2}(t+\varepsilon)^d}\int_{\mathbf{R}^d}\exp[\frac{1}{2(s+\varepsilon)}|w|_s^2-\frac{s+\varepsilon}{2(t+\varepsilon)^2}|y|^2]\,dy=h(s+\varepsilon,w_s)\ (a.s.).$$

Corollary 3. *If $0<s<T<\infty$, then*

$$P\{s<\tau(s)\le T\}+E\frac{1}{\beta(T)}h(T,w_T)I_{\tau(s)>T}=$$

$$\frac{\alpha^d(s)}{\beta(s)s^d}\kappa_d+E\int_s^T h(r,w_r)\gamma(r)I_{\tau(s)>r}\,dr, \tag{1}$$

where κ_d is $(2\pi)^{-d/2}$ times the volume of the unit sphere.

Indeed, $Em_\varepsilon(s)I_{\tau(s)>s}=Em_\varepsilon(\tau(s)\wedge T)I_{\tau(s)>s}$, which means that

$$E\frac{h(s+\varepsilon,w_s)}{\beta(s+\varepsilon)}I_{\tau(s)>s}+E\int_s^{\tau(s)\wedge T}h(r+\varepsilon,w_r)\gamma(r+\varepsilon)\,dr=$$

$$E\frac{h(\tau(s)\wedge T+\varepsilon,w_{\tau(s)\wedge T})}{\beta(\tau(s)\wedge T+\varepsilon)}I_{\tau(s)>s}.$$

It remains only to let here $\varepsilon\downarrow 0$ and apply the dominated convergence theorem along with the observations that $|w_r|\le\sup_{r\le T}\alpha(r)$ for $s<r\le\tau(s)\wedge T$,

$$E\frac{h(s,w_s)}{\beta(s)}I_{\tau(s)>s}=E\frac{h(s,w_s)}{\beta(s)}I_{|w_s|<\alpha(s)}=$$

$$\frac{1}{\beta(s)s^d(2\pi)^{d/2}}\int_{|x|<\alpha(s)}dx=\frac{\alpha^d(s)}{\beta(s)s^d}\kappa_d=\kappa_d[\frac{\alpha^2(s)}{s}]^{d/2}\exp[-\frac{\alpha^2(s)}{2s}], \tag{2}$$

$$\frac{h(\tau(s)\wedge T,w_{\tau(s)\wedge T})}{\beta(\tau(s)\wedge T)}I_{\tau(s)>s}=\frac{1}{\beta(T)}h(T,w_T)I_{\tau(s)>T}+$$

$$\frac{1}{\beta(\tau(s))}h(\tau(s),w_{\tau(s)})I_{T\ge\tau(s)>s}=\frac{1}{\beta(T)}h(T,w_T)I_{\tau(s)>T}+I_{T\ge\tau(s)>s}.$$

By letting $T\to\infty$ in (1) and by applying the monotone convergence theorem, relations (2) and the fact that $\{\tau(s)>T\}\subset\{|w_T|<\alpha(T)\}$ we immediately get the first assertion of the following lemma

Lemma 4. *Define $f(t)=\kappa_d[\frac{\alpha^2(t)}{t}]^{d/2}\exp[-\frac{\alpha^2(t)}{2t}]$, let $\gamma(t)\ge 0$ for large t and let*

$$\lim_{t\to\infty}\frac{\alpha(t)}{\sqrt{t}}=\infty\ (\lim_{t\to\infty}f(t)=0). \tag{3}$$

Then for any $s>0$

$$P\{s<\tau(s)<\infty\}=f(s)+\int_s^\infty Eh(r,w_r)\gamma(r)I_{\tau(s)>r}\,dr; \tag{4}$$

$$\int^\infty r^{-d/2}\gamma(r)\,dr=\infty\implies\limsup_{t\to\infty}[|w_t|-\alpha(t)]\ge 0\ \text{(a.s.)}; \tag{5}$$

$$\int^\infty \alpha^d(r)r^{-d}\gamma(r)\,dr<\infty\implies\limsup_{t\to\infty}[|w_t|-\alpha(t)]\le 0\ \text{(a.s.)}. \tag{6}$$

Proof. To prove (5) notice that $h(r,w_r) \geq r^{-d/2}$, so that

$$\int_s^\infty Eh(r,w_r)\gamma(r)I_{\tau(s)>r}\,dr \geq \int_s^\infty r^{-d/2}\gamma(r)\,dr\,P\{\tau(s)=\infty\},$$

and under the condition in (5) we have $P\{\tau(s)=\infty\}=0$. It remains only to observe that

$$\{\omega : \limsup_{t\to\infty}[|w_t| - \alpha(t)] < 0\} \subset \bigcup_{n=1}^{\infty}\{\omega : \tau(n)=\infty\}.$$

To prove (6) we use (2) and that $\{\tau(s)>r\} \subset \{|w_r| < \alpha(r)\}$ if $r > s$. Then

$$\int_s^\infty Eh(r,w_r)\gamma(r)I_{\tau(s)>r}\,dr \leq \kappa_d \int_s^\infty \alpha^d(r)r^{-d}\gamma(r)\,dr,$$

and from (4) we see that under the condition in (6), $P\{s<\tau(s)<\infty\}\to 0$ as $s\to\infty$. Finally, for any $s>0$

$$\{\omega : \limsup_{t\to\infty}[|w_t| - \alpha(t)] > 0\} \subset \{\omega : |w_s| \geq \alpha(s)\} \cup \{\omega : s<\tau(s)<\infty\},$$

$$P\{\limsup_{t\to\infty}[|w_t| - \alpha(t)] > 0\} \leq \lim_{s\to\infty} P\{|w_s| \geq \alpha(s)\} = \lim_{s\to\infty} P\{|w_1| \geq \frac{\alpha(s)}{\sqrt{s}}\} = 0.$$

The lemma is proved.

Proof of Theorem 1. Take $\varepsilon \in [0,1)$ and define $\alpha(t) = ((1+\varepsilon)2t\log\log t)^{1/2}$ if $t \geq 10$ and for $t < 10$ define $\alpha(t)$ in any way just to get a positive differentiable function on $[0,\infty)$. Then $\beta(t) = t^{-d/2}(\log t)^{1+\varepsilon}$, $\alpha(t)/\sqrt{t} \to \infty$ for any ε, and as easy to check

$$\int^\infty r^{-d/2}\,d\frac{1}{\beta(r)} = \infty \text{ if } \varepsilon = 0, \quad \int^\infty \alpha^d(r)r^{-d}\,d\frac{1}{\beta(r)} < \infty \text{ if } \varepsilon > 0;$$

from Lemma 4 it follows that

$$\limsup_{t\to\infty}\frac{|w_t|}{(2t\log\log t)^{1/2}} \geq 1, \quad \limsup_{t\to\infty}\frac{|w_t|}{(2t\log\log t)^{1/2}} \leq 1+\varepsilon \text{ (a.s.) if } \varepsilon > 0.$$

The theorem is proved.

Next observe that

$$\int^\infty \frac{\alpha^d(r)}{r^d}\gamma(r)\,dr = \int^\infty \frac{\alpha^d(r)}{r^d}\,d(r^{d/2}\exp[-\frac{\alpha^2(r)}{2r}]) =$$

$$\frac{d}{2}\int^\infty \frac{\alpha^d(r)}{r^{d/2}}\exp[-\frac{\alpha^2(r)}{2r}]\frac{dr}{r} - \frac{1}{2}\int^\infty \frac{\alpha^d(r)}{r^{d/2}}\exp[-\frac{\alpha^2(r)}{2r}]\,d\frac{\alpha^2(r)}{r},$$

and under condition (3) the integral

$$\int^\infty \frac{\alpha^d(r)}{r^{d/2}}\exp[-\frac{\alpha^2(r)}{2r}]\,d\frac{\alpha^2(r)}{r} = \int^\infty x^{d/2}e^{-x/2}\,dx$$

is finite, so that the integrals

$$\int^\infty \frac{\alpha^d(r)}{r^d}\gamma(r)\,dr, \quad \int^\infty \frac{\alpha^d(r)}{r^{d/2}}\exp[-\frac{\alpha^2(r)}{2r}]\frac{dr}{r}$$

converge or diverge simultaneously.

Now we see that the statement (6) in Lemma 4 implies the second statement in the following Kolmogorov–Petrovskii criterion.

Theorem 5. *Assume that $t^{d/2}\exp[-\alpha^2(t)/(2t)]$ and $t\alpha^{-1}(t)$ increase for large t, and assume (3). Then*

$$\int^{\infty}\frac{\alpha^d(r)}{r^{d/2}}\exp[-\frac{\alpha^2(r)}{2r}]\frac{dr}{r}=\infty \implies \limsup_{t\to\infty}\frac{|w_t|}{\alpha(t)}\geq 1 \text{ (a.s.)}; \tag{7}$$

$$\int^{\infty}\frac{\alpha^d(r)}{r^{d/2}}\exp[-\frac{\alpha^2(r)}{2r}]\frac{dr}{r}<\infty \implies \limsup_{t\to\infty}\frac{|w_t|}{\alpha(t)}\leq 1 \text{ (a.s.)}.$$

Proof. We only need to prove (7), and to do this we come back to statement (4) of Lemma 4 but analyze it slightly more carefully. Notice that for any given $r>0$ the process $w_t - w_r t/r$, $t\in[0,r]$, and the random vector w_r are independent. Therefore for $r>s$

$$Eh(r,w_r)I_{\tau(s)>r}=\frac{1}{(2\pi r)^{d/2}}\int_{|x|<\alpha(r)}e^{-|x|^2/(2r)}h(r,x)P\{\tau(s)>r|w_r=x\}\,dx=$$

$$\frac{1}{(2\pi)^{d/2}r^d}\int_{|x|<\alpha(r)}P\{\tau(s)>r|w_r=x\}\,dx=$$

$$\frac{1}{(2\pi)^{d/2}r^d}\int_{|x|<\alpha(r)}P\{\sup_{t\in[s,r]}|w_t-\frac{t}{r}w_r+\frac{t}{r}x|\alpha^{-1}(t)<1\}\,dx=$$

$$\frac{\alpha^d(r)}{(2\pi)^{d/2}r^d}\int_{|x|<1}P\{\sup_{t\in[s,r]}|w_t-\frac{t}{r}w_r+\frac{t}{r}\alpha(r)x|\alpha^{-1}(t)<1\}\,dx.$$

From (4) it now follows that if $\int^{\infty}\alpha^d(r)r^{-d}\gamma(r)\,dr=\infty$, then for any $s>0$

$$\liminf_{r\to\infty}\int_{|x|<1}P\{\sup_{t\in[s,r]}|w_t-\frac{t}{r}w_r+\frac{t}{r}\alpha(r)x|\alpha^{-1}(t)<1\}\,dx=0. \tag{8}$$

Now observe that for $\varepsilon\in(0,1)$, $|x|\leq\varepsilon$ and s so large that $t\alpha^{-1}(t)$ increases for $t\geq s$, we have $(t/r)\alpha(r)|x|\alpha^{-1}(t)\leq\varepsilon$,

$$P\{\sup_{t\in[s,r]}|w_t-\frac{t}{r}w_r+\frac{t}{r}\alpha(r)x|\alpha^{-1}(t)<1\}\geq$$

$$P\{\sup_{t\in[s,r]}|w_t-\frac{t}{r}\alpha(r)[w_r\alpha^{-1}(r)]|\alpha^{-1}(t)<1-\varepsilon\}\geq$$

$$P\{\sup_{t\in[s,r]}|w_t-\frac{t}{r}\alpha(r)[w_r\alpha^{-1}(r)]|\alpha^{-1}(t)<1-\varepsilon,\ |w_r|\alpha^{-1}(r)\leq\varepsilon\}\geq$$

$$P\{\sup_{t\in[s,r]}\frac{|w_t|}{\alpha(t)}<1-2\varepsilon\}-P\{|w_r|>\varepsilon\alpha(r)\}.$$

Since $\alpha(r)/\sqrt{r}\to\infty$, the last probability tends to zero as $r\to\infty$, and from (8) it follows that for any $\varepsilon\in(0,1)$ we have

$$P\{\sup_{t\geq s}\frac{|w_t|}{\alpha(t)}<1-2\varepsilon\}\leq\lim_{r\to\infty}P\{\sup_{t\in[s,r]}\frac{|w_t|}{\alpha(t)}<1-2\varepsilon\}=0.$$

This obviously yields the first assertion of our theorem, which is thus proved.

Remark 6. From the zero-one law it follows easily that for one-dimensional Wiener process B_t

$$\limsup_{t\to\infty}\frac{B_t}{\alpha(t)}\leq(\geq)1 \text{ (a.s.)} \implies \limsup_{t\to\infty}\frac{|w_t|}{\alpha(t)}\leq(\geq)1 \text{ (a.s.)}.$$

Therefore, consideration of arbitrary d does not yield any advantage, though it actually might happen that for $d = 1$ the integral in (7) converges and, say for $d = 2$ diverges. In this case lim sup in (7) simply equals 1 (a. s.).

Acknowledgment The author is sincerely grateful to P.-A. Meyer and M. Yor for comments and discussions related to the article.

References

1. D. Revuz, M. Yor, *Continuous martingales and Brownian motion*, Springer, New York etc., 1991.

127 Vincent Hall, University of Minnesota, Minneapolis, MN, 55455
E-mail address: krylov@math.umn.edu

Intertwining of Markov semi-groups, some examples.

Philippe Biane

Abstract: We give a general group theoretic scheme for obtaining intertwining relations between Markov semi-groups. This is applied to the Heisenberg group and yields an intertwining relation between the Ornstein-Uhlenbeck and the Yule semi-groups.

1. Introduction.

In this paper we would like to point out the relevance of group theory in the problem of finding intertwining relations between Markov semi-groups, which has been considered recently in [C-P-Y].

Let $(P_t)_{t\in\mathbb{R}_+}$ and $(Q_t)_{t\in\mathbb{R}_+}$ be two Markov semi-groups on the spaces $(E,\mathcal{E})$ and $(F,\mathcal{F})$ respectively. We say that a Markov kernel $\Lambda : (E,\mathcal{E}) \to (F,\mathcal{F})$ intertwines Q and P if one has the relation

$$Q_t\Lambda = \Lambda P_t$$

for all $t > 0$.

The intertwining of two Markov semi-groups is not a symmetric relation; indeed, if Λ intertwines Q and P, there is in general no Markov kernel $\Gamma : (F,\mathcal{F}) \to (E,\mathcal{E})$ such that

$$\Gamma Q_t = P_t\Gamma$$

for all $t > 0$. We refer to [C-P-Y] for a further discussion of the probabilistic properties of the intertwining relation between semi-groups.

There is a general framework which will allow us to build many examples of such intertwining relations, and which we will describe below. The important point in this approach is that we will have to enlarge the notion of Markov semi-group to a non-commutative context, in order to obtain in a natural way intertwinings between purely "commutative" Markov semi-groups.

Let A be a C^*-algebra, and let $(T_t)_{t\in\mathbb{R}_+}$ be a semi-group of completely positive linear contractions of A (recall that a linear T map between two C^*-algebras A and B is said to be completely positive if for any $n \geq 1$ the map $T\otimes Id : A\otimes M_n(\mathbb{C}) \to B\otimes M_n(\mathbb{C})$ is positive). Such a semi-group is a non-commutative analogue of a (sub)-markovian semi-group, indeed if A is a commutative C^*-algebra then it is isomorphic to the algebra $C_0(spec\ A)$ and the maps T_t come from a semi-group of sub-markovian kernels on the topological space *spec* A. In the quantum probabilist litterature, semi-groups of completely positive contractions are often called quantum dynamical semi-groups (see e.g. [P]).

Let now B and C be abelian C^*-subalgebras of A which are invariant under the semi-group $(T_t)_{t\in\mathbb{R}_+}$. The restrictions of this semi-group to these subalgebras define thus semi-groups of kernels $(P_t)_{t\in\mathbb{R}_+}$ and $(Q_t)_{t\in\mathbb{R}_+}$ on the spectra of B and C respectively. Denoting by i_B and i_C the inclusions of B and C in A respectively, we have the relations $T_t \circ i_B = i_B \circ P_t$ and $T_t \circ i_C = i_C \circ Q_t$.

In many instances it turns out that there exists a completely positive projection $\pi : A \to C$ such that $\pi \circ T_t = Q_t \circ \pi$ for all $t > 0$. Taking the composition $\pi \circ i_B$

we get a completely positive map $B \to C$ which is given by a kernel $\Lambda : spec\, B \to spec\, C$ verifying the intertwining relation

$$Q_t\Lambda = \Lambda P_t$$

between the semi-groups $(P_t)_{t\in\mathbb{R}_+}$ and $(Q_t)_{t\in\mathbb{R}_+}$.

In the following section we will give some constructions from group theory which will provide us with many examples of such intertwining relations. In section 3 we will treat explicitly an interesting case, related to the Heisenberg group. In this study we will obtain an intertwining relation between the Ornstein-Uhlenbeck semi-group and the Yule semi-group (the Yule process is a special type of Galton Watson process cf [A-N]).

2. Some groupe theoretic constructions.

2.1 Let G be a locally compact group, there are several algebras commonly associated to G. The convolution algebra $L^1(G)$ is a Banach $*$-algebra. Any weakly continuous unitary representation π of G gives rise to a $*$-representation of $L^1(G)$, and if we let $\|f\| = \sup_\pi |\pi(f)|$, where the supremum is over all weakly continuous unitary representations of G, we obtain a C^*-norm on $L^1(G)$. The completion of $L^1(G)$ for this norm is the C^*-algebra of G, called $C^*(G)$. Of some use is also the enveloping von Neumann algebra of G denoted by $U(G)$. As a Banach space, it is the bidual of $C^*(G)$.

Let φ be a continuous positive definite function on G, with $\varphi(e) \leq 1$. Then the map $\varphi \mapsto \varphi f$ (this is the pointwise product of functions on G) is a completely positive contraction of $L^1(G)$, which extends to a completely positive contraction of $C^*(G)$. Taking the bidual, we get also a normal completely positive contraction on $U(G)$.

If ψ is a continuous, conditionnally positive definite, function with $\psi(e) \leq 0$, then $(e^{t\psi})_{t\in\mathbb{R}_+}$ is a multiplicative semi-group of continuous, positive definite, functions on G and hence gives rise to a semi-group of completely positive contractions of $C^*(G)$. Some properties of such semi-groups have been discussed in [B].

2.2 Let G and ψ be as in 2.1, and denote by $(T_t)_{t\in\mathbb{R}_+}$ the semi-group of completely positive contractions on $C^*(G)$ (or on $U(G)$) generated by ψ.

Let $H \subset G$ be a closed abelian subgroup then there are natural inclusions $C^*(U) \subset U(H) \subset U(G)$, and the semi-group $(T_t)_{t\in\mathbb{R}_+}$ leaves these subalgebras invariant. In fact, its restrition to $C^*(H)$ (or $U(H)$) is the semi-group associated to the conditionnally positive definite function ψ restricted to H.

Let K be a compact group of automorphisms of G, it gives rise to a group of automorphisms of $C^*(G)$. Suppose that the subalgebra of elements of $C^*(G)$ fixed by K is commutative, and let us call $C^*_K(G)$ this subalgebra. If the function ψ is invariant under the automorphisms of K then this subalgebra is invariant by $(T_t)_{t\in\mathbb{R}_+}$. Furthermore the map $\Pi : x \mapsto \int_K \theta(x)d\theta$ where $d\theta$ is the normalized Haar measure on K, is a completely positive projection of $C^*(G)$ onto $C^*_K(G)$, and it verifies $\Pi \circ T_t = Q_t \circ \Pi$ where Q_t is the restriction of T_t to $C^*_K(G)$.

More generally, let (G', K') be a Gelfand pair, so that K' is a compact subgroup of G', such that the algebra $L^1(K'\backslash G'/K')$ of K'-biinvariant functions on G' is abelian. Let again $C^*_K(G')$ be the C^*-algebra generated by $L^1(K'\backslash G'/K')$, and suppose also that ψ is bi-invariant by K', then the semi-group $(T_t)_{t\in\mathbb{R}_+}$ leaves $C^*_K(G')$ invariant, and there is a completely positive projection of $C^*(G')$ onto $C^*_K(G')$ given by $\Pi : x \mapsto \int_{K'}\int_{K'} kxk'dkdk'$, verifiying again $\Pi\circ T_t = Q_t\circ\Pi$. This projection extends to a projection of the corresponding von Neumann algebras $\Pi : U(G') \to U_K(G')$, verifying the same identity.

The preceding example is a special case of this one since one can form the semi-direct product $K \times G$ and $(K \times G, K)$ is then a Gelfand pair.

2.3 Summarizing the preceding section, we see that we obtain an intertwining relation between Markov semi-groups as soon as we have the following data: a Gelfand pair (G, K), and a K-bi-invariant, continuous, conditionnally positive definite, function ψ, on G, with $\psi(e) \leq 0$. In this case, the restrictions of the semi-group T_t, generated by ψ, to the algebras $C^*(H)$, where H is an abelian subgroup of G, and $C^*_K(G)$ are given by submarkovian semi-groups of kernels on the spectra of these C^*-algebras, and the composition of the inclusion $i_H : C^*(H) \to U(G)$ and of the projection onto $U_K(G)$ is given by a kernel which intertwins the two semi-groups.

A simple example is obtained (in the notation of sect. 2.2) by taking $G = \mathbb{R}^d$, $H = \mathbb{R}^k$, K is the orthogonal group $O(d)$, and $\psi(x) = -\frac{1}{2}|x|^2$. In this case, the semi-group T_t is the brownian semi-group on $\mathbb{R}^d$ (considered here as the dual group of G), the semi-group obtained by restriction to $C^*_K(G)$ is the Bessel semi-group of dimension d, the restriction to $C^*(H)$ is the k-dimensional brownian semi-group and, we thus obtain an intertwining between the k-dimensional brownian motion and the d-dimensional Bessel process. We leave the detailed computation to the reader.

3. An example.

We will use the general framework developped in the preceding section to work out an explicit intertwining relation between Markov semi-groups.

Recall that the Heisenberg group is $H_1 = \mathbb{C} \times \mathbb{R}$ with the group law

$$(z, w) \star (z', w') = (z + z', w + w' + \Im m\, z\bar{z}')$$

The function $\psi(z, w) = iw - \frac{1}{2}|z|^2$ is a conditionnally positive definite function on H_1 and the semi-group of completely positive contractions of $C^*(H_1)$ that it induces is a non-commutative analogue of the heat semi-group. It has been studied in [B]. We denote it by $(T_t)_{t \in \mathbb{R}_+}$ in the rest of this section.

Let $\mathbb{R} \times \mathbb{R} \subset \mathbb{C} \times \mathbb{R}$ which is a maximal abelian subgroup of H_1. The restriction of the semi-group $(T_t)_{t \in \mathbb{R}_+}$ to the corresponding group C^*-algebra (as defined in 2.2) was seen in [B] to be the heat semi-group on $\mathbb{R} \times \mathbb{R}$.

There is a group of automorphisms of H_1 of the form $(z, w) \mapsto (\xi z, w)$ for $\xi \in \mathbb{C}$, $|\xi| = 1$. The subalgebra of $C^*(H_1)$ consisting of elements fixed by this group is abelian and the semi-group obtained by restriction of $(T_t)_{t \in \mathbb{R}_+}$ to this subalgebra was called "non-commutative Bessel semi-group" in [B], where the corresponding transition kernels were also computed. In view of sections 2.2, we can deduce an intertwining relation between this semi-group and the heat semi-group on $\mathbb{R} \times \mathbb{R}$. Instead of carrying out the computations for these semi-groups, we shall do it for related semi-groups obtained by a transformation analogous to the usual transformation yielding Ornstein-Uhlenbeck process from brownian motion. Recall that if $(B_t)_{t \in \mathbb{R}_+}$ is a brownian motion then the processes $(X_s = e^{\pm \frac{s}{2}} B_{e^{\mp s}})_{s \in \mathbb{R}_+}$ are Ornstein-Uhlenbeck processes. The corresponding construction for the Heisenberg group was effected in [B], using the scaling automorphisms of H_1 defined as $\alpha_c(z, w) = (cz, c^2 w)$ for all $c > 0$. Let us recall this construction.

Let $\pi_\pm$ be the two unitary representations of H_1 on $L^2(\mathbf{R})$ defined by

$$\pi_\pm(z,w)f(x) = f(x+p)e^{\pm i(w+pq+2qx)} \tag{3.1}$$

where $z = p + iq$. These representations extend to representations $\pi_\pm : C^*(H_1) \to \mathcal{B}(L^2(\mathbf{R}))$. In [B] we proved the following

3.1 Proposition. *There exists two semi-groups, $(R_s^+)_{s\in\mathbf{R}_+}$ and $(R_s^-)_{s\in\mathbf{R}_+}$, of completely positive contractions on $\mathcal{B}(L^2(\mathbf{R}))$, such that, for all $s \in \mathbf{R}_+$,*

$$\pi_+ \circ \alpha_{e^{-\frac{s}{2}}} \circ T_{1-e^{-s}} = R_s^+ \circ \pi_+ \tag{3.2}$$

$$\pi_- \circ \alpha_{e^{\frac{s}{2}}} \circ T_{e^s-1} = R_s^- \circ \pi_- \tag{3.3}$$

Let us consider the subalgebra of $\mathcal{B}(L^2(\mathbf{R}))$ consisting of multiplication operators by functions in $C_0(\mathbf{R})$. This algebra is invariant by the semi-groups $R^\pm$, indeed one has

3.2 Proposition. *For any $v \in C_0(\mathbf{R})$*

$$R_s^+v(x) = \sqrt{\frac{2}{\pi}}\int_{\mathbf{R}} v(y)\exp(-\frac{2(y-e^{-\frac{s}{2}}x)^2}{1-e^{-s}})\frac{dy}{\sqrt{1-e^{-s}}}$$

$$R_s^-v(x) = \sqrt{\frac{2}{\pi}}\int_{\mathbf{R}} v(y)\exp(-\frac{2(y-e^{\frac{s}{2}}x)^2}{e^s-1})\frac{dy}{\sqrt{e^s-1}}$$

Proof Let v be of the form $v(x) = \int_{\mathbf{R}} \eta(q)e^{2iqx}dq$ for some integrable function η. By (3.1) and (3.2) we have, for $f \in L^2(\mathbf{R})$

$$(\int_{\mathbf{R}} \eta(q)\pi_+(iq,0)f dq)(x) = v(x)f(x)$$

and so

$$R_s^+(v)f(x) = \int_{\mathbf{R}} e^{2iqe^{-\frac{s}{2}}x}e^{-\frac{1}{2}(1-e^{-s})|q|^2}\eta(q)f(x)dq$$

Using the formula

$$e^{-\frac{1}{2}(1-e^{-s})|q|^2} = \sqrt{\frac{2}{\pi}}\int_{\mathbf{R}} e^{2iqy}e^{-\frac{2y^2}{1-e^{-s}}}\frac{dy}{\sqrt{1-e^{-s}}}$$

and using Fubini's theorem we obtain that R_s^+v is the multiplication operator by the function

$$\sqrt{\frac{2}{\pi}}\int_{\mathbf{R}} v(e^{-\frac{s}{2}}x+y)e^{-\frac{2y^2}{1-e^{-s}}}\frac{dy}{\sqrt{1-e^{-s}}}$$

and a change of variable gives the required conclusion for R^+. The formula for general $v \in C_0(\mathbf{R})$ is obtained by taking uniform limits, and the case of R^- is treated similarly.

We recognize the restrictions of the two semi-groups $(R_s^\pm)_{s\in\mathbf{R}_+}$ to $C_0(\mathbf{R})$ as the usual Ornstein-Uhlenbeck semi-groups on $\mathbf{R}$, given by the following kernels with respect to

Lebesgue measure

$$p_s^+(x,y) = \frac{\sqrt{2}}{\sqrt{\pi(1-e^{-s})}} \exp(-(\frac{2(y-e^{-\frac{s}{2}}x)^2}{1-e^{-s}}))$$

and

$$p_s^-(x,y) = \frac{\sqrt{2}}{\sqrt{\pi(e^{s}-1)}} \exp(-(\frac{2(y-e^{\frac{s}{2}}x)^2}{e^{s}-1}))$$

We now turn to the Yule semi-group, which can be obtained from $(R_s^{\pm})_{s\in\mathbf{R}_+}$ by restriction to a suitable algebra.

Let us introduce the functions

$$\mathcal{E}(\alpha)(x) = (\frac{2}{\pi})^{\frac{1}{4}} e^{-2\alpha x - \frac{\alpha^2}{2} - x^2}$$

for any complex α. One has $\mathcal{E}(\alpha) \in L^2(\mathbf{R})$, and

$$\mathcal{E}(\alpha) = \sum_{n=0}^{\infty} (-\alpha)^n \frac{u_n}{n!}$$

where $u_n(x) = h_n(x)(\frac{2}{\pi})^{\frac{1}{4}} e^{-x^2}$ and the h_n are the Hermite polynomials. One has $|u_n|_{L^2}^2 = n!$. The representations $\pi_{\pm}$ are given by

$$\pi_+(z,w)\mathcal{E}(\alpha) = \mathcal{E}(\alpha+\bar{z})e^{iw-\alpha z-\frac{|z|^2}{2}} \tag{3.4}$$

$$\pi_-(z,w)\mathcal{E}(\alpha) = \mathcal{E}(\alpha+z)e^{-iw-\alpha\bar{z}-\frac{|z|^2}{2}} \tag{3.5}$$

The number operator, N, of $\pi_{\pm}$ is defined by the formula

$$Nu_n = nu_n$$

or alternatively, for all $\omega \in \mathbf{R}$,

$$e^{i\omega N}\mathcal{E}(\alpha) = \mathcal{E}(e^{i\omega}\alpha)$$

It turns out that the C^*-algebra generated by the spectral projections of the number operator is the image by $\pi_{\pm}$ of the algebra of elements of $C^*(H_1)$ invariant under the automorphisms $(z,w) \mapsto (\xi z, w)$, for $\xi \in U(1)$ and so it is stable under the semi-groups $R^{\pm}$. In [B] we computed the corresponding semi-groups of kernels on the spectrum of this algebra which is exactly the spectrum of N, the set $\mathbf{N}$. We obtained in [B] (corollaire 3.3.3) the following result.

3.3 Proposition. *The semi-groups of kernels on the spectrum of N induced by $R^{\pm}$ are given by the formulas*

$$q_s^+(k,l) = \frac{k!}{l!(k-l)!} e^{-ls}(1-e^{-s})^{k-l} \quad \text{for } 0 \le l \le k$$

$$q_s^-(k,l) = \frac{l!}{(l-k)!k!} e^{-ks}(1-e^{-s})^{l-k} \text{for } 0 \le k \le l$$

(beware that in [B] the formula for q^- is shifted by 1).

The semi-group of kernels q^- is the Yule semi-group (cf [A-N]).

Let us now show that there is an an intertwining between the kernels p^+ and q^+ on one hand and between p^- and q^- on the other, and compute explicitly this kernel. This is done in the following way.

3.4 Theorem. *There exists a kernel $\Lambda : \mathbf{R} \to \mathbf{N}$ which satisfies the interwining relations*

$$q_s^+ \circ \Lambda = \Lambda \circ p_s^+$$

and

$$q_s^- \circ \Lambda = \Lambda \circ p_s^-$$

This kernel is given by the following formula

$$\Lambda(n, dx) = \frac{1}{n!}\sqrt{\frac{2}{\pi}} h_n^2(x) e^{-2x^2} dx$$

Proof To any bounded operator A on $L^2(\mathbf{R})$ let us associate the diagonal operator $D(A)$ such that $D(A)(u_n) = \frac{<Au_n, u_n>}{n!} u_n$, alternatively, one has $D(A) = \sum_n \Pi_n A \Pi_n$ where the Π_n are the orthogonal projections on the lines $\mathbb{C}.u_n$. This defines a completely positive projection D from $\mathcal{B}(L^2(\mathbf{R}))$ onto the von Neumann algebra generated by the spectral projections of N and hence a completely positive map $D : \mathcal{B}(L^2(\mathbf{R})) \to l^\infty(spec\ N)$. We have the following

3.5 Lemma. *For any $s \in \mathbf{R}_+$ one has*

$$D \circ R_s^\pm = R_s^+ \circ D$$

Proof. The map D is also given by the formula

$$D(A) = \frac{1}{2\pi} \int_0^{2\pi} e^{i\omega N} A e^{-i\omega N} d\omega$$

From formulas (3.4) and (3.5) we deduce that, for all $(z, w) \in H_1$,

$$e^{i\omega N} \pi_\pm(z, w) e^{-i\omega N} = \pi_\pm(e^{\pm i\omega} z, w)$$

This implies that

$$\begin{aligned}
D \circ R_s^+(\pi_+(z, w)) &= \frac{1}{2\pi} \int_0^{2\pi} e^{i\omega N} \pi_+(ze^{-\frac{s}{2}}, we^{-s}) e^{-i\omega N} e^{(1-e^{-s})(iw - \frac{1}{2}|z|^2)} d\omega \\
&= \frac{1}{2\pi} \int_0^{2\pi} \pi_+(e^{i\omega} ze^{-\frac{s}{2}}, we^{-s}) e^{(1-e^{-s})(iw - \frac{1}{2}|z|^2)} d\omega \\
&= \frac{1}{2\pi} \int_0^{2\pi} \pi_+(e^{i\omega} ze^{-\frac{s}{2}}, we^{-s}) e^{(1-e^{-s})(iw - \frac{1}{2}|e^{i\omega} z|^2)} d\omega \\
&= R_s^+ \circ D(\pi_+(z, w))
\end{aligned}$$

The result follows then easily for R^+ and a similar computation works for R^-.

Let $i : C_0(\mathbb{R}) \to \mathcal{B}(L^2(\mathbb{R}))$ be the embedding by multiplication operators. It follows from lemma 3.5 that the composition $D \circ i$ is the required map from $C_0(\mathbb{R})$ to $l^\infty(\mathbb{N})$, given by the kernel Λ. So we obtain immediately for $v \in C_0(\mathbb{R})$.

$$\int_{\mathbb{R}} v(x)\Lambda(n, dx) = \frac{< i(v)u_n, u_n >}{n!}$$
$$= \frac{1}{n!}\sqrt{\frac{2}{\pi}} \int_{\mathbb{R}} v(x) h_n^2(x) e^{-2x^2} dx$$

Finally we get the right formula for $\Lambda(n, dx)$.

References.

[A-N] K. B. Athreya and P. E. Ney, *Branching processes, Springer verlag, Berlin, Heidelberg, New york, 1972.*

[B] P. Biane, *Quelques propriétés du mouvement brownien non-commutatif, preprint, Laboratoire de Probabilités, Université Paris VI, 1994.*

[C-P-Y] P. Carmona, F. Petit, M. Yor, *Beta-gamma random variables and intertwining relations between certain Markov processes, preprint, Laboratoire de Probabilités, Université Paris VI, 1994.*

[P] K. R. Parthasarathy, *An Introduction to Quantum Stochastic Calculus, Monographs in Mathematics, Vol 85, Birkhäuser, Basel, 1992.*

C.N.R.S. Laboratoire de Probabilités
Tour 56, 3e étage, Université Paris VI
4, place Jussieu, 75252, PARIS cedex 05
FRANCE

The author acknowledges partial support from the HUMAN CAPITAL AND MOBILITY programme, CONTRACT NUMBER: ERBCHRXCT930094

Some remarks on perturbed reflecting Brownian motion

Wendelin Werner

C.N.R.S. and University of Cambridge

0. Introduction

Let B denote a one-dimensional Brownian motion started from 0 and L its local time process at level 0. For fixed $\mu > 0$, the perturbed reflecting Brownian motion X is defined for all $t \geq 0$ by

$$X_t = |B_t| - \mu L_t.$$

It has aroused some interest in the last few years (see Le Gall-Yor [7], Yor [13], chapters 8 and 9, Carmona-Petit-Yor [2], Perman [8]). We are going to make a few remarks concerning this process and give short elementary proofs of some known results, such as the generalized Ray-Knight Theorems for X. Let us just stress that none of the results derived here is new, and that our modest aim is to shed a new light on them, which we hope can improve our understanding of these identities.

We now recall a few relevant facts: For all $a \in R$, $T_a = \inf\{t \geq 0;\ X_t = a\}$ will denote the hitting time of a by X. Except when $\mu = 1$, X is not Markovian; however, for $a > 0$, T_{-a} is the hitting time of a/μ by L and hence a stopping time for B. The strong Markov property then yields that the processes $(X_t, t \geq 0)$ and $(a + X_{t+T_{-a}}, t \geq 0)$ have the same law. We will refer to this property as the 'strong Markov property' for X.

Note also that for $\mu = 1$, Lévy's identity (that is: if $S_t = \sup_{s<t} B_s$, then the processes $(S, S - B)$ and $(L, |B|)$ have the same law) shows that X is in fact a Brownian motion.

1. A hitting time property

In [11], we used the following result: For all $a > 0$, $b > 0$,

$$P(T_{-a} < T_b) = \left(\frac{b}{a+b}\right)^{1/\mu}. \tag{1}$$

This is a generalization of the classical hitting time property for Brownian motion (which is in fact (1) for $\mu = 1$):

$$P(\sigma_{-a} < \sigma_b) = \frac{b}{a+b}, \tag{2}$$

where $\sigma_x = \inf\{t > 0,\ B_t = x\}$.

In [11], we derived (1) from the explicit law of L_{T_1} derived by Carmona-Petit-Yor [2] (corollary 3.4.1 there) (one has $P(T_{-a} < T_b) = P(L_{T_b} > a/\mu)$). As briefly pointed out in [2], the law of L_{T_1} (and therefore (1)) is in fact also a direct consequence of the explicit solution to Skorokhod's problem by Azéma and Yor [1] (see also exercise (5.9) chapter VI in Revuz-Yor [10]) in a very special case: One just has to compute the right-hand side of (5.9) in [10] for an affine function γ and then use Lévy's identity.
We now give an alternative elementary short proof of (1): First, for all $x \geq 1$ we put

$$g(x) = P(T_{1-x} < T_1).$$

For $x > 1$ and $y > 1$, one has immediately $T_{1-x} < T_{1-xy}$. The 'strong Markov property' at time T_{1-x} and the scaling property imply that

$$g(xy) = P(T_{1-x} < T_1)P(T_{x-xy} < T_x) = g(x)g(y).$$

Moreover, g is continuous decreasing on $[1,\infty)$ and $g(1) = 1$. Hence, for some fixed $c = c(\mu)$,

$$g(x) = x^{-c}. \tag{3}$$

It now remains to show that $c = 1/\mu$: We look at the asymptotic behaviour of

$$f(x) = P(T_{-1} > T_x) = 1 - g(1 + 1/x)$$

as $x \to \infty$. (3) implies that $f(x) = 1 - (1 + 1/x)^{-c} \sim c/x$ as $x \to \infty$. On the other hand, Lévy's identity implies that

$$P(\sigma_{-(x+1/\mu)} < \sigma_{1/\mu}) \leq f(x) \leq P(\sigma_{-x} < \sigma_{1/\mu}),$$

and consequently (using (2)), $f(x) \sim 1/(\mu x)$ as $x \to \infty$, and (1) follows.

2. The generalized second Ray-Knight Theorem as a consequence of (1)

In [2] (see also Yor [13], chapter 9), Carmona-Petit-Yor have derived a generalized second Ray-Knight Theorem for the local times of X (Theorem 3.3 in [2], Theorem 9.1 in [13]; we refer to Yor [13], chapter 3 or Revuz-Yor [10], Chapter XI for the Ray-Knight Theorems for Brownian motion). They then derive the law of L_{T_1} (which

implies (1)) as a consequence of this Theorem. We now briefly point out how this generalized second Ray-Knight Theorem for X can in fact be derived 'backwards', as a consequence of (1), using a general result of Lamperti [5] on semi-stable Markov processes we first recall.
Suppose $(Y_t, t \geq 0)$ is a non-deterministic continuous Markov process in $[0, \infty)$, started from $x \in [0, \infty)$ under the probability measure P_x. Suppose furthermore that Y is semi-stable of index 1 (in the sense of [5]), that is $(c^{-1}Y_{ct}, t \geq 0)$ under P_x and $(Y_t, t \geq 0)$ under $P_{x/c}$ have the same law for all $c > 0$. Then, Theorem 5.1 in Lamperti [5] implies that Y is a multiple of a squared Bessel process (of index $\delta \in R$), which is either absorbed or reflected at 0. This result is the key to our approach.

We now put down some notation and state the generalized second Ray-Knight Theorem. Let ℓ_t^a denote the local time of X at level a and time t, and let τ denote the right-continuous inverse process of ℓ^0. Then:

Theorem (Carmona-Petit-Yor). *The processes* $(\ell_{\tau_1}^a, a \geq 0)$ *and* $(\ell_{\tau_1}^{-a}, a \geq 0)$ *are independent and:*
(i) $(\ell_{\tau_1}^a, a \geq 0)$ *is a squared Bessel process of dimension 0 started from 1 and absorbed at 0.*
(ii) $(\ell_{\tau_1}^{-a}, a \geq 0)$ *is a squared Bessel process of dimension* $2 - 2/\mu$ *started from 0 and absorbed at 0.*

Let $A^+(t) = \int_0^t 1_{\{X_s > 0\}} ds$ and $A^-(t) = \int_0^t 1_{\{X_s < 0\}} ds$. Let also σ^+ (respectively σ^-) denote the right-continuous inverse of A^+ (resp. A^-). We put for all $u \geq 0$, $X_u^+ = X_{\sigma_u^+}$ and $X_u^- = X_{\sigma_u^-}$. In other words and loosely speaking: X^+ (resp. X^-) is obtained by glueing the positive (resp. negative) excursions of X together. Then:

Lemma *The two-processes* X^+ *and* X^- *are independent. Moreover* X^+ *is a reflected Brownian motion.*

There are various possible proofs of this lemma. Yor ([13], Chapter 8) indicates a proof based upon Knight's Theorem on orthogonal martingales. Mihael Perman suggested an excursion-theoretical approach The last part of this note provides yet another possible justification.

This lemma shows immediately that $(\ell_{\tau_1}^a, a \geq 0)$ and $(\ell_{\tau_1}^{-a}, a \geq 0)$ are independent; (i) then follows from the second Ray-Knight Theorem for Brownian motion (it actually also follows from (ii) with $\mu = 1$). It remains to show (ii).
The 'Markov property' for X and the lemma show that $(\ell_{\tau_1}^{-a}, a \geq 0)$ is a Markov process (one just has to apply the Lemma to $(a + X_{T_{-a}+t}, t \geq 0)$). As X is a continuous semi-martingale, Theorem (1.7) in Chapter VI of Revuz-Yor [10] yields that $(\ell_{\tau_1}^{-a}, a \geq 0)$ is continuous. The scaling property for B (which is also the scaling property for X) implies that $(\ell_{\tau_1}^{-a}, a \geq 0)$ is a semi-stable Markov process of index 1

in the sense of Lamperti [5]. Hence, Lamperti's result mentionned at the beginning of this section shows that $(\ell_{\tau_1}^{-a}, a \geq 0)$ is a multiple of a squared Bessel process Y. Let δ denote its dimension and $y = Y_0$. Y is absorbed at 0 since otherwise, $\ell_{\tau_1}^{-a}$ is not identically 0 for all sufficiently large a. It remains to identify δ and y, which can be done using section 1: As ℓ^0 increases,

$$P(T_{-a} < T_b) = P(\ell_{T_{-a}}^0 < \ell_{T_b}^0).$$

But $\ell_{T_b}^0$ depends only on X^+ whereas $\ell_{T_{-a}}^0$ depends only on X^-; hence, these two random variables are independent. It is well-known that $\ell_{T_b}^0$ is an exponential random variable of parameter $1/2b$ (see e.g. Proposition (4.6), Chapter 6 in Revuz-Yor [10]). Consequently, if ξ denotes an exponential random variable of parameter $\lambda = 1/(2b)$, if ρ denotes the hitting time of 0 by Y and Z_γ a Gamma-random variable of index $\gamma > 0$ (that is with density $z^{\gamma-1}e^{-z}/\Gamma(\gamma)$ on R_+),

$$\begin{aligned} E(e^{-\lambda/\rho}) &= P(1/\rho < \xi) = P(\inf(X_s, s \leq \tau_1) < -\frac{1}{\xi}) \\ &= P(\ell_{T_{-1}}^0 < \ell_{T_b}^0) = \left(\frac{b}{1+b}\right)^{1/\mu} = (1+2\lambda)^{-1/\mu} \end{aligned} \tag{4}$$

which is the Laplace transform of $(2Z_{1/\mu})$. Hence, ρ has the same law as $1/(2Z_{1/\mu})$. On the other hand, if $\rho(\alpha, x)$ is the hitting time of 0 by a squared Bessel process of dimension $2 - 2\alpha$ started from $x > 0$, then it is well-known that $\rho(\alpha, x)$ has the same law as $x^2/(2Z_\alpha)$ (one can for instance compare the Laplace transforms, using the results of Kent [4], and equation (15) section 6.22 in Watson [12]; alternatively, one can note by time-reversal that $\rho(\alpha, x)$ is the last passage time at x by a Bessel process of index $2 + 2\alpha$ started from 0, and use the results of Getoor [3], see also Yor [14]). Hence, $y = 1$ and $\delta = 2 - 2/\mu$, which completes the proof of the theorem.

3. The generalized first Ray-Knight Theorem

We now briefly point out how the same approach also yields the generalization of the first Ray-Knight Theorem for perturbed reflecting Brownian motion derived by Le Gall-Yor [6] (see also Yor [13], Section 3.3). However, in this case, the original proofs are shortish anyway.

Theorem (Le Gall-Yor). *$(\ell_{T_{-1}}^{-1+a}, 0 \leq a \leq 1)$ is a squared Bessel process of dimension $2/\mu$ started from 0 and reflected at 0.*

By time-reversal (since $(B_t, t \leq T_{-1})$ and $(B_{T_{-1}-t}, t \leq T_{-1})$ have the same law), one can consider the process $\tilde{X}_u = |B_u| + \mu L_u$ and its local times taken at infinite time: $\tilde{\ell}^a = \ell_\infty^a(\tilde{X})$, for $a \geq 0$, and remark that $(\ell_{T_{-1}}^{-1+a}, 0 \leq a \leq 1)$ and $(\tilde{\ell}^a, 0 \leq a \leq 1)$ have the same law. As previously, $\tilde{\ell}$ is a continuous Markov process, which is self-similar of index 1 because of the scaling property of $\tilde{X}$. $\tilde{\ell}$ is henceforth (using again

Theorem 5.1 in Lamperti [5]) a multiple of a squared Bessel process β of dimension δ: $\tilde{\ell} = \alpha\beta$, with $\alpha > 0$. This time, β has to be reflected at 0, since almost surely, for all rational $a > 0$, $\tilde{\ell}^a \neq 0$. We now identify δ and α using (4). On the one hand, one has for all $\lambda > 0$ (see e.g. Revuz-Yor [10], line before Corollary (1.4) in Chapter XI):

$$E(e^{-\lambda\alpha\beta_1}) = (1+2\lambda\alpha)^{-\delta/2}.$$

On the other hand, (4) implies that (using the same notations ξ, λ as in (4)),

$$E(e^{-\lambda\alpha\beta_1}) = P(\xi > \alpha\beta_1) = P(\ell^0_{T_{1/(2\lambda)}} > \ell^0_{T_{-1}}) = P(T_{1/(2\lambda)} > T_{-1}) = (1+2\lambda)^{-1/\mu}$$

and the Theorem follows.

4. The discrete approach

We now mention an approximation of X by a random walk, which converges towards perturbed reflecting Brownian motion as the simple random walk does towards Brownian motion. We define $(S_n, n \geq 0)$ as follows: We fix $\mu > 0$ and we put $q = (1+\mu)^{-1} \in (0,1)$.

Let $I_n = \min\{S_0, S_1, \ldots, S_n\}$ and $S_0 = 0$. By induction, for all $n \geq 0$, if $S_0, \ldots, S_n$ are defined, then the law of S_{n+1} is the following:

$$P(S_{n+1} = S_n + 1) = P(S_{n+1} = S_n - 1) = 1/2 \text{ if } S_n \neq I_n$$

and

$$P(S_{n+1} = S_n + 1) = q, \; P(S_{n+1} = S_n - 1) = 1 - q \text{ if } S_n = I_n.$$

Using for instance Lévy's identity and Proposition 2 page 137-138 in Révész [4], one can show that the processes

$$(n^{-1/2} S_{[nt]}, t \in [0,1])$$

converge weakly towards $(X_t, t \in [0,1])$ as $n \to \infty$, where $[x]$ denotes the integer part of x. With little extra work, this approach provides another possible proof of the Lemma, since the independence of the positive and negative parts of the random walk $(S_n, n \geq 0)$ is trivial. Equation (1) can also be deduced, since (if $N_p = \inf\{n \geq 0, S_n = p\}$), $P(N_{-1} < N_p)$ and consequently $P(N_{-p'} < N_p)$ and $P(N_{-[ap]} < N_{[bp]})$ can be very easily explicitely computed, when $p > 0$, $p' > 0$, $a > 0$, $b > 0$: Indeed, it is a good undergraduate exercise to see that

$$P(N_{-1} < N_p) = (1-q) \sum_{k \geq 0} \left(\frac{q(p-1)}{p} \right)^k = (1 + 1/(\mu p))^{-1},$$

and consequently as $P(N_{-p'} < N_p) = P(N_{-1} < N_p) P(N_{-1} < N_{p+1}) \ldots P(N_{-1} < N_{p+p'-1})$,

$$\log P(N_{-[ap]} < N_{[bp]}) = - \sum_{k=[bp]}^{k=[ap]+[bp]-1} \log(1 + 1/(\mu k)) \sim \frac{1}{\mu} \log \left(\frac{b}{a+b} \right)$$

as $p \to \infty$, which yields (1).

Aknowledgements. I owe many thanks to Mihael Perman and Marc Yor for important references, remarks and enlightening discussions. I also thank Davar Khoshnevisan for pointing out reference [9].

Note added in proof. I would like to mention the two recent preprints by Burgess Davis [15] and Darryl Nester [16], which are very closely related with Section 4 of this note.

References

[1] Azéma, J., Yor, M.: Une solution simple au problème de Skorokhod, in: Séminaire de Probabilités XIII, Lecture Notes in Mathematics 721, Springer, pages 90-115 (1979)

[2] Carmona, P., Petit, F., Yor, M.: Some extensions of the Arcsine law as partial consequences of the scaling property for Brownian motion, Probab. Theory Relat. Fields **100**, 1-29 (1994)

[3] Getoor, R.K.: The Brownian escape process, Ann. Prob. **7**, 864-867 (1979)

[4] Kent, J.: Some probabilistic properties of Bessel functions, Ann. Probab. **6**, 760-770 (1978)

[5] Lamperti, J.: Semi-stable Markov processes. I, Z. Wahrscheinlichkeitstheorie verw. Geb. **22**, 205-225 (1972)

[6] Le Gall, J.F., Yor, M.: Excursions browniennes et carrés de processus de Bessel, C. R. Acad. Sci. Paris, Série I, **303**, 73-76 (1986)

[7] Le Gall, J.F., Yor, M.: Enlacements du mouvement brownien autour des courbes de l'espace, Trans. Amer. Math. Soc. **317**, 687-722 (1990)

[8] Perman, M.: An excursion approach to Ray-Knight theorems for perturbed reflecting Brownian motion, preprint (1994).

[9] P. Révész: Local time and invariance, in: Analytical methods in Probability (D. Dugué, E. Lukacs, V.K. Rohatgi ed.), Lecture Notes in Math. 861, 128-145 (1981)

[10] Revuz, D., Yor, M.: Continuous martingales and Brownian motion, 2nd edition, Springer, 1994.

[11] Shi, Z., Werner, W.: Asymptotics for occupation times of half-lines by stable processes and perturbed reflecting Brownian motion, Stochastics, to appear (1995)

[12] Watson, G.N.: Theory of Bessel functions, Cambridge University Press, 1922.

[13] Yor, M.: Some aspects of Brownian motion, Part I: Some special functionals, Lecture Notes ETH Zürich, Birkhäuser, Basel, 1992.

[14] Yor, M.: Sur certaines fonctionnelles exponentielles du mouvement brownien réel, J. Appl. Prob. **29**, 202-208 (1992)

[15] Davis, B.: Path convergence of random walk partly reflected at extrema, Technical report #94-22, Departement of Statistics, Purdue University (1994)

[16] Nester, D.K.: Random walk with partial reflection of repulsion at both extrema, preprint (1994)

Current address: D.M.I., E.N.S., 45 rue d'Ulm, F-75230 PARIS cedex 05, France
wwerner@dmi.ens.fr

Onsager–Machlup Functionals for Solutions of Stochastic Boundary Value Problems

Mireille Chaleyat-Maurel[(1)] and David Nualart[(2)]

(1) Université de Paris VI, Laboratoire de Probabilités, 4, place Jussieu, tour 46-56, 75252 Paris, France.

(2) Facultat de Matemàtiques, Universitat de Barcelona, Gran Via 585, 08007 Barcelona, Spain.

Abstract. The purpose of this paper is to compute the asymptotic probability that the solution of a stochastic differential equation with boundary conditions belongs to a small tube of radius $\epsilon > 0$ centered around the solution of the deterministic equation without drift.

1 Introduction

Let $\{X_t, 0 \le t \le 1\}$ be a continuous stochastic process defined on the Wiener space $(\Omega, \mathcal{F}, P)$. Given a smooth function ϕ belonging to the support of the law of X, we are interested in the asymptotic behaviour as ϵ tends to zero of the probability that X belongs to a tube of radius ϵ around ϕ.

If X is the standard d-dimensional Wiener process, and ϕ has a square integrable derivative, we know that the probability $P(\|X-\phi\|_\infty < \epsilon)$ is equivalent to

$$c_1 e^{-\lambda_1/\epsilon^2} \exp\left(-\frac{1}{2}\int_0^1 \dot{\phi}_s^2 ds\right),$$

as ϵ tends to zero, where c_1 is a constant, and λ_1 is the first eigenvalue of the Laplacian in the unit ball. More generally, if X is the Wiener process with drift b, then we have

$$P(\|X-\phi\|_\infty < \epsilon) \sim c_1 e^{-\lambda_1/\epsilon^2} \exp\left(\int_0^1 L(\dot{\phi}_s, \phi_s) ds\right), \quad \epsilon \downarrow 0, \tag{1.1}$$

where

$$L(x,y) = -\frac{1}{2}|b(y)-x|^2 - \frac{1}{2}(\operatorname{div} b)(y).$$

This is true if the drift b belongs to $C_b^2(\mathbf{R}^d)$, and ϕ is an arbitrary function in the Cameron–Martin space (cf. Ikeda and Watanabe [3]) and [10]).

The functional $L(\dot{\phi}, \phi)$ appearing in (1.1) is called the Onsager–Machlup functional of the process X. In [1] we have computed this functional for the Brownian motion with drift, and assuming that the initial value is a random functional of the Brownian path. In this paper our aim is to compute the functional $L(\dot{\phi}, \phi)$ when the process

X satisfies a stochastic differential equation with boundary conditions, of the type studied in the references [7] and [8]. Like in [1], the main ingredient in computing the Onsager–Machlup functional will be a noncausal version of Girsanov theorem, due to Kusuoka (cf. [4]). The next two sections will be devoted to the computation of the Onsager–Machlup functional for first order and second order stochastic differential equations, respectively.

2 Onsager–Machlup functional for stochastic differential equations with boundary conditions

Let $\omega = \{\omega_t, 0 \leq t \leq 1\}$ be a d-dimensional Brownian motion defined on the canonical probability space $(\Omega, \mathcal{F}, P)$, where $\Omega = C_0([0,1], \mathbf{R}^d)$. Suppose that $f, g : \mathbf{R}^d \to \mathbf{R}^d$ are two continuous functions. Consider the following stochastic differential equation:

$$\left\{ \begin{array}{l} X_t = X_0 - \int_0^t f(X_s)ds + \omega_t \quad , \quad 0 \leq t \leq 1 \\ X_0 = g(X_1 - X_0) \end{array} \right. \tag{2.1}$$

That means, instead of giving the initial condition X_0, we impose some nonlinear relation between X_0 and X_1. This functional relation could have been written in the more general form $h(X_0, X_1) = 0$, but the above condition is more suitable for our purpose. The existence and uniqueness of a solution to equation (2.1), and the Markov property of this process have been studied in [7].

Using Theorem 2.3 below, it is easy to prove that the support of the law of X is the set of continuous functions u in $C([0,1], \mathbf{R}^d)$ satisfying the boundary condition $u_0 = g(u_1 - u_0)$. Then, smooth functions in the support are of the form $\phi - g(\phi_1)$ where ϕ belongs to the Cameron Martin space H^1. We recall that H^1 is the subspace of functions $\phi_t = \int_0^t \dot{\phi}_s ds$, and $\dot{\phi} \in L^2([0,1], \mathbf{R}^d)$.

Given an arbitrary function ϕ in H^1, we are interested in the evaluation of the asymptotic behaviour as ϵ tends to zero of the probability that X belongs to a tube of radius ϵ around a smooth functions of the support, namely

$$J^\epsilon(\phi) = P\left(\|X - \phi - g(\phi_1)\|_\infty < \epsilon\right), \tag{2.2}$$

where $\|\cdot\|_\infty$ denotes the supremum norm in Ω. This will provide the computation of the Onsager–Machlup functional.

In order to compute the asymptotic behaviour of $J^\epsilon(\phi)$, we will apply a generalized version of Girsanov theorem that has been used in [7] to study of the Markov property. For this purpose we introduce the process $Y = \{Y_t, 0 \leq t \leq 1\}$ which is the solution to equation (2.1) when f is zero. That means,

$$\left\{ \begin{array}{l} Y_t = Y_0 + \omega_t, \quad ,0 \leq t \leq 1 \\ Y_0 = g(Y_1 - Y_0) \end{array} \right. \tag{2.3}$$

Clearly, $Y_t = \omega_t + g(\omega_1)$.

We define the transformation $T : \Omega \to \Omega$ by

$$\begin{aligned} T(\omega)_t &= \omega_t + \int_0^t f(\omega_s + g(\omega_1))ds \\ &= \omega_t + \int_0^t f(Y_s(\omega))ds. \end{aligned} \tag{2.4}$$

We will assume the following conditions.

(h1) $f, g : \mathbf{R}^d \to \mathbf{R}^d$ are continuously differentiable functions.

(h2) The transformation T given by (2.4) is bijective.

(h3) We have

$$\det(I - \Phi_1 g'(\omega_1) + g'(\omega_1)) \neq 0,$$

a.s., where $\{\Phi_t, 0 \le t \le 1\}$ is the fundamental solution of the linear system

$$\begin{cases} d\Phi_t = -f'(Y_t)\Phi_t dt \\ \Phi_0 = I \end{cases}$$

In [7] some sufficient conditions on the functions f and g are given for the hypotheses (h1) to (h3) to hold. An example of such conditions would be:

a) f is a monotone function of class C^1 such that

$$\lim_{a\uparrow\infty} \frac{1}{a} \sup_{|x|\le a} |f(x)| = 0.$$

b) g is a monotone function of class C^1 and there exists a constant $C > 0$ such that $|g(x)| \le C(1 + |x|)$.

The bijective property of T is equivalent to the existence and uniqueness of the solution to equation (2.1).

As it is proved in [7, Theorem 3.6], under hypotheses **(h1)** to **(h3)** one can apply a generalized version of Girsanov theorem due to Kusuoka [4]:

Proposition 2.1 *Suppose that hypotheses* **(h1)** *to* **(h3)** *hold. There exists a probability Q on $(\Omega, \mathcal{F})$ such that $Q \circ T^{-1} = P$ (namely, $T(\omega)$ is a Brownian motion under Q), and Q is given by*

$$\frac{dQ}{dP} = |\det(I - \Phi_1 g'(\omega_1) + g'(\omega_1))| \times \exp\left\{\frac{1}{2}\int_0^1 \operatorname{Tr} f'(Y_t)dt - \int_0^1 f(Y_t) \circ d\omega_t - \frac{1}{2}\int_0^1 |f(Y_t)|^2 dt\right\},$$

where $\int_0^1 f(Y_t) \circ d\omega_t$ denotes the extended Stratonovich integral (cf. [6]).

We can state now the main result of this section.

Theorem 2.1 *Suppose that the above hypotheses* **(h1)** *to* **(h3)** *hold, and, in addition, the following conditions are satisfied:*

(h4) $f \in C_b^2(\mathbf{R}^d; \mathbf{R}^d)$.

(h5) $\det(I - g'(x)) \neq 0$, *a.e.*

(h6) *Either g is linear ($g(x) = Gx$) or ϕ is of class C_b^2.*

Then, for any $\phi \in H^1$, *we have*

$$
\begin{aligned}
P(\|X-\phi-g(\phi_1)\|_\infty<\epsilon) &\sim L_1(\phi_1)e^{-\frac{\lambda_1}{\epsilon^2}} \\
&\times |\det(I-\Phi_1(\phi)g'(\phi_1)+g'(\phi_1))| \\
&\times \exp\left\{\frac{1}{2}\int_0^1 \operatorname{Tr} f'(\phi_t+g(\phi_1))dt-\frac{1}{2}\int_0^1 |f(\phi_t+g(\phi_1))+\dot{\phi}_t|^2dt\right\},
\end{aligned}
$$

as $\epsilon \downarrow 0$, *where* λ_1 *is the first eigenvalue of the Laplacian in the unit ball,* f_1 *is the associated eigenfunction, and*

$$
L_1(\phi_1)=\int_{\{|g'(\phi_1)u|<1,|(I+g'(\phi_1))u|<1\}} f_1(g'(\phi_1)u)f_1((I+g'(\phi_1))u)du.
$$

Proof: Fix a function $\phi \in H^1$. Consider the transformation $T^\phi : \Omega \to \Omega$ defined by

$$
\begin{aligned}
T^\phi(\omega)_t &= \omega_t+\phi_t+\int_0^t f(\omega_s+\phi_s+g(\omega_1+\phi_1))ds \qquad (2.5)\\
&= \omega_t+\phi_t+\int_0^t f(Y_s(\omega+\phi))ds.
\end{aligned}
$$

In other words, $T^\phi(\omega)=T(\omega+\phi)$. Applying Proposition 2.2 and the ordinary Girsanov theorem one can show that there exists a probability Q^ϕ on $(\Omega,\mathcal{F})$ such that $Q^\phi \circ (T^\phi)^{-1}=P$ (namely, $T^\phi(\omega)$ is a Brownian motion under Q^ϕ) and

$$
\begin{aligned}
\frac{dQ^\phi}{dP} &= |\det(I-\Phi_1(\omega+\phi)g'(\omega_1+\phi_1)+g'(\omega_1+\phi_1))| \\
&\times \exp\left\{\frac{1}{2}\int_0^1 \operatorname{Tr} f'(Y_t(\omega+\phi))dt-\int_0^1 f(Y_t(\omega+\phi))\circ d\omega_t\right. \\
&\left.-\int_0^1 \dot{\phi}_t d\omega_t-\frac{1}{2}\int_0^1 |f(Y_t(\omega+\phi))+\dot{\phi}_t|^2dt\right\}.
\end{aligned}
$$

The process $Y(\omega+\phi)$ satisfies the equation

$$
\left\{
\begin{array}{ll}
Y_t(\omega+\phi)=Y_0(\omega+\phi)+T^\phi(\omega)_t-\int_0^t f(Y_s(\omega+\phi))ds \quad , & 0\le t\le 1\\
Y_0=g(Y_1-Y_0) &
\end{array}
\right. \qquad (2.6)
$$

Therefore, the law of the process X under the original probability P coincides with the law of the process $Y(\omega+\phi)$ under the probability Q^ϕ. This allows us to write the functional $J^\epsilon(\phi)$ in the following form

$$
\begin{aligned}
J^\epsilon(\phi) &= P(\|X-\phi-g(\phi_1)\|_\infty<\epsilon) \\
&= Q^\phi(\|Y(\omega+\phi)-\phi-g(\phi_1)\|_\infty<\epsilon) \\
&= Q^\phi(\|\omega+g(\omega_1+\phi_1)-g(\phi_1)\|_\infty<\epsilon).
\end{aligned}
$$

Define the set

$$
G_\epsilon=\{\|\omega+g_\phi(\omega_1)\|_\infty<\epsilon\},
$$

where $g_\phi(x)=g(x+\phi_1)-g(\phi_1)$. Then $J^\epsilon(\phi)$ can be written as

$$
\begin{aligned}
J^\epsilon(\phi) &= E\left(\frac{dQ^\phi}{dP}|G_\epsilon\right)P(G_\epsilon) \\
&= a_1(\epsilon)a_2(\epsilon).
\end{aligned}
$$

In order to prove the theorem it suffices to prove the following facts:

(C1)

$$\lim_{\epsilon\downarrow 0} a_1(\epsilon) = |\det(I - (\Phi_1(\phi) - I)g'(\phi_1)|$$
$$\times \exp\left\{\frac{1}{2}\int_0^1 \operatorname{Tr} f'(Y_t(\phi)dt - \frac{1}{2}\int_0^1 |f(Y_t(\phi)) + \dot{\phi}_t|^2 dt\right\}.$$

(C2) $\lim_{\epsilon\downarrow 0} e^{\frac{\lambda_1}{\epsilon^2}} a_2(\epsilon) = L_1(\phi_1)$.

Proof of **(C1)**: Consider the following Wiener functionals:

$$A_1(\omega) = |\det(I - (\Phi_1(\omega) - I)g'(\omega_1)|$$
$$\times \exp\left\{\frac{1}{2}\int_0^1 \operatorname{Tr} f'(Y_t(\omega)dt - \frac{1}{2}\int_0^1 |f(Y_t(\omega)) + \dot{\phi}_t|^2 dt\right\}.$$
$$A_2(\omega) = \exp\left\{-\int_0^1 f(Y_t(\omega+\phi)) \circ d\omega_t\right\}$$
$$A_3(\omega) = \exp\left\{-\int_0^1 \dot{\phi}_t d\omega_t\right\}.$$

We have

$$|E[(a_1(\epsilon) - A_1(\phi))|G_\epsilon]|$$
$$= |E[(A_1(\omega+\phi)A_2(\omega)A_3(\omega) - A_1(\phi)) \mid G_\epsilon]|$$
$$\leq \left(E\left[(A_1(\omega+\phi) - A_1(\phi))^2 \mid G_\epsilon\right]\right)^{1/2} \left(E[A_2(\omega)^2 A_3(\omega)^2 \mid G_\epsilon]\right)^{1/2}$$
$$+ A_1(\phi)\,|E[A_2(\omega)A_3(\omega) \mid G_\epsilon] - 1|.$$

Therefore, the proof of the convergence **(C1)** reduces to establish the following three facts:

$$\lim_{\epsilon\downarrow 0} E\left[(A_1(\omega+\phi) - A_1(\phi))^2 \mid G_\epsilon\right] = 0, \tag{2.7}$$
$$\sup_{\epsilon>0} E[(A_2A_3)^2 \mid G_\epsilon] < \infty, \tag{2.8}$$
$$\lim_{\epsilon\downarrow 0} E[A_2A_3 \mid G_\epsilon] = 1. \tag{2.9}$$

The convergence (2.7) is immediate because A_1 is a continuous functional of ω, and on the set G_ϵ we have $|g(\omega_1+\phi_1) - g(\phi_1)| < \epsilon$ and, therefore, $\|\omega\|_\infty < 2\epsilon$. We are going to prove the relation (2.9) and the proof of (2.8) would follow the same lines. Using the arguments of Ikeda and Watanabe ([3], page 449), which are still true if we condition by the set G_ϵ instead of $\{\|\omega\|_\infty < \epsilon\}$, in order to prove (2.9) it suffices to show that for any c in $\mathbf{R}$, we have

$$\limsup_{\epsilon\downarrow 0} E\left[\exp\left(c\int_0^1 f(Y_t(\omega+\phi)) \circ d\omega_t\right) \mid G_\epsilon\right] \leq 1, \tag{2.10}$$

$$\limsup_{\epsilon\downarrow 0} E\left[\exp\left(c\int_0^1 \dot{\phi}_t d\omega_t\right) \mid G_\epsilon\right] \leq 1. \tag{2.11}$$

The inequality (2.11) follows from the results of Shepp and Zeitouni (cf. [10]) if we assume that the function g is linear. In fact, in this case the set G_ϵ is convex and symmetric, and we can use the arguments based on the correlation inequalities. On the other hand, this inequality is also true by an integration by parts argument if the function ϕ is of class C_b^2.

In order to show the inequality (2.10) we will use the same arguments as in [1]. We write

$$f(Y_t(\omega+\phi)) = f(\phi_t + g(\omega_1+\phi_1)) + f'(\phi_t + g(\omega_1+\phi_1))\omega_t + R_t,$$

and hypotheses (h4) implies that $|R_t| \leq C\epsilon^2$ on the set G_ϵ. Then, using again the arguments of Ikeda and Watanabe, it suffices to show that

$$\limsup_{\epsilon\downarrow 0} E\left[\exp\left(c\int_0^1 f(\phi_t + g(\omega_1+\phi_1)) \circ d\omega_t\right) \mid G_\epsilon\right] \leq 1, \tag{2.12}$$

$$\limsup_{\epsilon\downarrow 0} E\left[\exp\left(c\int_0^1 f'(\phi_t + g(\omega_1+\phi_1))\omega_t \circ d\omega_t\right) \mid G_\epsilon\right] \leq 1, \tag{2.13}$$

and

$$\limsup_{\epsilon\downarrow 0} E\left[\exp\left(c\int_0^1 R_t \circ d\omega_t\right) \mid G_\epsilon\right] \leq 1, \tag{2.14}$$

for all $c \in \mathbf{R}$. The proof of these inequalities would follow the same lines as the proof of Theorem 2.1 in [1]. For this reason we omit the details of this proof and we just indicate the method of proof. The inequality (2.12) is obtained using the integration by parts formula of the nonadapted Stratonovich calculus (see [6]). The inequality (2.13) follows by the same arguments as in Ikeda and Watanabe ([3], page 451) with the help of the Lévy area and exponential inequalities. In order to handle inequality (2.14) one uses uniform exponential estimates and the susbstitution theorem for the Stratonovich integral (see [6]).

Proof of (**C2**): We have to estimate the probability

$$a_2(\epsilon) = P\left(\sup_{0\leq t\leq 1} |\omega_t + g_\phi(\omega_1)| < \epsilon\right)$$

when ϵ goes to zero. Note that g_ϕ is continuously differentiable and $g_\phi(0) = 0$, $g'_\phi(0) = g'(\phi_1)$. Using an idea of Le Gall we can express $a_2(\epsilon)$ in terms of the exit time of Brownian motion. More precisely, let us denote by $B_a(r)$ the open ball in $\mathbf{R}^d$ with center a and radius $r > 0$. Then, if $T_{a,r}$ is the exit time of this ball for the Brownian motion starting at zero, the above probability can be written as

$$a_2(\epsilon) = P\left(T_{-g_\phi(\omega_1),\epsilon} > 1\right). \tag{2.15}$$

For any $b \in \mathbf{R}^d$, $|b| < \epsilon$ and any Borel set $A \subset B_b(\epsilon)$ we can write

$$P\left(T_{b,\epsilon} > 1, \omega_1 \in A\right) = \int_A \psi_{b,\epsilon}(x)dx,$$

where the function $\psi_{b,\epsilon}$ is obtained as follows. Let Q_t be the semigroup whose generator is $-\frac{1}{2}\Delta$ on the ball $B_b(\epsilon)$ with Dirichlet boundary conditions. Then, for any bounded measurable function f on $B_b(\epsilon)$, and for any $x \in B_b(\epsilon)$, we have

$$E_x\left(f(\omega_t)1_{\{T_{b,\epsilon}>t\}}\right) = Q_t f(x).$$

Therefore,

$$P(T_{b,\epsilon} > 1, \omega_1 \in A) = Q_1 \mathbf{1}_A(0).$$

If we denote by $q_t(x,y)$ the kernel of Q_t, we have $\psi_{b,\epsilon}(x) = q_1(0,x)$.

Suppose that $0 < \lambda_1 < \lambda_2 \leq \cdots \leq \lambda_n \leq \cdots$ are the eigenvalues, and $\{f_n, n \geq 1\}$ an orthonormal sequence of associated eigenfunctions of the problem

$$\begin{cases} \frac{1}{2}\Delta f_n + \lambda_n f_n = 0 \\ f_n|_{\partial B_0(1)} = 0 \end{cases} \tag{2.16}$$

Then,

$$q_t(x,y) = \sum_{n=1}^{\infty} e^{-\frac{\lambda_n t}{\epsilon^2}} \epsilon^{-d} f_n\left(\frac{x-b}{\epsilon}\right) f_n\left(\frac{y-b}{\epsilon}\right), \tag{2.17}$$

and

$$\psi_{b,\epsilon}(x) = \sum_{n=1}^{\infty} e^{-\frac{\lambda_n}{\epsilon^2}} \epsilon^{-d} f_n\left(-\frac{b}{\epsilon}\right) f_n\left(\frac{x-b}{\epsilon}\right). \tag{2.18}$$

It holds that

$$P\left(T_{-g_\phi(\omega_1),\epsilon} > 1\right) = \int_{\{|g_\phi(x)|<\epsilon, |x+g_\phi(x)|<\epsilon\}} \psi_{-g_\phi(x),\epsilon}(x)dx. \tag{2.19}$$

In fact, we have that

$$\psi_{b,\epsilon}(x) = P(T_{b,\epsilon} > 1 | \omega_1 = x) f_{\omega_1}(x), \tag{2.20}$$

where f_{ω_1} is the density of ω_1. Moreover, $\psi_{b,\epsilon}(x)$ is a continuous function of the variables (b,ϵ) on the set $\{|b| < \epsilon, |x-b| < \epsilon\}$ (this follows from (2.17)). Consequently, we can substitute b by $-g_\phi(x)$ in (2.19) and we obtain

$$\begin{aligned} \psi_{-g_\phi(x),\epsilon}(x) &= P\left(T_{-g_\phi(x),\epsilon} > 1 | \omega_1 = x\right) f_{\omega_1}(x) \\ &= P\left(T_{-g_\phi(\omega_1),\epsilon} > 1 | \omega_1 = x\right) f_{\omega_1}(x), \end{aligned}$$

which implies (2.18).

Finally, from (2.15), (2.18) and (2.17) we get

$$a_2(\epsilon) = \sum_{n=1}^{\infty} e^{-\frac{\lambda_n}{\epsilon^2}} \epsilon^{-d} \int_{\{|g_\phi(x)|<\epsilon, |x+g_\phi(x)|<\epsilon\}} f_n\left(\frac{g_\phi(x)}{\epsilon}\right) f_n\left(\frac{x+g_\phi(x)}{\epsilon}\right) dx. \tag{2.21}$$

Set

$$L_n^\epsilon(\phi_1) = \epsilon^{-d} \int_{\{|g_\phi(x)|<\epsilon, |x+g_\phi(x)|<\epsilon\}} f_n\left(\frac{g_\phi(x)}{\epsilon}\right) f_n\left(\frac{x+g_\phi(x)}{\epsilon}\right) dx.$$

Using the change of variables $x = u\epsilon$ we can write

$$L_n^\epsilon(\phi_1) = \int_{\{|\frac{1}{\epsilon}g_\phi(u\epsilon)|<1, |u+\frac{1}{\epsilon}g_\phi(u\epsilon)|<1\}} f_n\left(\frac{1}{\epsilon}g_\phi(u\epsilon)\right) f_n\left(u + \frac{1}{\epsilon}g_\phi(u\epsilon)\right) du.$$

Taking into account that the functions f_n are bounded we deduce that the limit

$$L_n(\phi_1) = \lim_{\epsilon \downarrow 0} L_n^\epsilon(\phi_1)$$

exists and it is expressed as

$$L_n(\phi_1) = \int_{\{|g'_\phi(0)u|<1, |(I+g'_\phi(0))u|<1\}} f_n(g'_\phi(0)u) f_n((I+g'_\phi(0))u)du.$$

It is well-known that f_1 is strictly positive (see [2]). Consequently, $L_1(\phi_1) > 0$ and from (2.21) we obtain the exact asymptotic behaviour

$$a_2(\epsilon) \sim L_1(\phi_1) e^{-\frac{\lambda_1}{\epsilon^2}},$$

and the proof of the theorem is complete. □

Remarks:

1. When $g \equiv 0$ we get

$$L_1(\phi_1) = f_1(0) \int_{\{|u|<1\}} f_1(u)du,$$

which is the usual constant for the Onsager-Machlup functional with given initial value.

2. The periodic boundary condition $X_0 = X_1$ is not covered by the equation $X_0 = g(X_1 - X_0)$. Nevertheless with minor modifications and using the ideas of [7] one can also obtain the Onsager-Machlup functional for this and related cases.

3 Onsager-Machlup functional for second order stochastic differential equations

In this section we will assume that $d = 1$, namely, Ω is the space of real valued functions on $[0,1]$ which vanish at zero. Let $f : \mathbf{R} \to \mathbf{R}$ be a continuous function and consider the equation

$$\begin{cases} \dot{X}_t = \dot{X}_0 - \int_0^t f(X_s)ds + \omega_t & ,0 \le t \le 1 \\ X_0 = X_1 = 0 \end{cases} \tag{3.1}$$

This equation has been studied in [8] and [9]. There exists a unique solution for each $\omega \in \Omega$ provided the function f is continuously differentiable and $f' \le 0$.

Let us denote by $\{Y_t(\omega), 0 \le t \le 1\}$ the solution of the above equation when $f \equiv 0$, that is,

$$\begin{cases} \dot{Y}_t = \dot{Y}_0 + \omega_t & ,0 \le t \le 1 \\ Y_0 = Y_1 = 0 \end{cases} \tag{3.2}$$

Clearly, $Y_t = \int_0^t \omega_s ds - t\int_0^1 \omega_s ds$ and $\dot{Y}_t = \omega_t - \int_0^1 \omega_s ds$.

From Proposition 3.1 below it follows that the support of the law of X is the set of continuously differentiable functions on $[0,1]$ which vanish at 0 and 1. Given a function ϕ in the Cameron-Martin space H^1, the path

$$Y_t(\phi) = \int_0^t \phi_s ds - t\int_0^1 \phi_s ds$$

belongs to the support of the law of X and we are interested in the asymptotic behaviour as ϵ tends to zero of

$$J^\epsilon(\phi) = P(\|X - Y(\phi)\|_{1,\infty} < \epsilon). \tag{3.3}$$

In the above expression we have used the seminorm $\|\cdot\|_{1,\infty}$ defined by

$$\|y\|_{1,\infty} = \sup_{t\in[0,1]} |\dot{y}(t)|,$$

for $y \in C^1([0,1])$, instead of the usual supremum norm. This seminorm is well fitted to the process X because $\dot{X}_t$ behaves as a Brownian motion.

As in the previous section the computation of the asymptotic behaviour of $J^\epsilon(\phi)$ is based on the application of the extended Girsanov theorem to a suitable transformation T^ϕ on the Wiener space. This transformation will be defined as follows

$$T^\phi(\omega)_t = \omega_t + \phi_t + \int_0^t f(Y_s(\omega+\phi))ds. \tag{3.4}$$

From the results of [8] and [9] we have the following result:

Proposition 3.1 *Suppose that f is continuously differentiable and $f' \le 0$. Fix $\phi \in H^1$. Then the transformation T^ϕ defined by (3.4) is bijective and there exists a probability Q^ϕ such that $Q^\phi \circ (T^\phi)^{-1} = P$ and*

$$\begin{aligned}\frac{dQ^\phi}{dP} = |Z_1(\omega+\phi)| \exp\Big\{ &- \int_0^1 f(Y_t(\omega+\phi)) \circ d\omega_t - \int_0^1 \dot{\phi}_t d\omega_t \\ &- \frac{1}{2}\int_0^1 |f(Y_t(\omega+\phi)) + \dot{\phi}_t|^2 dt\Big\},\end{aligned}$$

where Z_1 is the solution at time $t=1$ of the second order differential equation

$$\left\{\begin{array}{l} \ddot{Z}_t + f'(Y_t)Z_t = 0, \quad ,0 \le t \le 1 \\ Z_0 = 0, \dot{Z}_1 = 0 \end{array}\right. \tag{3.5}$$

The process $Y(\omega+\phi)$ satisfies the equation

$$\dot{Y}_t(\omega+\phi) - \dot{Y}_0(\omega+\phi) = T^\phi(\omega)_t - \int_0^t f(Y_s(\omega+\phi))ds.$$

As a consequence, the law of the process X solution to (3.1) coincides with the law of $Y(\omega+\phi)$ under Q^ϕ. This allows to write the functional $J^\epsilon(\phi)$ in the following form

$$\begin{aligned} J^\epsilon(\phi) &= P(\|X - Y(\phi)\|_{1,\infty} < \epsilon) \\ &= Q^\phi(\|Y(\omega+\phi) - Y(\phi)\|_{1,\infty} < \epsilon) \\ &= Q^\phi(\|\omega - \int_0^1 \omega_t dt\|_\infty < \epsilon). \end{aligned}$$

In the sequel we will use the notation

$$H_\epsilon = \{\|\omega - \int_0^1 \omega_t dt\|_\infty < \epsilon\}.$$

Now we can state the main result of this section.

Theorem 3.1 *Suppose that f is continuously differentiable and $f' \leq 0$. For any $\phi \in H^1$ we have, as ϵ tends to zero,*

$$P(\|X - Y(\phi)\|_{1,\infty} < \epsilon) \sim P(H_\epsilon) Z_1(\phi) \exp\left\{-\frac{1}{2}\int_0^1 |f(Y_t(\phi)) + \dot{\phi}_t|^2 dt\right\}.$$

Proof: We have

$$J^\epsilon(\phi) = E\left(\frac{dQ^\epsilon}{dP}|H_\epsilon\right) P(H_\epsilon).$$

So in order to prove the theorem it suffices to show that

$$\lim_{\epsilon\downarrow 0} E\left(\frac{dQ^\epsilon}{dP}|H_\epsilon\right) = |Z_1(\phi)| \exp\left\{-\frac{1}{2}\int_0^1 |f(Y_t(\phi)) + \dot{\phi}_t|^2 dt\right\}.$$

Consider the following Wiener functionals

$$B_1(\omega) = |Z_1(\omega)| \exp\left\{-\frac{1}{2}\int_0^1 |f(Y_t(\omega)) + \dot{\phi}_t|^2 dt\right\},$$

$$B_2(\omega) = \exp\{-\int_0^1 f(Y_t(\omega+\phi)) \circ d\omega_t\}$$

$$B_3(\omega) = \exp\{-\int_0^1 \dot{\phi}_t d\omega_t\}.$$

We want to show that

$$\lim_{\epsilon\downarrow 0} E\left(\frac{dQ^\epsilon}{dP}|H_\epsilon\right) = B_1(\phi).$$

Using the same technique as in the previous paragraph we have

$$\begin{aligned}
&\left|E\left[\frac{dQ^\epsilon}{dP} - B_1(\phi)|H_\epsilon\right]\right| \\
&= |E\left[B_1(\omega+\phi)B_2(\omega)B_3(\omega) - B_1(\phi) \mid H_\epsilon\right]| \\
&\leq \left(E\left[(B_1(\omega+\phi) - B_1(\phi))^2 \mid H_\epsilon\right]\right)^{1/2} \left(E[B_2^2 B_3^2 \mid G_\epsilon]\right)^{1/2} \\
&+ B_1(\phi)\left|E[B_2(\omega)B_3(\omega) \mid H_\epsilon] - 1\right|,
\end{aligned}$$

and it suffices to show that

$$\lim_{\epsilon\downarrow 0} E\left[(B_1(\omega+\phi) - B_1(\phi))^2 \mid H_\epsilon\right] = 0, \tag{3.6}$$

$$\sup_{\epsilon>0} E[(B_2B_3)^2 \mid H_\epsilon] < \infty, \tag{3.7}$$

$$\lim_{\epsilon\downarrow 0} E[B_2B_3 \mid H_\epsilon] = 1. \tag{3.8}$$

The convergence (3.6) is immediate because B_1 is a continuous functional of ω, and on the set H_ϵ we have $\|\omega\|_\infty < 2\epsilon$. Using the arguments of Ikeda and Watanabe

([3], page 449), in order to prove (3.7) and (3.8) it suffices to show that for any c in $\mathbf{R}$, we have

$$\limsup_{\epsilon\downarrow 0} E\left[\exp\left(c\int_0^1 f(Y_t(\omega+\phi))\circ d\omega_t\right) \mid H_\epsilon\right] \leq 1, \tag{3.9}$$

$$\limsup_{\epsilon\downarrow 0} E\left[\exp\left(c\int_0^1 \dot\phi_t d\omega_t\right) \mid H_\epsilon\right] \leq 1. \tag{3.10}$$

The inequality (2.11) follows from the results of Shepp and Zeitouni because the set H_ϵ is convex and symmetric with respect to the origin.

In order to show the inequality (3.9) we will use the same arguments as in the proof of (2.10)

The following lemma, whose proof was given to us by Wenbo Li (cf. [5]), provides the asymptotic behaviour of the probability of the set H_ϵ as ϵ tends to zero.

Lemma 3.1 *We have*

$$\lim_{\epsilon\downarrow 0} \epsilon^2 \log P\left(\sup_{0\leq t\leq 1} |\omega_t - \int_0^1 \omega_s ds| < \epsilon\right) = -\frac{\pi^2}{8}.$$

Proof: Set $p_\epsilon = P(H_\epsilon)$. We have

$$\begin{aligned} p_\epsilon &\geq P\left(\sup_{0\leq t\leq 1} |\omega_t - \int_0^1 \omega_s ds| < \epsilon, |\int_0^1 \omega_s ds| < \epsilon^3\right) \\ &\geq P\left(\sup_{0\leq t\leq 1} |\omega_t| < (1-\epsilon^2)\epsilon, |\int_0^1 \omega_s ds| < \epsilon^3\right) \\ &\geq P\left(\sup_{0\leq t\leq 1} |\omega_t| < (1-\epsilon^2)\epsilon\right) P\left(|\int_0^1 \omega_s ds| < \epsilon^3\right), \end{aligned}$$

where the last step is due to the correlation inequalities (see Shepp and Zeitouni [10]). Finally, we deduce the inequality

$$p_\epsilon \geq C\epsilon^3 e^{-\frac{\pi^2}{8\epsilon^2}}, \tag{3.11}$$

for some constant $C > 0$ and for ϵ small enough.

To obtain an upper bound we write

$$\begin{aligned} p_\epsilon &= \sum_{|k|\leq[\frac{1}{2}\epsilon^{-2}]} P\left(\sup_{0\leq t\leq 1} |\omega_t - \int_0^1 \omega_s ds| < \epsilon, |\int_0^1 \omega_s ds - 2k\epsilon^3| < \epsilon^3\right) \\ &\leq \sum_{|k|\leq[\frac{1}{2}\epsilon^{-2}]} P\left(\sup_{0\leq t\leq 1} |\omega_t - 2k\epsilon^3| < (1+\epsilon^2)\epsilon, |\int_0^1 \omega_s ds - 2k\epsilon^3| < \epsilon^3\right) \\ &\leq \sum_{|k|\leq[\frac{1}{2}\epsilon^{-2}]} P\left(\sup_{0\leq t\leq 1} |\omega_t - 2k\epsilon^3| < (1+\epsilon^2)\epsilon\right) \\ &\leq (2[\frac{1}{2}\epsilon^{-2}]+1) P\left(\sup_{0\leq t\leq 1} |\omega_t| < (1+\epsilon^2)\epsilon\right). \end{aligned}$$

This implies that

$$p_\epsilon \leq C\epsilon^{-2}e^{-\frac{\pi^2}{8\epsilon^2}}, \tag{3.12}$$

for some constant $C > 0$ and for ϵ small enough. □

We conjecture that the estimates (3.11) and (3.12) can be improved in the sense that the polynomial term in ϵ can be replaced by a constant.

References

[1] M. Chaleyat-Maurel and D. Nualart: The Onsager–Machlup functional for a class of anticipating processes. *Probab. Theory Rel. Fields* **94**, 247-270 (1992).

[2] I. Chavel: *Eigenvalues in Riemannian Geometry.* Academic Press, 1984.

[3] N. Ikeda and S. Watanabe: *Stochastic differential equations and Diffusion processes.* Amsterdam, Oxford, New York: North–Holland, 1981.

[4] S. Kusuoka: The non linear transformation of Gaussian measure on Banach space and its absolute continuity. *(I). J. Fac. Sci. Tokyo Univ.* **32** Sec. IA, 567–597 (1985)

[5] Wenbo Li: Private communication.

[6] D. Nualart and E. Pardoux: Stochastic calculus with anticipating integrands. *Probab. Theory Rel. Fields* **78**, 535–581 (1988).

[7] D. Nualart and E. Pardoux: Boundary value problems for stochastic differential equations. *Annals of Probability* **19**, 1118-1144 (1991).

[8] D. Nualart and E. Pardoux: Second order stochastic differential equations with Dirichlet boundary conditions. *Stochastic Processes and Their Applications* **39**, 1-24 (1991).

[9] D. Nualart and E. Pardoux: Stochastic differential equations with boundary conditions. In: *Stochastic Analysis and Appl.* ed. A.B. Cruzeiro and J.C. Zambrini. Birkhauser 1991, 155-175.

[10] L.A. Shepp and O. Zeitouni: A note on conditional exponential moments and Onsager Machlup functionals. *Annals of Probability* **20**, 652-654 (1992).

The work of D. Nualart was done during his staying at the Laboratoire de Probabilités, Univ. Paris VI.

SUR QUELQUES FILTRATIONS

ET TRANSFORMATIONS BROWNIENNES

S. Attal, K. Burdzy, M. Émery et Y. Hu

Cette étude est issue de questions posées par Marc Yor, que nous remercions pour de nombreuses conversations, une correspondance fournie et son inlassable disponibilité.

Une filtration étant donnée sur un espace probabilisé, est-il possible de reconnaître si elle est engendrée par un mouvement brownien sans construire explicitement un tel processus ? Ce problème difficile, auquel nous ignorons la réponse, a fait récemment des progrès grâce à Dubins, Feldman, Smorodinsky & Tsirelson [3]. Nous n'allons pas l'attaquer ici, mais seulement, en exhibant un mouvement brownien générateur (et même plusieurs), vérifier que la filtration de Goswami-Rao, qui décrit la connaissance d'un mouvement brownien à un facteur ± 1 près, est brownienne; nous nous livrerons ensuite à des variations sur ce thème et sur les méthodes employées.

Tous les mouvements browniens considérés seront issus de l'origine et à valeurs réelles. L'espace de Wiener (unidimensionnel) sera désigné par $(\Omega, \mathcal{A}, \mathbb{P})$, sa filtration canonique par $\mathcal{F} = (\mathcal{F}_t)_{t\geqslant 0}$ et le mouvement brownien canonique des coordonnées sur Ω par B. Si $\mathcal{B}$ est une sous-tribu de $\mathcal{A}$ et R une relation d'équivalence sur Ω telle que le saturé pour R de tout élément de $\mathcal{B}$ soit encore dans $\mathcal{B}$,[1] on notera $\mathcal{B}/R$ la sous-tribu de $\mathcal{B}$ formée des événements de $\mathcal{B}$ saturés pour R, ou plutôt la complétée pour $(\mathcal{A}, \mathbb{P})$ de cette tribu : ici et dans la suite, les sous-tribus sur Ω contiennent implicitement tous les événements négligeables. De même, on notera $\mathcal{F}/R$ la filtration prenant à l'instant t la valeur $\mathcal{F}_t/R$ si chaque $\mathcal{F}_t$ est stable par saturation pour R.

1. Cette condition n'est pas nécessaire pour définir $\mathcal{B}/R$, mais lorsqu'elle n'est pas remplie, la notation $\mathcal{B}/R$ est vraiment abusive !

Si l'on prend pour R la relation d'équivalence telle que

$$\omega_1 \, R \, \omega_2 \quad \Longleftrightarrow \quad [\, \omega_1 = \omega_2 \quad \text{ou} \quad \omega_1 = -\omega_2 \,] \, ,$$

elle vérifie cette condition et la filtration quotient $\mathcal{G} = \mathcal{F}/R$, appelée[2] *filtration de Goswami-Rao,* peut être ainsi caractérisée : les variables aléatoires mesurables pour $\mathcal{G}_t$ sont exactement les variables aléatoires mesurables pour $\mathcal{F}_t$ qui sont paires, c'est-à-dire qui prennent p. s. la même valeur en ω et $-\omega$.

Pour chaque t, il est clair que la tribu $\mathcal{G}_t$ est aussi engendrée par le processus $(B_s \operatorname{sgn} B_t \, , \; s \in [0,t])$. Goswami & Rao ont remarqué dans [4] qu'elle l'est aussi par les variables aléatoires $|B_u - B_v|$ quand u et v décrivent $[0,t]$; en effet, ces variables aléatoires sont paires et réciproquement $B_s \operatorname{sgn} B_t$ est le produit de $|B_s| = |B_s - B_0|$ par

$$\operatorname{sgn}(B_s B_t) = \begin{cases} 1 & \text{si } |B_s - B_t| = \big||B_s| - |B_t|\big| ; \\ -1 & \text{si } |B_s - B_t| = |B_s| + |B_t| . \end{cases}$$

Si l'on se fixe t_1 et t_2 dans $[0,t]$ tels que $t_1 \neq t_2$, la tribu $\mathcal{G}_t$ est aussi engendrée par les processus $|B_s - B_{t_1}|$ et $|B_s - B_{t_2}|$, où s décrit $[0,t]$. Ceci peut se vérifier en retrouvant d'abord, à une symétrie par rapport à l'origine près, le couple (B_{t_1}, B_{t_2}) à partir des trois distances $|B_{t_1}|$, $|B_{t_2}|$, $|B_{t_1} - B_{t_2}|$, puis, à cette même symétrie près, B_s à partir de ses distances à B_{t_1} et B_{t_2}.

Pour $0 < s \leqslant t$, $\mathcal{F}_t$ est engendrée par la tribu $\mathcal{G}_t$ et la variable aléatoire $\operatorname{sgn} B_s$ qui est indépendante de $\mathcal{G}_t$; on peut donc dire, en un certain sens, que la filtration $\mathcal{F}$ est obtenue en grossissant $\mathcal{G}$ par un bit d'information indépendant. Mais, puisque $\mathcal{F}_0$ est égale à $\mathcal{G}_0$ (et triviale), ce supplément d'information nécessaire pour passer de $\mathcal{G}$ à $\mathcal{F}$ semble apparaître par magie à l'instant 0+ (bien que $\mathcal{F}_{0+}$ soit aussi triviale); ceci évoque la situation de l'exemple de Tsirelson (voir Revuz & Yor [10], Yor [11] et surtout Le Gall & Yor [6], sur lequel nous reviendrons plus bas).

DÉFINITION. — *Une filtration sera dite* brownienne *si c'est la filtration naturelle d'un mouvement brownien (réel, issu de zéro).*

Dans une filtration brownienne $\mathcal{H}$, engendrée par un mouvement brownien X, tout mouvement brownien Y est de la forme $\int H \, dX$ où H est de carré 1, donc inversible. Il en résulte que Y hérite de X la propriété de représentation prévisible relativement à $\mathcal{H}$; en particulier, deux mouvements browniens pour $\mathcal{H}$ sont toujours intégrale stochastique l'un par rapport à l'autre.

Mais la réciproque est fausse : il existe des filtrations qui ne sont pas browniennes, bien qu'elles contiennent des mouvements browniens et que tous les mouvements browniens qu'elles contiennent y aient la propriété de représentation prévisible. C'est par exemple le cas de la filtration canonique sur l'espace de Wiener muni d'une probabilité $\mathbb{Q}$ équivalente à $\mathbb{P}$ convenablement choisie (Dubins, Feldman, Smorodinsky et Tsirelson ont montré dans [3] qu'il est possible de choisir $\mathbb{Q}$ de sorte que, sur $(\Omega, \mathcal{A}, \mathbb{Q})$, $\mathcal{F}$ ne soit pas une filtration brownienne).

DÉFINITION. — *Une sous-filtration $\mathcal{K}$ d'une filtration brownienne $\mathcal{H}$ sera dite* représentable dans $\mathcal{H}$ *si elle est engendrée par un processus qui est un mouvement brownien (réel, issu de zéro) à la fois pour $\mathcal{H}$ et pour $\mathcal{K}$.*

---

2. Par Revuz & Yor, exercice (3.22), chapitre V de la 2e édition de [10] (1994).

Ce nom vient de ce que, si X est n'importe quel mouvement brownien pour $\mathcal{H}$, $\mathcal{K}$ est engendrée par un mouvement brownien de la forme $\int H\,dX$, où H est un processus prévisible pour $\mathcal{H}$ et de carré 1. Il est évident que toute sous-filtration représentable de $\mathcal{H}$ est elle-même brownienne. La réciproque est fausse; il est facile de caractériser, parmi les sous-filtrations browniennes de $\mathcal{H}$, celles qui sont représentables dans $\mathcal{H}$; c'est l'objet de la proposition suivante.

PROPOSITION 1. — *Soit $\mathcal{K}$ une sous-filtration brownienne d'une filtration brownienne $\mathcal{H}$. Les cinq assertions suivantes sont équivalentes.*

(i) *$\mathcal{K}$ est représentable dans $\mathcal{H}$;*

(ii) *toute martingale pour $\mathcal{K}$ en est aussi une pour $\mathcal{H}$;*

(iii) *tout mouvement brownien pour $\mathcal{K}$ en est aussi un pour $\mathcal{H}$;*

(iv) *il existe un mouvement brownien pour $\mathcal{K}$ qui en soit aussi un pour $\mathcal{H}$ et ait la propriété de représentation prévisible dans $\mathcal{H}$;*

(v) *tout mouvement brownien pour $\mathcal{K}$ en est aussi un pour $\mathcal{H}$ et a la propriété de représentation prévisible dans $\mathcal{H}$;*

(vi) *pour tout t, les tribus $\mathcal{H}_t$ et $\mathcal{K}_\infty$ sont conditionnellement indépendantes étant donnée $\mathcal{K}_t$.*

DÉMONSTRATION. — (i) $\Rightarrow$ (iv) : Il existe par hypothèse un mouvement brownien X engendrant $\mathcal{H}$ et un mouvement brownien Y engendrant $\mathcal{K}$ tel que $Y = \int H\,dX$ avec $H^2 = 1$; et toute martingale pour $\mathcal{H}$ est de la forme $\int K\,dX$, c'est-à-dire $\int KH\,dY$.

(iv) $\Rightarrow$ (v) : Puisque deux mouvements browniens pour $\mathcal{K}$ sont toujours intégrale stochastique l'un de l'autre dans $\mathcal{K}$, si l'un vérifie l'hypothèse (iv), il en va de même de l'autre; d'où (v).

(v) $\Rightarrow$ (iii) est trivial.

(iii) $\Rightarrow$ (ii) : Toute martingale pour $\mathcal{K}$, intégrale stochastique par rapport à un mouvement brownien pour $\mathcal{K}$, donc pour $\mathcal{H}$, est aussi une martingale pour $\mathcal{H}$.

(ii) $\Rightarrow$ (i) : Martingale pour $\mathcal{K}$, un mouvement brownien engendrant $\mathcal{K}$ en est aussi une pour $\mathcal{H}$, donc, grâce à la caractérisation de Lévy, un mouvement brownien pour $\mathcal{H}$.

(ii) $\Leftrightarrow$ (vi) : Cette équivalence ne nécessite pas que les filtrations considérées soit browniennes. Elle est mentionnée dans [10] (exercice (4.16) du chapitre V) et résulte facilement du critère suivant : *Si $\mathcal{B}$, $\mathcal{C}$ et $\mathcal{D}$ sont des sous-tribus telles que $\mathcal{D} \subset \mathcal{B} \cap \mathcal{C}$, $\mathcal{B}$ et $\mathcal{C}$ sont conditionnellement indépendantes étant donnée $\mathcal{D}$ si et seulement si leurs opérateurs d'espérance conditionnelle vérifient* $\mathbb{E}^{\mathcal{B}}\mathbb{E}^{\mathcal{C}} = \mathbb{E}^{\mathcal{D}}$. Ceci est laissé au lecteur, qui n'éprouvera aucune difficulté à le vérifier directement, ou à le déduire du théorème II-51 de Meyer [8]. ∎

La condition (vi) implique en particulier que $\mathcal{K}_t = \mathcal{K}_\infty \cap \mathcal{H}_t$. En conséquence, dans une filtration brownienne, une sous-filtration représentable $\mathcal{K}$ est déterminée de façon unique par sa tribu terminale $\mathcal{K}_\infty$. (Ceci est aussi un cas particulier du lemme 1 ci-dessous.) Mais nous ne voyons pas comment caractériser, parmi les sous-tribus de $\mathcal{H}_\infty$, les tribus terminales des sous-filtrations représentables.

Une autre conséquence de (ii) est la transitivité : si $\mathcal{K}$ est représentable dans $\mathcal{H}$ et $\mathcal{L}$ représentable dans $\mathcal{K}$, $\mathcal{L}$ est représentable dans $\mathcal{H}$. Réciproquement, si $\mathcal{K}$ et $\mathcal{L}$

sont deux sous-filtrations représentables d'une même filtration brownienne et si $\mathcal{L}$ est incluse dans $\mathcal{K}$, alors $\mathcal{L}$ est représentable dans $\mathcal{K}$.

PROPOSITION 2. — *La filtration de Goswami-Rao $\mathcal{G}$ est représentable dans $\mathcal{F}$; c'est en particulier une filtration brownienne.*

D'après la remarque qui précède, ceci entraîne par exemple que toute martingale pour $\mathcal{G}$ peut être écrite comme une intégrale stochastique par rapport au mouvement brownien de Lévy $\beta = \int \operatorname{sgn} B\, dB$. (La filtration naturelle de ce dernier est strictement incluse dans $\mathcal{G}$, puisque c'est $\mathcal{F}/\tilde{R}$, où la relation d'équivalence $\tilde{R}$, donnée par

$$\omega_1 \tilde{R}\, \omega_2 \iff |\omega_1| = |\omega_2| ,$$

est strictement moins fine que R. Par exemple, la variable aléatoire $\operatorname{sgn} B_{t/2} \operatorname{sgn} B_t$ est mesurable pour $\mathcal{G}_t$ mais n'est pas une fonctionnelle de β.)

Nous allons donner de cette proposition trois démonstrations, consistant chacune à exhiber un processus H prévisible pour $\mathcal{F}$, de carré 1, tel que le mouvement brownien $\int H\, dB$ engendre $\mathcal{G}$. Nous regroupons auparavant dans des lemmes les éléments communs à ces trois démonstrations.

LEMME 1. — *Soit X un mouvement brownien pour une filtration $\mathcal{H}$. Si le processus X est adapté à une sous-filtration $\mathcal{K}$ de $\mathcal{H}$ et engendre la tribu $\mathcal{K}_\infty$, il engendre la filtration $\mathcal{K}$.*

Ce résultat est faux si, au lieu de supposer que le mouvement brownien X est une martingale pour $\mathcal{H}$, on le suppose seulement adapté à $\mathcal{H}$. Un contre-exemple est donné par $\mathcal{K} = \mathcal{H} = \mathcal{F}$ et $X_t = 2B_{t/4}$; un autre contre-exemple, à la filtration beaucoup plus intéressante, est le processus

$$X_t = B_t - \int_0^t \frac{B_s}{s}\, ds$$

étudié par Jeulin et Yor dans [5] et par Meyer dans [9].

DÉMONSTRATION DU LEMME 1. — La filtration naturelle $\tilde{\mathcal{K}}$ de X est incluse dans $\mathcal{K}$; il s'agit de montrer l'inverse. Fixons $A \in \mathcal{K}_t$. Le processus $M_s = \mathbb{P}[A|\tilde{\mathcal{K}}_s]$ est une martingale pour $\tilde{\mathcal{K}}$, donc une intégrale stochastique par rapport à X, donc une martingale pour $\mathcal{H}$. Comme $M_\infty = \mathbb{P}[A|\tilde{\mathcal{K}}_\infty] = \mathbb{P}[A|\mathcal{K}_\infty] = \mathbb{1}_A$ est mesurable pour $\mathcal{K}_t$, donc pour $\mathcal{H}_t$, on a $M_t = M_\infty = \mathbb{1}_A$ et A est donc dans $\tilde{\mathcal{K}}_t$. ∎

LEMME 2. — *Sur l'espace de Wiener, soit H un processus prévisible impair, c'est-à-dire tel que $H(-\omega) = -H(\omega)$, ne prenant que les valeurs 1 et −1.*

a) *Le mouvement brownien $\tilde{B} = \int H\, dB$ est adapté à $\mathcal{G}$.*

b) *Pour que $\tilde{B}$ engendre $\mathcal{G}$, il suffit qu'il existe une variable aléatoire ε sur $(\Omega, \mathcal{A})$ à valeurs dans $\{-1, 1\}$ et une fonction mesurable ϕ de $\Omega \times \{-1, 1\}$ dans Ω telles que $B = \phi(\tilde{B}, \varepsilon)$ p. s.*

DÉMONSTRATION DU LEMME 2. — a) La variable aléatoire $\tilde{B}_t = \int_0^t H_s\,dB_s$ est mesurable pour $\mathcal{F}_t$; elle est paire car H peut être approché par des processus prévisibles simples impairs; $\tilde{B}$ est donc adapté à $\mathcal{G}$.

b) En appliquant le lemme 1, il suffit de vérifier que $\tilde{B}$ engendre la tribu $\mathcal{G}_\infty$, c'est-à-dire que toute variable aléatoire paire Z s'exprime en fonction de $\tilde{B}$. En remplaçant ω par $-\omega$ dans l'égalité p. s. $B(\omega) = \phi(\tilde{B}(\omega), \varepsilon(\omega))$, on obtient $-B(\omega) = \phi(\tilde{B}(\omega), \varepsilon(-\omega))$, d'où $\varepsilon(-\omega) \neq \varepsilon(\omega)$ puisque $B \neq -B$ p. s. Ainsi l'ensemble à deux éléments $\{B(\omega), -B(\omega)\}$ est égal à $\{\phi(\tilde{B}(\omega), 1), \phi(\tilde{B}(\omega), -1)\}$; et, Z étant paire, $Z(\omega) = Z \circ \phi(\tilde{B}(\omega), 1)$ p. s. ne dépend que de $\tilde{B}$. ∎

LEMME 3. — *Toute suite $u = (u_n)_{n\in\mathbb{Z}} \in \{-1,1\}^{\mathbb{Z}}$ s'exprime comme fonction borélienne de u_0 et de la suite v définie par $v_n = u_{n-1}u_n$.*

DÉMONSTRATION DU LEMME 3. — On a $u_n = u_0 \prod_{n<i\leqslant 0} v_i$ pour $n \leqslant 0$ et $u_n = u_0 \prod_{0<i\leqslant n} v_i$ pour $n \geqslant 0$. ∎

DÉFINITION. — *Nous appellerons ici* subdivision *toute famille $(t_n)_{n\in\mathbb{Z}}$ telle que $0 < t_n < t_{n+1}$ pour tout n et que* $\inf_n t_n = 0$, $\sup_n t_n = \infty$. (On pourrait sans inconvénient remplacer $\mathbb{Z}$ par $-\mathbb{N}$ dans cette définition; l'important est seulement l'accumulation des t_n vers zéro.)

PREMIÈRE DÉMONSTRATION DE LA PROPOSITION 2. — Une subdivision $(t_n)_{n\in\mathbb{Z}}$ étant fixée, on pose $\varepsilon_n = \operatorname{sgn}(B_{t_n} - B_{t_{n-1}})$ et on définit un processus prévisible impair de carré 1 par $H^1 = \sum_{n\in\mathbb{Z}} \varepsilon_n \mathbb{1}_{]t_n, t_{n+1}]}$. Pour établir que le mouvement brownien $B^1 = \int H^1\,dB$ engendre $\mathcal{G}$, nous allons appliquer le lemme 2 et reconstruire B à partir de ε_0 et B^1. Il suffit pour cela de reconstruire chacun des processus $Z^n = (B_{t_n+t} - B_{t_n},\ t \in [0, t_{n+1} - t_n])$. Mais, par définition de B^1, $Z^n = \varepsilon_n \tilde{Z}^n$ où $\tilde{Z}^n = (B^1_{t_n+t} - B^1_{t_n},\ t \in [0, t_{n+1} - t_n])$ est un morceau de B^1; il suffit donc d'exprimer ε_n en fonction de B^1 et de ε_0. La définition de B^1 entraîne que $\varepsilon_{n-1}\varepsilon_n$ est égal à $\operatorname{sgn}(B^1_{t_n} - B^1_{t_{n-1}})$ et le lemme 3 permet de conclure. ∎

REMARQUE. — Dans [6], Le Gall et Yor étudient le même couple (B, B^1), mais vu sous un autre angle : se donnant B^1 a priori, ils s'intéressent à la solution B de l'équation $dB = H^1(B)\,dB^1$. (Elle n'a pas de solution forte, puisque B^1 a une filtration strictement incluse dans celle de B.)

DEUXIÈME DÉMONSTRATION DE LA PROPOSITION 2. — La subdivision $(t_n)_{n\in\mathbb{Z}}$ restant fixée, on définit comme ci-dessus un processus prévisible impair de carré 1 par $H^2 = \sum_{n\in\mathbb{Z}} \eta_n \mathbb{1}_{]t_n, t_{n+1}]}$, où l'on a posé cette fois-ci $\eta_n = \operatorname{sgn} B_{t_n}$. Toujours grâce au lemme 2, il nous suffit de retrouver B à partir de $B^2 = \int H^2\,dB$ et de η_0. Comme précédemment, il suffit de retrouver les signes η_n, qui transforment les morceaux $(B^2_{t_n+t} - B^2_{t_n},\ t \in [0, t_{n+1} - t_n])$ du brownien B^2 en les morceaux Z^n de B; selon le lemme 3, les η_n s'expriment à l'aide de η_0 et des produits $\eta_{n-1}\eta_n$ et il suffit donc d'écrire ces produits à partir de B^2. Or

$$\begin{aligned}\eta_{n-1}\eta_n = \operatorname{sgn}(\eta_{n-1}B_{t_n}) &= \operatorname{sgn}\big(\eta_{n-1}B_{t_{n-1}} + \eta_{n-1}(B_{t_n} - B_{t_{n-1}})\big)\\ &= \operatorname{sgn}\big(|B_{t_{n-1}}| + (B^2_{t_n} - B^2_{t_{n-1}})\big)\,;\end{aligned}$$

il ne reste donc qu'à exprimer $|B_{t_n}|$ en termes du nouveau brownien B^2.

À cette fin, remarquons que $B_{t_n} = \eta_{n-1}\big((B^2_{t_n} - B^2_{t_{n-1}}) + |B_{t_{n-1}}|\big)$, donc

$$|B_{t_n}| = \big|(B^2_{t_n} - B^2_{t_{n-1}}) + |B_{t_{n-1}}|\big| = \Phi_1\big(B^2_{t_n} - B^2_{t_{n-1}}\,,\ |B_{t_{n-1}}|\big)\,,$$

où l'on a posé $\Phi_1(x,y) = |x+y|$. En définissant pour $p \geqslant 1$ des fonctions Φ_p par

$$\Phi_{p+1}(x_1,\ldots,x_p,x_{p+1},y) = \Phi_p\big(x_1,\ldots,x_p,\Phi_1(x_{p+1},y)\big)\,,$$

en posant $\Delta_n = B^2_{t_{n+1}} - B^2_{t_n}$ et en utilisant $|B_{t_{n-p}}| = \Phi_1(\Delta_{n-p-1}\,,\ |B_{t_{n-p-1}}|)$, on obtient de proche en proche

$$|B_{t_n}| = \Phi_p\big(\Delta_{n-1},\ldots,\Delta_{n-p}\,,\ |B_{t_{n-p}}|\big)\,.$$

Mais les fonctions Φ_p vérifient $\big|\Phi_p(x_1,\ldots,x_p,y) - \Phi_p(x_1,\ldots,x_p,0)\big| \leqslant |y|$ (cela se vérifie immédiatement par récurrence sur p). On peut donc écrire

$$\big||B_{t_n}| - \Phi_p(\Delta_{n-1},\ldots,\Delta_{n-p}\,,0)\big| \leqslant |B_{t_{n-p}}|$$

et il en résulte que $\Phi_p\big(\Delta_{n-1},\ldots,\Delta_{n-p}\,,0\big)$ tend, quand p tend vers l'infini, vers $|B_{t_n}|$, qui est donc une fonction mesurable du processus B^2. ∎

REMARQUES. — a) Ce résultat montre que l'analogie syntaxique entre la formule définissant B^2 et la transformation de Lévy est trompeuse : en un sens, B^2 ressemble bien plus à B^1 qu'à β et sa construction ne doit pas être interprétée comme une sorte de transformation de Lévy discrète.

b) Au vu des deux démonstrations qui précèdent, on pourrait espérer qu'en prenant n'importe quelles variables aléatoires σ_n de carré 1, impaires et respectivement mesurables pour $\mathcal{F}_{t_n}$ et en posant $H = \sum_n \sigma_n \mathbb{1}_{]t_n,t_{n+1}]}$ et $\tilde{B} = \int H\,dB$, on obtiendrait toujours un mouvement brownien engendrant $\mathcal{G}$. Cela ne marche pas, comme on peut s'en convaincre par exemple en itérant la transformation $T : B \mapsto B^1$ de l'espace de Wiener. Son carré $T \circ T$ est du type indiqué, avec

$$\sigma_n = \operatorname{sgn}(B^1_{t_n} - B^1_{t_{n-1}})\,\operatorname{sgn}(B_{t_n} - B_{t_{n-1}})\,,$$

mais la filtration obtenue, celle des processus dépendant de façon paire de B^1, est strictement incluse dans celle de B^1. Ceci suggère d'étudier la suite décroissante des filtrations obtenues à l'aide des puissances de la transformation T; nous y reviendrons avec la proposition 3.

TROISIÈME DÉMONSTRATION DE LA PROPOSITION 2. — Appelons H^3_t la variable aléatoire égale au signe de la plus longue excursion de B avant t, ou, plus précisément, avant l'instant $G_t = \sup\{s \in [0,t[\ : B_s = 0\}$. Le processus H^3 est prévisible (car adapté et continu à gauche) et impair; pour établir que $B^3 = \int H^3\,dB$ engendre $\mathcal{G}$, il suffit, grâce au lemme 2, de montrer que $B = \phi(B^3, H^3_1)$.

Puisque $H^3_t = H^3_{G_t}$, la formule du balayage donne $B^3 = \int H^3\,dB = H^3 B$, d'où $|B^3| = |B|$; ainsi B^3 a les mêmes zéros et, aux signes près, les mêmes excursions que B. Comme $B = H^3 B^3$, nous devons exprimer H^3 à l'aide de B^3 et de H^3_1.

Soit $]U_0, V_0[$ l'intervalle portant la plus longue excursion avant G_1; pour $n < 0$ on définit inductivement des variables aléatoires U_n et V_n en appelant $]U_n, V_n[$ l'intervalle qui porte la plus longue excursion de B avant U_{n+1}; pour $n > 0$ on appelle $]U_n, V_n[$ l'intervalle qui porte la première excursion plus longue que $V_{n-1} - U_{n-1}$ (donc plus longue que toutes les excursions antérieures). On notera e_n le signe de B sur $]U_n, V_n[$; avec ces notations, $H^3 = \sum_{n\in\mathbb{Z}} e_n \mathbb{1}_{]V_n,V_{n+1}]}$. Puisque les U_n et V_n

peuvent aussi bien être définis à partir de B^3 que de B, il nous suffit de retrouver les e_n à partir de B^3 et de $H_1^3 = e_0$. D'après le lemme 3, il suffit pour cela de savoir exprimer les produits $e_n e_{n+1}$ à l'aide de B^3. Mais sur l'intervalle $]U_{n+1}, V_{n+1}[$ on a d'une part $B = e_n B^3$ (c'est vrai sur tout l'intervalle $]V_n, V_{n+1}]$) et d'autre part $\operatorname{sgn} B = e_{n+1}$, donc aussi $\operatorname{sgn} B^3 = e_n e_{n+1}$, d'où le résultat. ∎

REMARQUES. — a) Une démonstration en tout point semblable à cette dernière permet d'établir le même résultat lorsque l'on prend pour H_t le signe de l'excursion la plus haute (en valeur absolue) avant G_t.

Mais on ne peut pas prendre pour H_t le signe de n'importe quelle excursion choisie canoniquement dans l'intervalle $[0, G_t]$. Par exemple, nous verrons plus bas (dans la démonstration de la proposition 4) que si l'on prend H_t égal au signe de la plus longue excursion entre $V(t)$ et G_t, où $V(t)$ désigne la fin de la plus longue excursion avant G_t, la filtration obtenue est strictement plus petite que $\mathcal{G}$.

b) Les deux premières démonstrations faisaient intervenir le choix arbitraire d'une subdivision; la troisième est plus intrinsèque. Plus précisément, pour tout $\lambda > 0$, la transformation $B \mapsto B^3$ de Ω dans lui-même commute avec la dilatation $B \mapsto \left(\frac{1}{\lambda} B_{\lambda^2 t}\right)_{t \geqslant 0}$, de sorte que celle-ci opère non seulement sur B, mais sur le couple (B, B^3). Bien entendu, pour $\lambda < 0$, cette commutation n'a plus lieu et est au contraire remplacée par une anticommutation.

DÉFINITIONS. — *Nous appellerons* transformation brownienne *toute application mesurable de Ω dans lui-même* (définie partout ou seulement presque partout) *qui préserve la mesure de Wiener* $\mathbb{P}$. Dans ce cas, $B \circ T$ est un mouvement brownien sur l'espace de Wiener.

Une transformation brownienne T sera dite adaptée *si le nouveau brownien $B \circ T$ est adapté à $\mathcal{F}$; elle sera dite* représentable *si $B \circ T$ est en outre un mouvement brownien pour $\mathcal{F}$.*

Dans ce dernier cas, T est caractérisée par le processus H prévisible pour $\mathcal{F}$, de carré 1, tel que $B \circ T = \int H\, dB$; nous le noterons H^T. La composée $U = T \circ S$ de deux transformations adaptées est adaptée (car $B \circ U$ est adapté à la filtration naturelle de $B \circ T$, qui est incluse dans $\mathcal{F}$); si en outre S et T sont toutes deux représentables, il en va de même de U, avec $H^U = H^S\, H^T \circ S$.

Le choix de l'adjectif représentable se justifie d'une part par le lien avec les filtrations représentables (explicité dans le paragraphe suivant) et d'autre part parce que ces transformations peuvent toutes être décrites à l'aide d'intégrales stochastiques d'opérateurs dans le cadre du calcul stochastique non commutatif (voir [2]).

Si T est une transformation brownienne adaptée (respectivement représentable), la filtration naturelle de $B \circ T$ est une sous-filtration brownienne (respectivement représentable) de $\mathcal{F}$, que nous noterons $\mathcal{F}^T$; les itérées T^n de T sont aussi des transformations browniennes adaptées (respectivement représentables) et les $\mathcal{F}^{T^n}$ forment donc une suite décroissante de sous-filtrations browniennes (respectivement représentables) de $\mathcal{F}$, strictement décroissante si $\mathcal{F}^T$ est une sous-filtration stricte de $\mathcal{F}$. Nous appellerons $\mathcal{F}^{T^\infty}$ la limite des $\mathcal{F}^{T^n}$ (elle est automatiquement continue à droite comme intersection de filtrations continues à droite). Cette filtration limite est parfois difficile à décrire : lorsque T est la transformation de Lévy (transformation représentable donnée par $H = \operatorname{sgn} B$), $\mathcal{F}^{T^\infty}$ n'est pas connue (on peut conjecturer que c'est la filtration triviale; ceci impliquerait l'ergodicité de la transformation de Lévy).

Toute sous-filtration représentable de $\mathcal{F}$ est de la forme $\mathcal{F}^T$ pour une transformation représentable T qui est, bien sûr, très loin d'être unique (le choix d'une telle T équivaut au choix de l'un des mouvements browniens qui engendrent cette sous-filtration). Les trois démonstrations de la proposition 2 ont consisté à exhiber trois transformations représentables T pour lesquelles $\mathcal{F}^T$ soit la filtration de Goswami-Rao $\mathcal{G}$. Pour la première et la troisième de ces trois transformations, la filtration limite $\mathcal{F}^{T^\infty}$ peut être décrite explicitement comme un quotient de $\mathcal{F}$ et est représentable; ceci fera l'objet des deux prochaines propositions.

PROPOSITION 3. — *Étant donnée une subdivision $(t_n)_{n\in\mathbb{Z}}$, si l'on note T la transformation représentable associée au processus prévisible*

$$H^1 = \sum_{n\in\mathbb{Z}} \operatorname{sgn}(B_{t_n}-B_{t_{n-1}})\, \mathbb{1}_{]t_n,t_{n+1}]}\,,$$

la filtration $\mathcal{F}^{T^\infty}$ n'est autre que $\mathcal{F}/R'$, où R' est la relation qui lie ω_1 et ω_2 si et seulement si, pour chaque n, les trajectoires $\big(\omega_1(s)-\omega_1(t_n)\,,\ t_n \leqslant s \leqslant t_{n+1}\big)$ et $\big(\omega_2(s)-\omega_2(t_n)\,,\ t_n \leqslant s \leqslant t_{n+1}\big)$ sont les mêmes à un facteur ± 1 près. En outre, cette filtration est représentable dans $\mathcal{F}$.

Rappelons un lemme de Lindvall & Rogers [7] qui résulte de la préservation des martingales lors d'un grossissement indépendant.

LEMME 4. — *Étant donné un espace probabilisé, soient $\mathcal{B}$ et $\mathcal{C}$ deux sous-tribus indépendantes, $(\mathcal{B}^k)_{k\in\mathbb{N}}$ (respectivement $(\mathcal{C}^k)_{k\in\mathbb{N}}$) une suite décroissante de sous-tribus de $\mathcal{B}$ (respectivement $\mathcal{C}$) et $\mathcal{B}^\infty$ (respectivement $\mathcal{C}^\infty$) la limite $\bigcap_k \mathcal{B}^k$ (respectivement $\bigcap_k \mathcal{C}^k$). On a $\bigcap_k (\mathcal{B}^k \vee \mathcal{C}^k) = \mathcal{B}^\infty \vee \mathcal{C}^\infty$.*
En particulier, $\bigcap_k (\mathcal{B}^k \vee \mathcal{C}) = \mathcal{B}^\infty \vee \mathcal{C}$.

LEMME 5. — *Soient $E = (E_n)_{n\in\mathbb{Z}}$ une suite indépendante de variables aléatoires de même loi uniforme dans $\{-1,1\}$ et τ l'application de $\{-1,1\}^{\mathbb{Z}}$ dans lui-même qui transforme $(e_n)_{n\in\mathbb{Z}}$ en $(e'_n)_{n\in\mathbb{Z}}$, où $e'_n = e_{n-1}e_n$. La suite décroissante des tribus $\sigma(\tau^k \circ E)$ a une intersection dégénérée.*

DÉMONSTRATION DU LEMME 5. — En itérant l'opération τ, on voit facilement, par récurrence, que lorsque k est une puissance de 2, τ^k transforme E_n en $E_{n-k}E_n$. En conséquence, pour un tel k, la tribu $\mathcal{B}^k = \sigma(\tau^k \circ E)$ est engendrée par les variables aléatoires $E_{n-k}E_n$, et consiste en les fonctionnelles de la suite E qui sont invariantes par chacune des k transformations consistant à changer en bloc le signe de toute une sous-suite $(E_{kn+i})_{n\in\mathbb{Z}}$. Cette tribu $\mathcal{B}^k$ est donc indépendante de $\sigma(E_n, n \in A)$ pour tout A n'ayant dans chacune de ces sous-suites qu'un élément au plus. Donc, si A est une partie finie de $\mathbb{Z}$, la variable aléatoire $E_A = \prod_{n\in A} E_n$ est indépendante de $\mathcal{B}^k$ lorsque k est une puissance de 2 assez grande (supérieure au diamètre de A). Ceci entraîne que $\mathbb{E}[E_A|\mathcal{B}^k] = \mathbb{E}[E_A]$ et à la limite $\mathbb{E}[E_A|\bigcap_k \mathcal{B}^k] = \mathbb{E}[E_A]$. Le lemme en découle puisque les E_A forment une base orthonormée de $\mathcal{B}^0$. ∎

DÉMONSTRATION DE LA PROPOSITION 3. — Pour chaque n, à partir du mouvement brownien $Z^n = (B_{t_n+s} - B_{t_n}\,,\ s \in [0, t_{n+1} - t_n])$, la proposition 2 permet de construire un mouvement brownien $X^n = (X^n_s\,,\ s \in [0, t_{n+1} - t_n])$ qui engendre la filtration de Goswami-Rao de Z^n et qui soit de la forme $X^n_s = \int_0^s H^n_u\, dZ^n_u$ où H^n est prévisible pour Z^n. Les X^n étant, comme les Z^n, indépendants, on peut définir un nouveau mouvement brownien B' par $B'_t - B'_{t_n} = X^n_{t-t_n}$ pour $t_n < t \leqslant t_{n+1}$; il vérifie $B' = \int H'\, dB$, où le processus H' prévisible pour $\mathcal{F}$ est défini par $H'_t = H^n_{t-t_n}$ pour $t_n < t \leqslant t_{n+1}$. Par construction de B', sa filtration naturelle $\mathcal{G}'$ est $\mathcal{F}/R'$, qui est donc une sous-filtration représentable de $\mathcal{F}$. Il reste à établir que $\mathcal{F}^{T^\infty} = \mathcal{G}'$. Par le lemme 1 appliqué à $\mathcal{K} = \mathcal{F}^{T^\infty}$, $\mathcal{H} = \mathcal{F}$ et $X = B'$, il suffit pour cela de vérifier que $\mathcal{F}^{T^\infty}_\infty = \mathcal{G}'_\infty$.

La tribu $\mathcal{A}$ sur l'espace de Wiener est engendrée par les processus Z^n ; elle est donc aussi engendrée par les X^n et la suite ε des signes $\varepsilon_n = \mathrm{sgn}(B_{t_n} - B_{t_{n-1}})$; ces ingrédients sont indépendants. La tribu $\mathcal{B} = \sigma(\varepsilon)$ est donc indépendante de $\mathcal{G}'_\infty$ et telle que $\mathcal{A} = \mathcal{G}'_\infty \vee \mathcal{B}$.

La transformation T laisse invariante toute la tribu $\mathcal{G}'_\infty$ et opère sur la suite ε par la transformation τ qui remplace chaque ε_n par $\varepsilon_{n-1}\varepsilon_n$. Les tribus $\mathcal{B}^k = \sigma(\tau^k \circ \varepsilon)$ forment une filtration inverse, indépendante de $\mathcal{G}'_\infty$; on a $\mathcal{F}^{T^k}_\infty = \mathcal{G}'_\infty \vee \mathcal{B}^k$ et à la limite $\mathcal{F}^{T^\infty}_\infty = \bigcap_k (\mathcal{G}'_\infty \vee \mathcal{B}^k)$. Le lemme 4 dit que cette tribu est engendrée par $\mathcal{G}'_\infty$ et par la tribu asymptotique $\bigcap_k \mathcal{B}^k$; cette dernière étant dégénérée par le lemme 5, $\mathcal{F}^{T^\infty}_\infty$ est égale à $\mathcal{G}'_\infty$. ∎

REMARQUE. — Si dans cette proposition on remplace H^1 par le processus $H^2 = \sum_n \mathrm{sgn}(B_{t_n}) \mathbb{1}_{]t_n, t_{n+1}]}$, il est clair que la filtration asymptotique $\mathcal{F}^{T^\infty}$ contient $\mathcal{G}'$ (qui est invariante par T) mais nous ne savons pas s'il y a égalité.

En revanche, pour H^3 (introduit dans la troisième démonstration de la proposition 2), la filtration asymptotique peut être caractérisée. Reprenons des notations introduites lors de cette démonstration : G_t est le dernier zéro de B avant t ; $]U_0, V_0[$ est l'intervalle de la plus longue excursion de B avant G_1 ; pour $n < 0$, $]U_n, V_n[$ porte la plus longue excursion de B avant U_{n+1} ; pour $n > 0$, $]U_n, V_n[$ porte la première excursion plus longue que $V_{n-1} - U_{n-1}$; e_n est le signe de B sur $]U_n, V_n[$.

PROPOSITION 4. — *Si l'on note S la transformation représentable associée au processus $H^3 = \sum_{n \in \mathbb{Z}} e_n \mathbb{1}_{]V_n, V_{n+1}]}$ (prévisible car H^3_t est le signe de la plus longue excursion de B sur l'intervalle $[0, G_t]$), la filtration $\mathcal{F}^{S^\infty}$ n'est autre que $\mathcal{F}/R''$, où la relation d'équivalence R'' relie ω_1 et ω_2 si et seulement si $|\omega_1| = |\omega_2|$ et, pour chaque n, les trajectoires $(\omega_1(s)\,,\ V_n(\omega_1) \leqslant s \leqslant V_{n+1}(\omega_1))$ et $(\omega_2(s)\,,\ V_n(\omega_2) \leqslant s \leqslant V_{n+1}(\omega_2))$ sont les mêmes à un facteur ± 1 près. En outre, cette filtration est représentable dans $\mathcal{F}$.*

Remarquer que la condition $|\omega_1| = |\omega_2|$ implique que ω_1 et ω_2 ont les mêmes zéros, donc $V_n(\omega_1) = V_n(\omega_2)$ et $V_{n+1}(\omega_1) = V_{n+1}(\omega_2)$.

DÉMONSTRATION. — Notons $\mathcal{G}''$ la filtration $\mathcal{F}/R''$. Pour $t > 0$, appelons $]U(t), V(t)[$ l'intervalle qui porte la plus longue excursion de B sur $[0, G_t]$, $]U'(t), V'(t)[$ celui qui porte la plus longue excursion sur $[V(t), G_t]$ et H''_t le signe de B sur $]U'(t), V'(t)[$. Le processus H'' est prévisible (et même continu à gauche) ;

puisque $H''_t = H''_{G_t}$, le mouvement brownien $B'' = \int H''\,dB$ est aussi égal à $H''B$ (formule du balayage). Si ω_1 et ω_2 sont liés par R'', ils ont la même trajectoire à un facteur ± 1 près sur l'intervalle $[V(t), t]$ donc $B''_t(\omega_1) = H''_t(\omega_1)B_t(\omega_1) = H''_t(\omega_2)B_t(\omega_2) = B''_t(\omega_2)$; ceci indique que B'' est adapté à $\mathcal{G}''$; il est en particulier indépendant de la suite $e = (e_n)_{n\in\mathbb{Z}}$. Réciproquement, nous allons voir que B'' engendre la filtration $\mathcal{G}''$; en appliquant le lemme 1 à $\mathcal{K} = \mathcal{G}''$, $\mathcal{H} = \mathcal{F}$ et $X = B''$, il suffit de vérifier que B'' engendre la tribu $\mathcal{G}''_\infty = \mathcal{A}/R''$. Mais la formule $B = H''B''$ montre que $\mathcal{A}$ est engendrée par B'', les V_n et e; comme B'' a les mêmes zéros que B, les V_n sont eux-mêmes fonction de B'' et $\mathcal{A}$ est engendrée par B'' et e; toute variable aléatoire est donc de la forme $Z = \phi(B'', e)$ p. s. Si Z est mesurable pour $\mathcal{G}''_\infty$, elle est constante sur les classes d'équivalence pour R''; mais lorsque ω parcourt une classe d'équivalence, $B''(\omega)$ reste fixe et $e(\omega)$ décrit tout $\{-1,1\}^{\mathbb{Z}}$. En conséquence, en appelant μ la loi de e, on a aussi $Z = \int_{u\in\{-1,1\}^{\mathbb{Z}}} \phi(B'', u)\,\mu(du)$; comme ceci ne dépend que de B'', B'' engendre $\mathcal{G}''$.

Il reste à établir que $\mathcal{F}^{S^\infty} = \mathcal{G}''$. Appelons $M(t)$ et $M'(t)$ les milieux respectifs des intervalles $[U(t), V(t)]$ et $[U'(t), V'(t)]$, et Σ la transformation représentable $B \mapsto B''$, associée à H''. La composée $\Sigma\circ S$ est représentable, avec

$$\begin{aligned} H^{\Sigma\circ S}_t &= H''_t\circ S\; H^3_t = \operatorname{sgn}(B_{M'(t)}\circ S)\;\operatorname{sgn} B_{M(t)} = \operatorname{sgn} B^3_{M'(t)}\;\operatorname{sgn} B_{M(t)} \\ &= H^3_{M'(t)}\;\operatorname{sgn} B_{M'(t)}\;\operatorname{sgn} B_{M(t)} = \operatorname{sgn} B_{M'(t)} = H''_t\,; \end{aligned}$$

il en résulte que $\Sigma\circ S = \Sigma$, d'où $B''\circ S = B''$ et la filtration $\mathcal{G}''$, engendrée par B'', est invariante par S, donc incluse dans $\mathcal{F}^{S^\infty}$. Le mouvement brownien B'' est donc adapté à $\mathcal{F}^{S^\infty}$; pour montrer que sa filtration naturelle est $\mathcal{F}^{S^\infty}$, il suffit d'après le lemme 1 de vérifier qu'il engendre la tribu $\mathcal{F}^{S^\infty}_\infty$, c'est-à-dire que $S^{-k}(\mathcal{A})$ décroît vers $\mathcal{G}''_\infty$ quand k tend vers l'infini.

Mais $\mathcal{A}$ est engendrée par les tribus indépendantes $\mathcal{G}''_\infty$ et $\sigma(e)$; S laisse invariants les événements de $\mathcal{G}''_\infty$ et opère sur e par l'application τ du lemme 5. Le lemme 4 dit que $\bigcap_k S^{-k}(\mathcal{A}) = \mathcal{G}''_\infty \vee \bigcap_k \sigma(\tau^k\circ e)$; puisque $\bigcap_k \sigma(\tau^k\circ e)$ est dégénérée (lemme 5), $\bigcap_k S^{-k}(\mathcal{A}) = \mathcal{G}''_\infty$. ∎

La proposition 4 admet une variante, dans laquelle H^3_t est remplacé par le signe de la plus haute excursion de $|B|$ sur l'intervalle $[0, G_t]$; la démonstration, tout à fait semblable, conduit au même résultat (à condition de remplacer partout "plus longue excursion de B" par "plus haute excursion de $|B|$" dans la construction des variables aléatoires U_n et V_n).

La décomposition du mouvement brownien en les signes de ses accroissements sur des intervalles aléatoires et les tribus de Goswami-Rao de ces accroissements, que nous venons d'utiliser pour étudier les propriétés asymptotiques de certaines transformations représentables, peut aussi être mise à profit pour construire des transformations représentables de période finie. La proposition suivante apporte une réponse négative à une question formulée dans [1].

PROPOSITION 5. — *Pour tout entier $p > 1$, il existe une transformation représentable T de l'espace de Wiener qui commute avec les dilatations positives et qui vérifie $T \neq \mathrm{Id}$ et $T^p = \mathrm{Id}$.*

Lorsque p est pair, cet énoncé est trivialement satisfait avec $T(\omega) = -\omega$. Dans la suite, l'entier p est fixé et impair.

Nous noterons W l'espace produit $\{-1,1\}^{\mathbb{Z}}$; nous le munirons de la loi de Rademacher μ, pour laquelle les coordonnées sont indépendantes et de même loi uniforme sur $\{-1,1\}$.

LEMME 6. — *Il existe un entier $q > 0$, une famille $(f_n)_{n\in\mathbb{Z}}$, périodique de période q, d'applications de $\{-1,1\}^{-\mathbb{N}^*}$ dans $\{-1,1\}$ et une transformation mesurable θ de W dans lui-même, définie presque partout pour μ et préservant μ, vérifiant $\theta \neq \mathrm{Id}$ et $\theta^p = \mathrm{Id}$ et telle que, pour tout $n \in \mathbb{Z}$ et presque tout $u \in W$, le produit $u_n\,\theta(u)_n$ s'écrive $f_n\big((u_{n-k})_{k>0}\big)$.*

DÉMONSTRATION DU LEMME 6. — Nous allons construire une application $t \neq \mathrm{Id}$ de $\{-1,1\}^{-\mathbb{N}^*}$ dans lui-même (au lieu de $\{-1,1\}^{\mathbb{Z}}$), préservant la mesure, vérifiant $t^p = \mathrm{Id}$ et telle que $u_n\,t(u)_n$ soit de la forme $f_n\big((u_{n-k})_{k>0}\big)$ pour une suite périodique $(f_n)_{n<0}$, de période q. Le lemme s'en déduira en définissant θ sur W par la formule $\theta(u)_n = t\big((u_{aq+m})_{m<0}\big)_{n-aq}$, où a est assez grand pour que $n-aq$ soit négatif; la condition de périodicité assurera que ceci ne dépend pas du choix d'un tel a.

Remplaçons le groupe multiplicatif $\{-1,1\}$ par le groupe additif $\mathbb{Z}/2\mathbb{Z}$ qui lui est isomorphe. La seule modification à faire est de remplacer le produit par l'addition modulo 2, que nous noterons $\dot{+}$. En associant à tout $x \in [0,1]$ son développement dyadique $\sum_{n<0} u_n\,2^n$, identifions les espaces probabilisés $\{0,1\}^{-\mathbb{N}^*}$ et $[0,1]$ (les ambiguïtés sont négligeables). L'application τ de $[0,1]$ dans lui-même définie par $\tau(x) \equiv x + 1/p$ (modulo 1) préserve la mesure et vérifie $\tau^p = \mathrm{Id}$ et $\tau \neq \mathrm{Id}$; en la transférant à $\{0,1\}^{-\mathbb{N}^*}$, on obtient une application t ayant les mêmes propriétés.

Nous devons vérifier que $u_n \dot{+} t(u)_n$ ne dépend que des u_m pour $m < n$. Mais l'algorithme d'addition des nombres réels développés en base 2 permet d'écrire $t(u)_n = u_n \dot{+} \alpha_n \dot{+} r_{n-1}$, où les $\alpha_k \in \{0,1\}$ sont définis par $1/p = \sum_{k<0} \alpha_k\,2^k$ et où $r_{n-1} \in \{0,1\}$ est la retenue provenant des termes de rang $< n$; donc $u_n \dot{+} t(u)_n = \alpha_n \dot{+} r_{n-1}$ est une fonctionnelle des u_m pour $m < n$.

Enfin, puisque la suite des α_k est périodique (appelons q sa période), l'algorithme donnant u_n en fonction de $(u_{n-k})_{k\geqslant 0}$ dépend aussi de façon périodique de n. ∎

DÉFINITION. — *Appelons* subdivision optionnelle *tout ensemble aléatoire optionnel de la forme $\bigcup_{n\in\mathbb{Z}} [\![R_n]\!]$ pour des variables aléatoires R_n telles que, pour presque tout ω, la famille $\big(R_n(\omega)\big)_{n\in\mathbb{Z}}$ soit une subdivision. Une telle famille $(R_n)_{n\in\mathbb{Z}}$ sera appelée une* numérotation croissante *de l'ensemble.*

La subdivision optionnelle est seulement la réunion des graphes des R_n, et ne comporte pas l'information supplémentaire qui consisterait en la numérotation fournie par les R_n. (Pour avoir des définitions cohérentes, nous aurions dû, dans le cas déterministe, définir une subdivision comme l'*ensemble* des t_n au lieu de la famille; ceci n'aurait pas changé grand chose.)

On n'exige pas que les R_n soient des temps d'arrêt; il peut arriver qu'une subdivision optionnelle soit impossible à numéroter par des temps d'arrêt strictement croissants. C'est le cas, par exemple, de la subdivision optionnelle union des graphes des V_n considérés plus haut : c'est l'ensemble Γ de toutes les fins des excursions qui sont plus longues que toute excursion antérieure. Le comportement de cette subdivision optionnelle Γ sous les dilatations $(D_\lambda\omega)(t) = \lambda^{-1}\omega(\lambda^2 t)$ donne lieu à une propriété d'invariance d'échelle : $\big(t, D_\lambda(\omega)\big) \in \Gamma$ si et seulement si $(\lambda^2 t, \omega) \in \Gamma$.

LEMME 7. — *Soit $q \geqslant 1$ un entier. Il existe, sur l'espace de Wiener pourvu de la filtration des zéros browniens, deux subdivisions optionnelles Γ' et Γ'' incluses dans la subdivision optionnelle Γ définie ci-dessus, vérifiant la propriété d'invariance d'échelle, telles que $\Gamma'' \subset \Gamma'$ et que deux points consécutifs de Γ'' soient toujours séparés par exactement $q-1$ points de Γ'.*

Pour $q = 1$, ce lemme est trivial : prendre $\Gamma'' = \Gamma' = \Gamma$. Pour $q \geqslant 2$, il signifie que bien qu'il ne soit pas possible de numéroter les points de Γ' de façon adaptée (par des temps d'arrêt), on peut, de façon adaptée, les numéroter cycliquement modulo q : affecter la valeur 0 aux points de Γ'', la valeur 1 aux points de Γ' qui suivent immédiatement un point de Γ'', etc.

DÉMONSTRATION DU LEMME 7. — Nous supposons $q \geqslant 2$. Reprenons des notations déjà utilisées : les intervalles portant les excursions plus longues que toute excursion antérieure sont appelés $]U_n, V_n[$, la numérotation étant choisie telle que $V_0 < 1 \leqslant V_1$; de plus, Γ est la réunion des graphes des V_n.

Les deux plus longues excursions de l'intervalle $[V_{n-1}, U_n]$ peuvent être classées par ordre de longueur ou par ordre chronologique; nous poserons $\xi_n = 1$ si ces deux ordres coïncident et $\xi_n = -1$ s'ils sont inverses. Pour chaque $n \in \mathbb{Z}$, la transformation consistant à retourner le temps sur l'intervalle $[V_{n-1}, U_n]$ préserve la mesure de Wiener; il en résulte que les variables aléatoires ξ_n sont indépendantes et uniformément distribuées sur $\{-1, 1\}$. Il existe donc presque sûrement des valeurs de n arbitrairement voisines de $-\infty$ et des valeurs de n arbitrairement voisines de $+\infty$ telles que $\xi_{n-q+1} = \ldots = \xi_n = 1$ (respectivement $\xi_n = -1$); ceci permet de définir la subdivision optionnelle

$$\Gamma'' = \{t > 0 : \exists n \in \mathbb{Z}\ t = V_n \text{ et } \xi_{n-q} = -1, \xi_{n-q+1} = \ldots = \xi_n = 1\} \subset \Gamma$$

qui possède la propriété d'invariance d'échelle et est telle que deux points consécutifs de Γ'' sont séparés par au moins $q-1$ points de Γ. Il ne reste qu'à définir Γ' comme l'ensemble des points de Γ qui comptent un point de Γ'' parmi leurs $q-1$ prédécesseurs immédiats dans Γ. ∎

DÉMONSTRATION DE LA PROPOSITION 5 — Étant donné p, le lemme 6 fournit θ et q et le lemme 7 donne Γ' et Γ''. Si $(R''_n)_{n\in\mathbb{Z}}$ est une numérotation croissante de Γ'', il existe une (unique) numérotation croissante $(R'_n)_{n\in\mathbb{Z}}$ de Γ' telle que, pour tout n, $R''_n = R'_{qn}$. Pour $0 \leqslant \ell < q$, l'ensemble

$$A_\ell = \bigcup_{n\in\mathbb{Z}}]\!]R'_{\ell+nq}, R'_{\ell+nq+1}]\!]$$

est formé des instants $t > 0$ tels que Γ' comporte exactement ℓ points entre $\sup(\Gamma'' \cap]0,t[)$ et t; cet ensemble A_ℓ dépend de Γ' et Γ'' et non du choix des R'_n et des R''_n. Fixant $\varepsilon > 0$, on aurait pu choisir les R''_n tels que R''_0 soit le début de $\Gamma'' \cap]\varepsilon, \infty[$; tous les R''_n pour $n \geqslant 0$ seraient alors des temps d'arrêt et $A_\ell \cap]\!]R''_0, \infty[\![$ serait donc prévisible; faisant maintenant tendre ε vers zéro, on voit que A_ℓ est prévisible comme union dénombrable de prévisibles. Il posssède en outre la propriété d'invariance d'échelle.

Chacun des instants R'_n est une fin d'excursion (car $\Gamma' \subset \Gamma$); soit $s_n = (\operatorname{sgn} B)_{R'_n-}$ le signe de cette excursion. Le processus S à valeurs dans $\{-1,1\}^{-\mathbb{N}^*}$ qui prend la valeur $(s_{n-k})_{k>0}$ sur $]\!]R'_{n-1}, R'_n]\!]$ est prévisible (on peut raisonner comme si les R'_n étaient des temps d'arrêt par le même argument que ci-dessus). La théorie des excursions entraîne que, conditionnellement aux zéros de B, les s_n sont indépendants et de même loi uniforme sur $\{-1,1\}$. En utilisant le lemme 6, on peut définir un processus prévisible

$$H = \sum_{\ell=0}^{q-1} \mathbb{1}_{A_\ell}\, f_\ell \circ S = \sum_{n\in\mathbb{Z}} s_n\, \theta(s)_n\, \mathbb{1}_{]\!]R'_{n-1},R'_n]\!]}$$

à valeurs dans $\{-1,1\}$. La transformation représentable T qui lui est associée commute avec les dilatations positives (en raison de l'invariance d'échelle); pour un ω fixé, elle change en bloc le signe de B sur chacun des intervalles $]R'_{n-1}(\omega), R'_n(\omega)]$ de façon que son opération sur les s_n soit donnée par θ; il en résulte que $T \neq \mathrm{Id}$ et $T^p = \mathrm{Id}$. ∎

REMARQUE. — Si l'on n'exige pas que T commute avec les dilatations, la démonstration de la proposition 5 peut être notablement simplifiée : on peut travailler avec une subdivision fixe $(t_n)_{n\in\mathbb{Z}}$ et prendre au lieu de s_n le signe de $B_{t_n} - B_{t_{n-1}}$; le lemme 6 se simplifie également, puisque l'on n'a plus besoin de la périodicité de l'algorithme fournissant θ. Ceci permet de ne pas traiter à part le cas où p est pair et fournit une transformation représentable T telle que $T^p = \mathrm{Id}$ et $T^{p'} \neq \mathrm{Id}$ pour tout $p' < p$.

Avec les propositions 3 et 4, nous avons étudié des suites décroissantes de filtrations browniennes obtenues par itération d'une même transformation. De façon analogue, on pourrait s'intéresser à des semi-groupes continus $(T_s)_{s\geqslant 0}$ de transformations browniennes donnant lieu à des familles strictement décroissantes de filtrations browniennes. Il est facile de donner des exemples de tels semigroupes; le plus simple est certainement le semi-groupe de dilatations défini par $T_s(\omega) = \omega'$ si $\omega'(t) = e^s\omega(te^{-2s})$; l'action de T_s sur la filtration est simplement l'homothétie de rapport e^{-2s} sur l'axe des temps.

En revanche, nous ignorons s'il existe un semi-groupe $(T_s)_{s\geqslant 0}$ de transformations browniennes *représentables* transformant $\mathcal{F}$ en une famille strictement décroissante de filtrations browniennes.

RÉFÉRENCES

[1] S. Attal. Thèse de doctorat. Université de Strasbourg, 1994.

[2] S. Attal. Représentation des endomorphismes de l'espace de Wiener qui préservent les martingales. *Ann. Inst. Henri Poincaré*, à paraître.

[3] L. E. Dubins, J. Feldman, M. Smorodinsky & B. Tsirelson. Decreasing Sequences of σ-Fields and a Measure Change for Brownian Motion. Prépublication, soumise à *Ann. Prob.*

[4] A. Goswami & B. V. Rao. Conditional Expectation of Odd Chaos given Even. *Stochastics and Stochastic Reports 35*, 213–214, 1991.

[5] T. Jeulin & M. Yor. Filtration des ponts browniens et équations différentielles stochastiques linéaires. *Séminaire de Probabilités XXIV*, Lecture Notes in Mathematics 1426, Springer 1990.

[6] J.-F. Le Gall & M. Yor. Sur l'équation stochastique de Tsirelson. *Séminaire de Probabilités XVII*, Lecture Notes in Mathematics 986, Springer 1983.

[7] T. Lindvall & L. C. G. Rogers. Coupling of Multidimensional Diffusions by Reflection. *Ann. Prob. 14*, 860–872, 1986.

[8] P.-A. Meyer. Probabilités et Potentiel. Hermann, 1966.

[9] P.-A. Meyer. Sur une transformation du mouvement brownien due à Jeulin et Yor. *Séminaire de Probabilités XXVIII*, Lecture Notes in Mathematics 1583, Springer 1994.

[10] D. Revuz & M. Yor. Continuous Martingales and Brownian Motion. *Grundlehren der mathematischen Wissenschaften*, Springer 1991.

[11] M. Yor. Tsirel'son's Equation in Discrete Time. *Probab. Theory Relat. Fields 91*, 135–152, 1992.

S. Attal et M. Émery
I.R.M.A., Université Louis Pasteur
Département de Mathématiques
7 rue René Descartes
67 084 Strasbourg Cedex
France

K. Burdzy(*)
Department of Mathematics
University of Washington
Seattle WA 98195
U. S. A.

Y. Hu
Université Pierre et Marie Curie
Laboratoire de Probabilités
4 place Jussieu
75 252 Paris Cedex 05
France

(*) Research supported in part by NSF grant DMS 91-00244 and AMS Centennial Research Fellowship.

Barycentres convexes et approximations des martingales continues dans les variétés

Marc Arnaudon

Institut de Recherche Mathématique Avancée

Université Louis Pasteur et CNRS

7, rue René Descartes

67084 Strasbourg Cedex

France.

Résumé

On majore le diamètre de l'ensemble des barycentres convexes d'une probabilité portée par un petit compact d'une variété avec connexion, par le moment d'ordre trois de la probabilité. Si le compact est un espace produit, on démontre que le projeté sur une composante de l'ensemble des barycentre convexes est l'ensemble des barycentres convexes de la loi marginale sur cette composante.

On utilise ces propriétés pour démontrer que les suites de martingales discrètes construites à partir de la valeur terminale d'une martingale continue convergent vers cette martingale continue lorsque le pas de la subdivision tend vers zéro, s'il existe une distance riemannienne convexe sur la variété. La convergence a lieu aussi dans tous les compacts suffisamment petits si la martingale continue a une variation quadratique dominée par un processus déterministe. On retrouve ainsi les résultats de convergence obtenus par Picard avec une définition différente des barycentres.

1. Introduction

Soient V un compact d'une variété W avec connexion ∇, et un espace probabilisé filtré $(\Omega, \mathcal{F}, (\mathcal{F}_t)_{0\leq t\leq 1}, P)$ vérifiant les conditions habituelles. Soit L une variable aléatoire à valeurs dans V. On sait (voir [E,M]), que si X est une martingale continue telle que $X_1 = L$, alors pour tout $t \leq 1$, X_t est dans l'ensemble de variables aléatoires $\mathbb{E}[L|\mathcal{F}_t]$, des espérances conditionnelles de L quand $\mathcal{F}_t$. Toute la difficulté du problème de construire X à partir de L réside dans le fait que cet ensemble n'est pas en général réduit à un singleton. Dans une première partie, on donne une majoration de la distance de deux éléments de cet ensemble, par le moment d'ordre 3 de L quand $\mathcal{F}_t$, sous certaines conditions de convexité de la variété.

Contrairement aux barycentres conditionnels de [P2] qui sont stables par produit, les produits d'espérances conditionnelles de [E,M] ne sont pas en général des espérances conditionnelles. On remédie à ce défaut en démontrant qu'avec des conditions de convexité sur une variété produit, si les lois conditionnelles par rapport aux sous-tribus de $\mathcal{F}$ existent, alors pour toute variable aléatoire (L, L') dans la variété produit, et pour tout élément X de $\mathbb{E}[L|\mathcal{F}_t]$, il existe X' dans $\mathbb{E}[L'|\mathcal{F}_t]$ tel que (X, X') soit dans $\mathbb{E}[(L, L')|\mathcal{F}_t]$. On utilise ce résultat pour démontrer que si L et L' sont proches, alors $\mathbb{E}[L|\mathcal{F}_t]$ et $\mathbb{E}[L'|\mathcal{F}_t]$ sont proches. En utilisant une méthode de Picard ([P2]), ce résultat nous conduit à démontrer que les suites de martingales discrètes construites à partir de la valeur terminale d'une martingale continue convergent vers la martingale continue lorsque le pas de la subdivision tend vers 0.

2. Majoration du diamètre des barycentres convexes, existence des barycentres convexes dans les variétés produits

2.1. Définitions et rappels

On considère toujours un compact V d'une variété W avec connexion ∇, et un espace probabilisé filtré $(\Omega, \mathcal{F}, (\mathcal{F}_t)_{0\leq t\leq 1}, P)$ vérifiant les conditions habituelles.

Une fonction f définie sur un ouvert W' de W sera dite convexe si pour toute géodésique $\gamma : I \to W'$ où I est un intervalle ouvert de $\mathbb{R}$, la fonction $f \circ \gamma$ est convexe sur I. Cette définition n'est intéressante que s'il existe suffisamment de géodésiques dans W', par exemple lorsque deux points de W' sont toujours reliés par une géodésique. Si au contraire W' n'est pas connexe, elle n'a pas beaucoup de sens.

On notera $C(V)$ l'ensemble des fonctions convexes définies sur un voisinage ouvert de V.

DÉFINITION 2.1. — *On dira que le compact V vérifie la condition (i) s'il possède un voisinage ouvert W' tel que deux points de W' soient reliés par une géodésique incluse dans W' et une seule qui dépend de façon C^∞ des deux points, et si la géodésique qui relie deux points de V est incluse dans V.*

Si la condition (i) est réalisée, pour tous x, y dans V, le vecteur $\overrightarrow{xy} \in T_xW$ désignera le vecteur vitesse en x de la géodésique passant en x au temps 0 et en y au temps 1. L'application qui à (x, y) associe $\overrightarrow{xy}$ est de classe C^∞.

Tout point d'une variété avec connexion possède un voisinage compact qui vérifie la condition (i). Dans une variété riemannienne, une boule géodésique régulière (voir [K1] théorème 1.7) vérifie aussi (i).

Les définitions qui suivent sont dues à Émery et Mokobodzki [E,M].

DÉFINITION 2.2. — *Si μ est une probabilité sur V, on dira que $x \in V$ est un barycentre convexe de μ si pour toute fonction f appartenant à $C(V)$, on a $f(x) \leq \mu(f)$. On notera $b(\mu)$ l'ensemble des barycentres convexes de μ.*

Si le compact V vérifie la condition (i), on appellera barycentre exponentiel d'une probabilité μ tout point e de V qui vérifie $\int_V \overrightarrow{ey}\mu(dy) = 0$.

D'après [E,M] proposition 2, tout barycentre exponentiel d'une probabilité μ est dans $b(\mu)$.

DÉFINITION 2.3. — *Si X est une variable aléatoire à valeurs dans V et si $\mathcal{G}$ est une sous-tribu de $\mathcal{F}$, on appellera espérance conditionnelle de X sachant $\mathcal{G}$ toute variable aléatoire $\mathcal{G}$-mesurable Y à valeurs dans V telle que $f \circ Y \leq \mathbb{E}[f \circ X|\mathcal{G}]$ pour toute fonction f appartenant à $C(V)$. On notera $\mathbb{E}[X|\mathcal{G}]$ l'ensemble (éventuellement vide) des espérances conditionnelles de X sachant $\mathcal{G}$.*

Si V vérifie (i), on appellera espérance conditionnelle exponentielle de X sachant $\mathcal{G}$ toute variable aléatoire $\mathcal{G}$-mesurable Y à valeurs dans V telle que $\mathbb{E}[\overrightarrow{YX}|\mathcal{G}] = 0$. On utilisera la notation $\mathcal{E}[X|\mathcal{G}]$ pour désigner une espérance conditionnelle exponentielle.

On dira que les espérances conditionnelles existent dans V si quelle que soit X variable aléatoire $\mathcal{F}$-mesurable à valeurs dans V, quelle que soit $\mathcal{G}$ sous-tribu de $\mathcal{F}$, l'ensemble $\mathbb{E}[X|\mathcal{G}]$ est non vide. On notera (i') cette propriété.

LEMME 2.4. — *Pour toute variable aléatoire X dans V, pour toute sous-tribu $\mathcal{G}$, toute espérance conditionnelle exponentielle $\mathcal{E}[X|\mathcal{G}]$ —s'il en existe— est un élément de $\mathbb{E}[X|\mathcal{G}]$.*

Démonstration. — La preuve est analogue à celle de [E,M] proposition 2. Il faut montrer que pour toute fonction f appartenant à $\mathcal{C}(V)$, on a $f(\mathcal{E}[X|\mathcal{G}]) \leq \mathbb{E}[f(X)|\mathcal{G}]$. Pour une telle fonction f, pour tout a dans V, on définit $\partial f(a)$ comme étant l'ensemble des $h \in T_a^*V$ tels que pour tout x dans V, on ait $f(x) - f(a) \geq h(\overrightarrow{ax})$. Nous pouvons remarquer que dans cette définition, on peut remplacer "pour tout x dans V" par "il existe un voisinage V_a de a, tel que pour tout x dans V_a", en vertu de la condition (i) et de la convexité de f sur les géodésiques. D'après [E,Z] corollaire 1, $\partial f(a)$ n'est pas vide. On définit ∂f comme étant la réunion des $\partial f(a)$, a parcourant V. L'ensemble ∂f est un borélien de T^*V dont on choisit une section borélienne $a \mapsto f'(a)$. On a alors pour tous a, x dans V, $f(x) - f(a) \geq f'(a)(\overrightarrow{ax})$, donc $f(x) - f(\mathcal{E}[X|\mathcal{G}]) \geq f'(\mathcal{E}[X|\mathcal{G}])\left(\overrightarrow{\mathcal{E}[X|\mathcal{G}]x}\right)$. Ceci implique que

$$\mathbb{E}[f(X)|\mathcal{G}] - f(\mathcal{E}[X|\mathcal{G}]) \geq \mathbb{E}\left[f'(\mathcal{E}[X|\mathcal{G}])\left(\overrightarrow{\mathcal{E}[X|\mathcal{G}]X}\right)|\mathcal{G}\right].$$

Par linéarité de $f'(\mathcal{E}[X|\mathcal{G}])(\cdot)$, le terme de droite est égal à

$$f'(\mathcal{E}[X|\mathcal{G}])\left(\mathbb{E}\left[\overrightarrow{\mathcal{E}[X|\mathcal{G}]X}|\mathcal{G}\right]\right).$$

Il est nul, et on a bien $\mathbb{E}[f(X)|\mathcal{G}] \geq f(\mathcal{E}[X|\mathcal{G}])$. □

DÉFINITION 2.5. — *On dira qu'une semi-martingale X à valeurs dans V est une $\mathcal{C}$-martingale si pour toute fonction f appartenant à $\mathcal{C}(V)$, le processus $f(X)$ est une sous-martingale réelle.*

D'après [E,Z] théorème 2, les martingales continues au sens usuel, que nous appellerons aussi ∇-martingales, sont des $\mathcal{C}$-martingales. La réciproque est fausse en général, et les $\mathcal{C}$-martingales ne sont pas toujours des processus continus. On les relie aux espérances conditionnelles en constatant qu'une semi-martingale est une $\mathcal{C}$-martingale si et seulement si pour tous $s, t \in [0,1]$ vérifiant $s \leq t$, on a $X_s \in \mathbb{E}[X_t|\mathcal{F}_s]$.

Soit L une variable aléatoire $\mathcal{F}_1$-mesurable à valeurs dans V. Lorsque la condition (i') est vérifiée, on peut construire des $\mathcal{C}$-martingales discrètes de valeur terminale donnée L, par rapport à toutes les subdivisions de $[0,1]$. Il suffit en effet, étant donnée une subdivision $(t_i)_{1\leq i\leq k}$, de prendre pour X_{t_i} un élément de $\mathbb{E}[X_{t_{i+1}}|\mathcal{F}_{t_i}]$, et de faire une récurrence descendante. Les $\mathcal{C}$-martingales discrètes sont des $\mathcal{C}$-martingales pour la filtration discrétisée, égale à $\mathcal{F}_{t_i}$ sur l'intervalle de temps $[t_i, t_{i+1}[$.

La condition (i') est réalisée dans les deux cas de figure présentés par les propositions 2.6 et 2.7.

PROPOSITION 2.6. — *Si le sous-ensemble compact V de W vérifie la condition (i), est de la forme $\{\phi \leq 0\}$, avec ϕ convexe de classe C^2 au voisinage de V, telle que $\{\varphi < 0\}$ ne soit pas vide, alors les espérances conditionnelles exponentielles existent dans V.*

Notons que Émery et Mokobodzki ont donné dans [E,M] une condition d'existence des barycentres exponentiels : il suffit que V soit de la forme $\{\varphi \leq 0\}$, avec φ convexe de classe C^2 au voisinage de V. La démonstration qui va suivre est inspirée de celle qui figure dans [E,M].

Démonstration. — Nous allons tout d'abord démontrer que le compact V est une variété à bord dont le bord est exactement égal à $\{\varphi = 0\}$. Pour cela, il suffit de démontrer que $d\varphi$ n'est pas nulle sur $\{\varphi = 0\}$. Soit o un point de V tel que $\varphi(o) < 0$. Si $x \in \{\varphi = 0\}$, alors la convexité de φ sur la géodésique reliant x et o permet d'écrire l'inégalité $\left\langle d\varphi, \overrightarrow{xo}\right\rangle \leq \varphi(o) - \varphi(x) < 0$. On en déduit que $d\varphi$ n'est pas nulle sur $\{\varphi = 0\}$. Nous allons maintenant établir que tout champ de vecteurs sur V qui est transverse au bord et dirigé vers l'extérieur admet au moins un zéro dans l'intérieur de V. La somme des indices aux points d'annulation des champs de vecteurs de V qui n'ont que des zéros isolés et qui sont transverses au bord, dirigés vers l'extérieur, est égale à la caractéristique d'Euler de V (théorème de Poincaré-Hopf [M] p. 35). Le champ de vecteurs $x \mapsto -\overrightarrow{xo}$ est égal à l'identité dans la carte exponentielle centrée en o, et est transverse au bord, dirigé vers l'extérieur. On en déduit que la caractéristique d'Euler de V est 1. Tout champ de vecteurs sur V, transverse au bord et dirigé vers l'extérieur, a donc au moins un zéro isolé s'il n'a pas de point d'accumulation de zéros.

Soient L une variable aléatoire à valeurs dans V, et $\mathcal{G}$ une sous-tribu de $\mathcal{F}$. On peut supposer que L prend ses valeurs dans l'intérieur de V, car dans le cas contraire, on fait la même démonstration avec tous les $V_\varepsilon = \{\varphi \leq \varepsilon\}$.

Pour x dans un sous-ensemble dénombrable dense d'un voisinage de V, on choisit $\omega \mapsto v_x(\omega)$ dans la classe de variables aléatoires $\mathbb{E}\left[\overrightarrow{xL}|\mathcal{G}\right]$ telle que sur un sous-ensemble Ω' de Ω de probabilité 1, le champ de vecteurs $x \mapsto v_x$ soit de classe C^1, avec des dérivées uniformément bornées, indépendamment de ω. Pour $\omega \in \Omega'$, on peut donc prolonger $v.(\omega)$ à un champ de vecteurs de classe C^1 sur V tout entier. Montrons que, presque sûrement, v est transverse à l'hypersurface $\{\varphi = 0\}$, dirigé vers l'intérieur de V. On note $(\omega, t, x) \mapsto U_t(x)(\omega)$ le groupe à un paramètre issu de $v.(\omega)$. Il vérifie $\dfrac{d}{dt}U_t(x) = v_{U_t(x)}$ et $U_0(x) = x$. Si $U_t(x) \in V$, on a

$$\frac{d}{dt}\varphi(U_t(x)) = \left\langle d\varphi(U_t(x)), \mathbb{E}\left[\overrightarrow{U_t(x)L}|\mathcal{G}\right]\right\rangle = \mathbb{E}\left[\langle d\varphi(U_t(x)), \overrightarrow{U_t(x)L}\rangle|\mathcal{G}\right].$$

Comme φ composée à la géodésique joignant $U_t(x)$ à L est convexe, on obtient

$$\frac{d}{dt}\varphi(U_t(x)) \leq \mathbb{E}[\varphi(L)|\mathcal{G}] - \varphi(U_t(x)),$$

ce qui donne pour $t = 0$ et $x \in \partial V$,

$$\langle d\varphi(x), v_x\rangle \leq \mathbb{E}[\varphi(L)|\mathcal{G}] < 0.$$

On a montré que $-v$ était transverse sur le bord et dirigé vers l'extérieur. On en déduit que $v(\omega)$ s'annule dans l'intérieur de V, pour tout $\omega \in \Omega'$. Définissons $\mathcal{E}_\mathcal{G} = \{(\omega, x) \in \Omega' \times V, v_x(\omega) = 0\}$. C'est un ensemble mesurable par rapport au produit de $\mathcal{G}$ et de l'ensemble des boréliens de V, et une section $\mathcal{G}$-mesurable $\omega \mapsto \mathcal{E}[L|\mathcal{G}](\omega)$ est une espérance conditionnelle exponentielle de L par rapport à $\mathcal{G}$. □

On obtient aussi la condition (i') avec des hypothèses géométriques plus faibles, mais en imposant des conditions sur $(\Omega, \mathcal{F}, P)$:

PROPOSITION 2.7. — *On suppose que pour toute probabilité μ sur V, l'ensemble des barycentres convexes de μ n'est pas vide, et qu'il existe des lois conditionnelles sur Ω relativement à n'importe quelle sous-tribu de $\mathcal{F}$.*

Alors les espérances conditionnelles existent dans V.

La démonstration de ce résultat figure dans [A]. Notons que dans la plupart des espaces canoniques, on a l'existence des lois conditionnelles par rapport à toutes les sous-tribus.

Lorsque deux points quelconques de V peuvent toujours être reliés par au moins une géodésique passant dans V, alors les barycentres convexes existent sur V. En effet, on commence par les construire pour des probabilités qui sont des sommes finies de masses de Dirac en remplaçant deux masses de Dirac par leur barycentre sur une géodésique, et en utilisant l'associativité. Si μ est une probabilité quelconque, on considère une suite $(\mu_n)_{n\in\mathbb{N}}$ de mesures du type précédent qui converge étroitement vers μ, et une suite $(x_n)_{n\in\mathbb{N}}$ de barycentres convexes associés à ces mesures. Une valeur d'adhérence de la suite (x_n) sera un barycentre convexe de μ.

En particulier, si V vérifie (i), alors les barycentres convexes existent sur V.

DÉFINITION 2.8. — *On notera (ii) la condition d'existence pour tout $(x, y) \in V \times V$ tel que $x \neq y$, d'une fonction $f \in \mathcal{C}(V)$ qui vérifie $f(x) < f(y)$.*

On notera (iii) la condition d'existence pour tous $a \in V$ et $\lambda \in T_a^*W$ d'une fonction f convexe de classe C^2 sur un voisinage de V telle que $df(a) = \lambda$ et Hess $f(a) = 0$.

Si deux points de V sont toujours reliés par une géodésique incluse dans V, alors la condition (iii) implique la condition (ii). En effet, si on choisit deux points différents x et y dans V, la condition (iii) permet de construire une fonction convexe nulle en x, dont la restriction à une géodésique qui va de x à y a une dérivée égale à 1 au point x. Cette fonction devra être strictement positive en y.

On notera (ii') la condition d'existence pour tout x dans V d'une fonction $f \in \mathcal{C}(V)$ qui vérifie $f(x) = 0$ et $f > 0$ sur $V \backslash \{x\}$. Cette condition est plus forte que (ii). Pour la réciproque, il suffit d'ajouter une condition sur les fonctions f de la condition (ii) :

PROPOSITION 2.9. — *S'il existe un voisinage compact V' de V tel que pour tout $(x,y) \in V \times V$ vérifiant $x \neq y$, il existe une fonction $f \in \mathcal{C}(V')$ qui vérifie $f(x) < f(y)$, alors le compact V vérifie la condition (ii').*

Démonstration. — On fixe une métrique sur W qui ne nous servira qu'à définir des fonctions lipschitziennes de rapport 1. La connexion utilisée est toujours celle du début. Montrons tout d'abord que pour toute fonction $f \in \mathcal{C}(V')$, il existe un voisinage de V' dans lequel f est lipschtzienne. D'après [E,Z] (proposition 1), pour toute carte locale de domaine inclus dans le domaine de définition de f, la fonction f lue dans cette carte locale est localement lipschitzienne. On peut donc recouvrir le compact V' par un nombre fini de cartes locales dans lesquelles f est lipschitzienne. Elle est donc lipschitzienne dans la réunion des domaines de ces cartes, qui contient V'.

Soit $x \in V$. Pour tout y dans V différent de x, il existe une fonction $f_y \in \mathcal{C}(V')$ telle que $0 = f_y(x) < f_y(y)$. Quitte à multiplier f_y par une constante, nous pouvons imposer à f_y d'être lipschtzienne de rapport inférieur à 1 sur un voisinage de V'. Ceci nous permet de démontrer que la fonction $f = \sup_{y \in V \backslash \{x\}} f_y$ est bien définie sur V', lipschtzienne de rapport 1 et convexe sur l'intérieur de V'. Elle appartient donc à $\mathcal{C}(V)$, et vérifie $f(x) = 0$, $f > 0$ sur $V \backslash \{x\}$. □

Rappelons les résultats dont la démonstration figure dans [A], et qui relient la convexité du compact V à des propriétés vérifiées par les $\mathcal{C}$-martingales.

PROPOSITION 2.10. — *Si toutes les martingales réelles de la filtration $(\mathcal{F}_t)$ sont continues et si le compact V vérifie la condition (ii), alors les $\mathcal{C}$-martingales de V sont continues.*

Si le compact V vérifie la condition (iii), alors les $\mathcal{C}$-martingales continues sont des ∇-martingales.

2.2. Les $\mathcal{C}$-martingales dans les variétés à géométrie convexe

DÉFINITION 2.11. — *On dira qu'une variété avec connexion W' a une géométrie convexe s'il existe une fonction convexe $\psi : W' \times W' \to \mathbb{R}$ de classe C^3, positive, qui s'annule exactement sur la diagonale, et telle que pour tout $a \in W'$, l'application $\psi_a = \psi(a, \cdot)$ ait une hessienne $\nabla d\psi_a$ strictement positive sur $W' \backslash \{a\}$.*

On dira qu'un compact V' d'une variété avec connexion a une géométrie convexe s'il possède un voisinage ouvert qui a une géométrie convexe.

Si la variété W' est suffisamment petite, alors d'après [E] lemme (4.59), elle a une géométrie convexe. D'autre part, Kendall a démontré (voir [K2] et [A]) que les boules géodésiques régulières dans les variétés riemanniennes ont une géométrie convexe.

PROPOSITION 2.12. — *Si le compact V a une géométrie convexe, alors il vérifie les conditions (ii) et (iii).*

Avant de démontrer la proposition, nous allons construire une famille de fonctions convexes sur V. Fixons une métrique riemannienne quelconque g sur W, et soit δ la distance riemannienne associée. La notation $\operatorname{Hess} f$ désignera ∇df (∇ est la connexion donnée au début, et n'est pas la connexion associée à la métrique g).

LEMME 2.13. — *On suppose que le compact V a une géométrie convexe. Alors il existe deux constantes strictement positives c_δ et C_δ, et une famille de fonctions $(h_a)_{a\in V}$ de classe C^2 sur V, telles que l'on ait pour tous a, x dans V,*

$$c_\delta\delta^3(a,x) \le h_a(x) \le C_\delta\delta^3(a,x) \quad et \quad c_\delta\delta(a,x)g(x) \le \operatorname{Hess} h_a(x) \le C_\delta\delta(a,x)g(x).$$

Démonstration du lemme 2.13. — On note Δ la diagonale de $V\times V$, et on définit une fonction η de $V\times V$ dans $\mathbb{R}$, positive, de classe C^∞ sur $V\times V\backslash\Delta$, et qui coïncide avec δ sur un voisinage ouvert $\mathcal{V}$ de Δ. On peut supposer, quitte à remplacer ψ par ψ^ν avec ν entier supérieur à 2, que pour tout $a\in V$, la hessienne $\operatorname{Hess}\psi_a(a)$ est nulle. Montrons que si A est un réel positif suffisamment grand, les fonctions $h_a(x)=\eta(a,x)^3 + A\psi(a,x)$ conviennent. Ce sont des fonctions de classe C^2 avec des dérivées jusqu'à l'ordre 2 qui s'annulent en a. Elles satisfont le premier encadrement. Notons η_a la fonction $\eta(a,\cdot)$.

Nous allons tout d'abord établir la minoration de $\operatorname{Hess} h_a(x)$. Quitte à réduire $\mathcal{V}$, on peut supposer qu'il existe une constante $c'>0$ telle que pour tout $(a,x)\in\mathcal{V}$, on ait $\operatorname{Hess}\eta_a^2(x)\ge c'g(x)$. Comme

$$\operatorname{Hess}\eta_a^3(x) = \frac{3}{2}\eta_a(x)\operatorname{Hess}\eta_a^2(x) + 3\eta_a(x)d\eta_a(x)\otimes d\eta_a(x)$$

lorsque $x\ne a$, on déduit que sur $\mathcal{V}$, on a $\operatorname{Hess}\eta_a^3(x)\ge\frac{3}{2}c'\delta(a,x)g(x)$, et donc pour tout A positif, $\operatorname{Hess}(\eta_a^3(x)+A\psi_a(x))\ge\frac{3}{2}c'\delta(a,x)g(x)$.

Pour obtenir la minoration sur le complémentaire de $\mathcal{V}$, il suffit sur ce compact de minorer δ par une constante strictement positive α, de minorer $\operatorname{Hess}\psi_a(x)$ par $\beta g(x)$ avec $\beta>0$, et $\operatorname{Hess}\eta_a^3(x)$ par $-Mg(x)$ avec $M\ge 0$. Si on choisit $A\ge\dfrac{\alpha+M}{\beta}$, on obtient l'inégalité désirée.

Nous allons établir la majoration de $\operatorname{Hess} h_a(x)$ avec une constante A vérifiant l'inégalité ci-dessus. La majoration de $\operatorname{Hess}\eta_a^3(x)$ découle de l'expression écrite plus haut. Pour la majoration de $\operatorname{Hess}\psi_a(x)$ il suffit de constater que pour tout $a\in V$, on a $\operatorname{Hess}\psi_a(a)=0$. Comme l'application $(a,x)\mapsto\psi_a(x)$ est de classe C^3, on déduit qu'il existe une constante C' telle que pour tout $(a,x)\in V\times V$, on ait $\operatorname{Hess}\psi_a(x)\le C'\delta(a,x)g(x)$. On obtient ensuite la majoration désirée pour $\operatorname{Hess} h_a(x)$. □

Démonstration de la proposition 2.12. — On suppose que V est un compact à géométrie convexe. Pour obtenir la condition (ii), il suffit d'associer au couple (x,y) la fonction ψ_x.

Démontrons que la condition (iii) est réalisée. Il existe un voisinage $\mathcal{V}$ de la diagonale Δ de $V \times V$, et une application $(a,x) \mapsto w_a(x) \in T_aW$ définie sur $V \times V$, de classe C^∞, tels que $(a,x) \mapsto \exp_a^{-1}(x)$ soit définie et de classe C^∞ sur $\mathcal{V}$, et les deux applications coïncident sur cet ensemble. Comme pour tout $a \in V$, l'application $x \mapsto w_a(x)$ a une hessienne nulle en a, il existe une constante K telle que pour tout (a,x), pour tout $\lambda \in T_a^*W$, on ait $-K\|\lambda\|\delta(a,x)g(x) \leq \text{Hess}(\lambda \circ w_a)(x)$. Pour obtenir une fonction f convexe telle que $df(a) = \lambda$ et $\text{Hess}\, f(a) = 0$, il suffit de poser

$$f(x) = (\lambda \circ w_a)(x) + \frac{K\|\lambda\|}{c_\delta} h_a(x).$$

On obtient ainsi la condition (iii). □

Contrairement aux conditions (ii) et (iii), qui sont réalisées dès que le compact V est suffisamment petit, la condition (i') d'existence des espérances conditionnelles peut ne pas être réalisée localement, par exemple s'il y a des petits trous dans le compact V. Au contraire, si cet ensemble est grand et s'il n'y a plus assez de fonctions convexes, alors (i') est toujours réalisée, l'ensemble des espérances conditionnelles est trop grand et il y a trop de $\mathcal{C}$-martingales (par exemple dans une variété compacte connexe sans bord, toutes les semi-martingales sont des $\mathcal{C}$-martingales).

2.3. Majoration du diamètre des barycentres convexes dans les variétés à géométrie convexe

On supposera dans cette partie que le compact V a une géométrie convexe, et qu'il vérifie la condition (i). On supposera aussi que toute probabilité μ sur V a un barycentre exponentiel e_μ. Notons que Kendall [K1] a démontré qu'il était unique en géométrie convexe.

PROPOSITION 2.14. — *Sous les conditions écrites ci-dessus, pour toute distance riemannienne δ sur V, il existe une constante C telle que pour toute probabilité μ sur V, pour tout v dans V, on ait l'inégalité*

$$|b(\mu)| \leq C \int_V \delta^3(v,y)\mu(dy),$$

où $|b(\mu)|$ désigne le diamètre de $b(\mu)$ pour la distance δ.

On pourrait démontrer, en se plaçant dans une petite boule géodésique d'une sphère et en prenant des probabilités portées par trois points, que cette estimation est optimale : si on remplace le moment d'ordre 3 dans l'inégalité par un moment d'ordre supérieur, l'inégalité devient fausse en général.

Démonstration de la proposition. — Considérons une métrique riemannienne g sur V, et la distance δ associée. On note S^*W l'ensemble des formes linéaires sur V de norme 1 pour g. Les applications exp et les vecteurs $\overrightarrow{xy}$ seront toujours calculés avec la connexion ∇ (et non pas avec g). En raison de la compacité de V, il existe une constante c_1 strictement positive, telle que pour tous $x, y \in V$, on ait l'inégalité $c_1\delta(x,y) \leq \sup_{\lambda \in S_x^*W} \lambda(\overrightarrow{xy})$.

Reprenant les notations du lemme 2.13, on pose toujours $h_a(x) = \eta(a,x)^3 + A\psi(a,x)$. En raison de la compacité de V, il existe une constante $C' > 0$ telle que pour tout $a \in V$, $\lambda \in S^*_a W$, les fonctions $x \mapsto \lambda(\overrightarrow{ax}) + C'h_a(x)$ soient convexes.

Soit μ une probabilité sur V. Nous allons tout d'abord démontrer la proposition avec $v = e_\mu$. Soient x un élément de $b(\mu)$ et $\lambda \in S^*_{e_\mu} W$. On a

$$\lambda(\overrightarrow{e_\mu x}) + C'h_{e_\mu}(x) \leq \int_V \mu(dy)\Big(\lambda(\overrightarrow{e_\mu y}) + C'h_{e_\mu}(y)\Big).$$

On a $\int_V \mu(dy)\left(\lambda(\overrightarrow{e_\mu y})\right) = \lambda\left(\int_V \mu(dy)\overrightarrow{e_\mu y}\right) = 0$ puisque e_μ est le barycentre exponentiel de μ. D'autre part, $C'h_{e_\mu}(x) \geq 0$, donc on obtient

$$\lambda(\overrightarrow{e_\mu x}) \leq C' \int_V \mu(dy) h_{e_\mu}(y).$$

L'inégalité est encore valable avec à gauche le supremum sur $\lambda \in S^*_{e_\mu} W$, ce qui permet ensuite d'obtenir $c_1 \delta(e_\mu, x) \leq C' \int_V \mu(dy)\left(h_{e_\mu}(y)\right)$. D'après la définition des h_a, pour tous a, x, on a $h_a(x) \leq C_\delta \delta^3(a,x)$. En posant $C = \dfrac{2C'C_\delta}{c_1}$, on obtient bien l'inégalité $|b(\mu)| \leq C \int_V \delta^3(e_\mu, y)\mu(dy)$.

Si v est un élément quelconque de V, il suffit d'établir qu'il existe une constante C' indépendante de μ telle que $\int_V \delta^3(e_\mu, y)\mu(dy) \leq C' \int_V \delta^3(v,y)\mu(dy)$. En raison de la convexité de h_v, on a $h_v(e_\mu) \leq \int_V h_v(y)\mu(dy)$, que l'on peut transformer en $\delta^3(v, e_\mu) \leq \frac{C_\delta}{c_\delta} \int_V \delta^3(v,y)\mu(dy)$, car $C_\delta \delta^3(a,x) \geq h_a(x) \geq c_\delta \delta^3(a,x)$ pour tous $a, x \in V$. De l'inégalité $\delta^3(e_\mu, y) \leq 4(\delta^3(e_\mu, v) + \delta^3(b,y))$ pour tout $y \in V$, on déduit

$$\int_V \delta^3(e_\mu, y)\mu(dy) \leq 4\frac{C_\delta}{c_\delta} \int_V \delta^3(v,y)\mu(dy) + 4\int_V \delta^3(v,y)\mu(dy).$$

Ceci achève la démonstration. □

PROPOSITION 2.15. — *Sous les conditions écrites au début de la partie 2.3, pour toute métrique riemannienne δ sur V, il existe une constante C, telle que quelles que soient la sous-tribu $\mathcal{G}$ de $\mathcal{F}$, la variable aléatoire L à valeurs dans V telle que $\mathcal{E}[L|\mathcal{G}]$ existe, quels que soient les éléments X et Y de $\mathbb{E}[L|\mathcal{G}]$, quelle que soit la variable aléatoire $\mathcal{G}$-mesurable Z à valeurs dans V on ait l'inégalité*

$$\delta(X,Y) \leq C\mathbb{E}[\delta^3(Z,L)|\mathcal{G}].$$

Démonstration. — Le principe est identique à celui de la démonstration précédente. On va tout d'abord démontrer que

$$\delta\left(\mathcal{E}[L|\mathcal{G}], Y\right) \leq C\mathbb{E}\left[\delta^3\left(\mathcal{E}[L|\mathcal{G}], L\right)|\mathcal{G}\right].$$

Pour tout λ appartenant à S^*V, de projeté a sur V, on a l'inégalité

$$\lambda\left(\overrightarrow{aY}\right) + C'h_a(Y) \leq \mathbb{E}\left[\lambda\left(\overrightarrow{aL}\right) + C'h_a(L)|\mathcal{G}\right].$$

Les applications $\lambda \mapsto \lambda\left(\overrightarrow{aY}\right) + C'h_a(Y)$ et $\lambda \mapsto \lambda\left(\overrightarrow{aL}\right) + C'h_a(L)$ sont uniformément lipschitziennes. L'inégalité ci-dessus est encore valable en remplaçant λ par un processus Λ étagé $\mathcal{G}$-mesurable, à valeurs dans S^*V, et en remplaçant a par le processus projeté de Λ. La propriété de Lipschitz permet, en passant à la limite sur les processus étagés, d'écrire l'inégalité avec tout processus Λ $\mathcal{G}$-mesurable, à valeurs dans S^*V, et en particulier avec les processus Λ de processus projeté sur V égal à $\mathcal{E}[L|\mathcal{G}]$. L'inégalité s'écrit alors

$$\Lambda\left(\overrightarrow{\mathcal{E}[L|\mathcal{G}]Y}\right) + C'h_{\mathcal{E}[L|\mathcal{G}]}(Y) \le \mathbb{E}\left[\Lambda\left(\overrightarrow{\mathcal{E}[L|\mathcal{G}]L}\right) + C'h_{\mathcal{E}[L|\mathcal{G}]}(L)|\mathcal{G}\right].$$

Le terme de droite du membre de gauche est positif, et le terme de gauche du membre de droite est nul puiqu'il est égal à $\Lambda\left(\mathbb{E}\left[\overrightarrow{\mathcal{E}[L|\mathcal{G}]L}|\mathcal{G}\right]\right)$, donc on obtient l'inégalité

$$\Lambda\left(\overrightarrow{\mathcal{E}[L|\mathcal{G}]Y}\right) \le C'\mathbb{E}\left[h_{\mathcal{E}[L|\mathcal{G}]}(L)|\mathcal{G}\right].$$

La suite de cette partie de la démonstration est identique au cas déterministe. On minore à gauche le supremum sur Λ par $c_1\delta(\mathcal{E}[L|\mathcal{G}], Y)$, et on majore $h_{\mathcal{E}[L|\mathcal{G}]}(L)$ par $C_\delta\delta^3(\mathcal{E}[L|\mathcal{G}], L)$.

Il reste à démontrer l'inégalité en remplaçant à droite $\mathcal{E}[L|\mathcal{G}]$ par Z quelconque $\mathcal{G}$-mesurable. Une fois que l'on aura prouvé que

$$h_Z(\mathcal{E}[L|\mathcal{G}]) \le \mathbb{E}[h_Z(L)|\mathcal{G}],$$

on pourra se ramener au cas déterministe. Cette dernière inégalité se prouve comme précédemment. Elle est vraie si on remplace Z par un processus étagé $\mathcal{G}$-mesurable à valeurs dans V, puis pour tout processus $\mathcal{G}$-mesurable car h est lipschitzienne en tant que fonction de deux variables. □

2.4. Existence de barycentres convexes dans les variétés produits

Dans cette partie, nous allons démontrer que le projeté de l'ensemble des barycentres convexes d'une probabilité sur un espace produit est exactement l'ensemble des barycentres de la loi marginale. L'utilité de cette propriété apparaîtra dans la partie 3 au cours du calcul de la distance entre une martingale continue et une C-martingale discrète.

La notation V' désignera un compact d'une variété W' munie d'une connexion ∇'. La variété $W \times W'$ sera munie de la connexion produit. Notons que si les compacts V et V' vérifient la condition (i), alors le compact $V \times V'$ vérifie aussi la condition (i).

PROPOSITION 2.16. — *On suppose que les ensembles V et V' vérifient la condition (i). Soit μ une probabilité sur $V \times V'$, de lois marginales μ^V et $\mu^{V'}$, et soit x un élément de $b(\mu^V)$. Alors il existe un élément y de $b(\mu^{V'})$, tel que (x, y) soit dans $b(\mu)$.*

Démonstration. — On écrit μ sous la forme $\mu^V(dv)Q(v,dv')$, et on considère le sous-ensemble $b(Q)$ de $V \times V'$ des barycentres convexes de Q, défini par

$$b(Q) = \left\{(v,v') \in V \times V', \forall f \in \mathcal{C}(V'), f(v') \leq \int_{V'} Q(v,du')f(u')\right\}.$$

Si $v \mapsto h(v)$ est une section borélienne de $b(Q)$, un élément y tel que (x,y) soit dans $b\left(\mu^V(dv)\delta_{h(v)}(dv')\right)$ répond à la question (δ_a désigne ici la masse de Dirac au point a). En effet, pour toute fonction f convexe sur un voisinage de $V \times V'$, on a $f(x,y) \leq \int_V \mu^V(dv)f(v,h(v))$ et $f(v,h(v)) \leq \int_{V'} Q(v,dv')f(v,v')$ donc $f(x,y) \leq \int_V \mu^V(dv)\int_{V'} Q(v,dv')f(v,v')$. Il suffit donc de démontrer la proposition avec μ de la forme $\mu^V(dv)\delta_{h(v)}(dv')$.

Dans un premier temps, nous supposerons que h est une application continue, et ensuite nous ferons la démonstration avec h quelconque. Nous utiliserons les notations de [E,M], que nous rappelons brièvement. Soit $n \in \mathbb{N}$. Si p est une probabilité sur $W^n = \{0,1\}^n$ et si y est une variable aléatoire définie sur W^n et à valeurs dans V, le barycentre géodésique itéré de (y,p) est le point $\beta_n(y,p)$ de V défini de la façon suivante. Pour $\varepsilon \in \{0,1\}$, on note p_ε la probabilité sur W^{n-1} égale à la loi conditionnelle de $(\omega_2,\ldots,\omega_n)$ sachant que $\omega_1 = \varepsilon$. Les variables y_ε sur W^{n-1} sont définies par $y_\varepsilon(\omega_1,\ldots,\omega_{n-1}) = y(\varepsilon,\omega_1,\ldots,\omega_{n-1})$.

Pour $n=0$, $\beta_0(y,1)$ est la valeur de la variable aléatoire constante y.

Pour $n > 0$, $\beta_n(y,p)$ est le point $\gamma(p\{\omega_1 = 1\})$ de V, où $\gamma : [0,1] \mapsto V$ est la géodésique telle que $\gamma(0) = \beta_{n-1}(y_0,p_0)$ et $\gamma(1) = \beta_{n-1}(y_1,p_1)$.

Notons que dans [E,M], les barycentres géodésiques itérés sont des ensembles, alors qu'ici ce sont des points. La différence provient du fait que notre hypothèse (i) est plus forte que celle de [E,M], et qu'elle garantit que les ensembles définis dans [E,M] sont des singletons.

Cas où h est continu. D'après [E,M] théorème 1, il existe une suite $(y_k,p_k)_{k\in\mathbb{N}}$, avec y_k une variable aléatoire définie sur un $W^{n(k)}$, à valeurs dans V, et p_k probabilité sur $W^{n(k)}$, telle que $\mu_k^V = y_k(p_k)$ converge étroitement vers μ^V et x soit le barycentre géodésique itéré de chaque (y_k,p_k). On définit pour chaque k la probabilité $\mu_k = \mu_k^V(dv)\delta_{h(v)}(dv')$. Elle est encore égale à $(y_k,h(y_k))(p_k)$. Le barycentre géodésique itéré de (y_k,p_k) est de la forme (x,z_k) avec z_k dans V', car les géodésiques de $V \times V'$ sont les produits des géodésiques de V et de V'. Par conséquent, (x,z_k) est un barycentre convexe de μ_k. La continuité de h implique que μ_k converge étroitement vers μ lorsque k tend vers l'infini. Cela implique qu'une valeur d'adhérence (x,y) de la suite $((x,z_k)_{k\in\mathbb{N}})$ soit un barycentre convexe de μ.

Cas où h est quelconque. Il existe une suite $(h_n)_{n\in\mathbb{N}}$ de fonctions continues telle que μ^V-presque sûrement, h_n converge vers h (c'est une conséquence du théorème de Lusin). On définit les probabilités μ_n sur $V \times V'$ par $\mu_n = \mu^V(dv)\delta_{h_n(v)}(dv')$. Il est facile de voir que les μ_n convergent étroitement vers μ. Pour chaque n, en utilisant le résultat du paragraphe précédent, on peut trouver $y_n \in V'$ tel que (x,y_n) soit un barycentre convexe de μ_n. Une valeur d'adhérence (x,y) de la suite $((x,y_n))_{n\in\mathbb{N}}$ est un barycentre convexe de μ. □

DÉFINITION 2.17. — *On notera* $(iv'(V, V'))$ *la condition suivante : pour tout couple de variables aléatoires* (L, L') *à valeurs dans* $V \times V$, *pour toute sous-tribu* $\mathcal{G}$ *de* $\mathcal{F}$, *pour tout élément* X *de* $\mathbb{E}[L|\mathcal{G}]$, *il existe un élément* X' *de* $\mathbb{E}[L'|\mathcal{G}]$ *tel que* (X, X') *appartienne à* $\mathbb{E}[(L, L')|\mathcal{G}]$

PROPOSITION 2.18. — *Si les compacts* V *et* V' *vérifient la condition* (i), *et s'il existe des lois conditionnelles sur* Ω *relativement à n'importe quelle sous-tribu de* $\mathcal{F}$, *alors la condition* $(iv'(V, V'))$ *de la définition 2.17 est réalisée.*

Démonstration. — D'après [A] lemme 2.5, il existe une suite $(f_n)_{n\in\mathbb{N}}$ de fonctions appartenant à $\mathcal{C}(V)$ telles que pour tout $\varepsilon > 0$ et toute fonction f de $\mathcal{C}(V)$, il existe $n \in \mathbb{N}$ tel que $\sup_V |f - f_n| < \varepsilon$.

Soit $Q(\omega, dvdv')$ la loi conditionnelle de (L, L') sachant $\mathcal{G}$. Alors $Q(\omega, dv) = \int_{V'} Q(\omega, dvdv')$ est la loi conditionnelle de L sachant $\mathcal{G}$. Puisque X est dans $\mathbb{E}[L|\mathcal{G}]$, pour toute fonction f dans $\mathcal{C}(V)$, on a presque sûrement $f(X) \leq \int_V Q(\omega, dv)f(v)$. Comme il existe une suite de fonctions de $\mathcal{C}(V)$ dense pour la norme uniforme, on a l'inégalité presque sûre pour toute f : il existe Ω' de probabilité 1 tel que $\forall\, \omega \in \Omega'$, $f(X(\omega)) \leq \int_V Q(\omega, dv)f(v)$, ce qui veut dire que $X(\omega)$ est dans $b(Q(\omega, dv))$. On en déduit d'après la proposition 2.16 qu'il existe $y \in V'$ tel que $(X(\omega), y) \in b(Q(\omega, dvdv'))$. Notons $b' = \{(\omega, y) \in \Omega' \times V', (X(\omega), y) \in b(Q(\omega, dvdv'))\}$. Nous venons de démontrer que cet ensemble a des coupes en ω non vides. D'après [A] lemme 2.5, il existe une suite $(g_n)_{n\in\mathbb{N}}$ d'éléments de $\mathcal{C}(V \times V')$ dense pour la topologie de la convergence uniforme. L'ensemble b' est aussi égal à

$$\left\{(\omega, y) \in \Omega' \times V', \forall n \in \mathbb{N},\ g_n(X(\omega), y) < \int_{V\times V'} Q(\omega, dvdv')g_n(v, v')\right\},$$

et est donc mesurable pour le produit de $\mathcal{G}$ et la tribu borélienne de V'. Une section $\mathcal{G}$-mesurable $\omega \mapsto X'(\omega)$ de b' répond à la question. □

3. Approximation des martingales continues par des $\mathcal{C}$-martingales discrètes

DÉFINITION 3.1. — *Soit* p *un réel supérieur ou égal à* 1. *On dira qu'une variété* W *a une géométrie* p*-convexe si elle a une géométrie convexe et s'il existe une fonction convexe* α *définie sur* $W \times W$, *de classe* C^∞ *sur* $(W \times W)\backslash\Delta$, *telle que pour une (donc pour toute) distance riemannienne* δ *sur* W, *il existe deux constantes* $c_{p,\delta}$ *et* $C_{p,\delta}$ *strictement positives, telles que pour tous* x, y *dans* V, *on ait*

$$c_{p,\delta}\delta^p(x, y) \leq \alpha(x, y) \leq C_{p,\delta}\delta^p(x, y).$$

On dira qu'un compact V *d'une variété* W *a une géométrie* p*-convexe s'il possède un voisinage ouvert qui a une géométrie* p*-convexe.*

Si W a une géométrie p-convexe et si $p' \geq p$, alors W a une géométrie p'-convexe (considérer $\alpha^{\frac{p'}{p}}$).

D'après [E 4.59], tout point d'une variété avec connexion possède un voisinage 2-convexe. Dans [K2], Kendall a démontré qu'une boule géodésique régulière a une géométrie $(2+\varepsilon)$-convexe pour tout $\varepsilon > 0$. Si la distance est convexe sur une variété riemannienne, alors tout compact a une géométrie 1-convexe. C'est le cas des compacts dans les variétés de Cartan-Hadamard.

La définition donnée ici est plus faible que celle de Picard. Une variété p-convexe au sens de [P2] est p-convexe selon notre définition. Je ne sais pas si la réciproque est vraie.

Rappelons la définition d'une C-martingale discrète (voir [A]).

DÉFINITION 3.2. — *Soient X une semi-martingale cadlag à valeurs dans V et $\sigma = \{0 = t_0 < \ldots < t_n = 1\}$ une subdivision de $[0,1]$. On dira que X est une σ-martingale discrète si pour tout $i \in \mathbb{N}_{n-1}$, pour tout $t \in [t_i, t_{i+1}[$, on a $X_t = X_{t_i}$ et $X_{t_i} \in \mathbb{E}[X_{t_{i+1}}|\mathcal{F}_{t_i}]$.*

On dira que X est une σ-martingale discrète exponentielle si c'est une σ-martingale discrète et pour tout i, $X_{t_i} = \mathcal{E}[X_{t_{i+1}}|\mathcal{F}_{t_i}]$.

On dira que X est une C-martingale discrète (resp. discrète exponentielle) s'il existe une subdivision σ telle que X soit une σ-martingale discrète (resp. discrète exponentielle).

Énonçons le résultat principal de cette partie.

PROPOSITION 3.3. — *On suppose que V a une géométrie convexe, vérifie la propriété (i), et que les espérances conditionnelles exponentielles existent. Soit X une martingale continue de valeur terminale L dans V. Soit $(\sigma_m)_{m\in\mathbb{N}}$ une suite de subdivisions de $[0,1]$ dont le pas tend vers 0. Pour chaque m, notons X^m la σ_m-martingale discrète exponentielle de valeur terminale L, et Y^m une σ_m-martingale discrète quelconque de valeur terminale L. On suppose que l'une des deux conditions (a) et (b) suivantes est réalisée.*

(a)-Le compact V a une géométrie 1-convexe.

(b)-Le compact V a une géométrie p-convexe pour un $p > 1$, et pour une, donc pour toute métrique riemannienne, il existe une fonction croissante continue $a(t)$ tel que la variation quadratique riemannienne $\langle X|X\rangle$ de X vérifie $d\langle X|X\rangle \leq da(t)$.

Alors X^m converge uniformément en probabilité vers X lorsque m tend vers l'infini.

De plus, si la condition (iv'(V,V)) est réalisée, alors Y^m converge uniformément en probabilité vers X lorsque m tend vers l'infini.

Démonstration. — Au cours de la démonstration, si σ est une subdivision de $[0,1]$, la notation X^σ désignera une σ-martingale discrète quelconque de valeur terminale L si (iv'(V,V)) est réalisée, et la σ-martingale discrète exponentielle de valeur terminale L sinon.

Nous allons utiliser la construction de Picard dans [P2]. Soit $\sigma = (0 = t_0 < \ldots < t_k = 1)$ une subdivision de $[0,1]$. Comme dans la démonstration de [P2] théorème 6.3, pour chaque $i \in \{1,\ldots,k\}$, on considère le processus en temps discret X_j^i, $0 \leq j \leq k$ relativement aux filtrations $\mathcal{F}_{t_j}$, $0 \leq j \leq k$, défini de la façon suivante. On pose pour tout j, $X_j^k = X_{t_j}^\sigma$. Lorsque le processus X^{i+1} est défini, on pose $X_j^i = X_{t_i}$

pour $j \geq i$, et avec une récurrence descendante, on choisit pour $j < i$ une variable aléatoire X^i_j mesurable par rapport à $\mathcal{F}_{t_j}$, telle que

$$(X^i_j, X^{i+1}_j) \in \mathbb{E}\left[(X^i_{j+1}, X^{i+1}_{j+1})|\mathcal{F}_{t_j}\right].$$

Notons que ce choix est possible d'après (iv'(V,V)), et si cette condition n'est pas réalisée, alors $X^{i+1}_j = \mathcal{E}\left[X^{i+1}_{j+1}|\mathcal{F}_{t_j}\right]$, et on choisit $X^i_j = \mathcal{E}[X^i_{j+1}|\mathcal{F}_{t_j}]$, ce qui nous donne aussi $(X^i_j, X^{i+1}_j) \in \mathbb{E}\left[(X^i_{j+1}, X^{i+1}_{j+1})|\mathcal{F}_{t_j}\right]$.

Ce choix a pour conséquence que pour tout $i < k$, le processus (X^i_j, X^{i+1}_j), $0 \leq j \leq i$ est une martingale discrète. On en déduit pour tout $j \leq i$, avec la fonction α de la définition 3.1, l'inégalité

$$\alpha((X^i_j, X^{i+1}_j)) \leq \mathbb{E}\left[\alpha(X^i_i, X^{i+1}_i)|\mathcal{F}_{t_j}\right],$$

et on remarque que $X^i_i(= X_{t_i})$ et X^{i+1}_i sont tous deux dans $\mathbb{E}\left[X_{t_{i+1}}|\mathcal{F}_{t_i}\right]$, ce qui conduit, d'après la proposition 2.15 à l'inégalité

$$\delta((X^i_i, X^{i+1}_i)) \leq C\mathbb{E}\left[\delta^3(X_{t_i}, X_{t_{i+1}})|\mathcal{F}_{t_i}\right].$$

De l'encadrement de α de la définition 3.1 et des inégalités précédentes, on déduit que

$$\delta^p((X^i_j, X^{i+1}_j)) \leq \frac{C^p C_{p,\alpha}}{c_{p,\alpha}} \mathbb{E}\left[\mathbb{E}\left[\delta^3(X_{t_i}, X_{t_{i+1}})|\mathcal{F}_{t_i}\right]^p|\mathcal{F}_{t_j}\right].$$

Or

$$\delta(X_{t_j}, X^\sigma_{t_j})(= \delta(X^j_j, X^k_j)) \leq \sum_{i=j}^{k-1} \delta(X^i_j, X^{i+1}_j),$$

donc

$$\delta(X_{t_j}, X^\sigma_{t_j}) \leq C\frac{C^p_{p,\alpha}}{c^p_{p,\alpha}} \sum_{i=j}^{k-1} \mathbb{E}\left[\mathbb{E}\left[\delta^3(X_{t_i}, X_{t_{i+1}})|\mathcal{F}_{t_i}\right]^p|\mathcal{F}_{t_j}\right]^{\frac{1}{p}}.$$

Comme les fonctions $x \mapsto \delta^3(a, x)$ sont majorées par les fonctions $\frac{1}{c_\delta} h_a$ du lemme 2.13, en utilisant la formule d'Itô, on obtient la majoration

$$\mathbb{E}\left[\delta^3(X_{t_i}, X_{t_{i+1}})|\mathcal{F}_{t_i}\right] \leq \frac{1}{2c_\delta} \mathbb{E}\left[\int_{t_i}^{t_{i+1}} \operatorname{Hess} h_{X_{t_i}}(X_s)(dX_s, dX_s)|\mathcal{F}_{t_i}\right],$$

ce qui donne, pour une constante C', en utilisant le lemme 2.13,

$$\mathbb{E}\left[\delta^3(X_{t_i}, X_{t_{i+1}})|\mathcal{F}_{t_i}\right] \leq C'\mathbb{E}\left[\int_{t_i}^{t_{i+1}} \delta(X_{t_i}, X_s)\langle dX_s|dX_s\rangle|\mathcal{F}_{t_i}\right].$$

La majoration de $\delta(X_{t_j}, X^\sigma_{t_j})$ devient

$$\delta(X_{t_j}, X^\sigma_{t_j}) \leq CC'\frac{C^p_{p,\alpha}}{c^p_{p,\alpha}} \sum_{i=j}^{k-1} \mathbb{E}\left[\mathbb{E}\left[\int_{t_i}^{t_{i+1}} \delta(X_{t_i}, X_s)\langle dX_s|dX_s\rangle|\mathcal{F}_{t_i}\right]^p|\mathcal{F}_{t_j}\right]^{\frac{1}{p}}.$$

Sous la condition (a), on peut choisir p égal à 1. On obtient

$$\delta(X_{t_j}, X^\sigma_{t_j}) \le CC' \frac{C^p_{p,\alpha}}{c^p_{p,\alpha}} \sum_{i=j}^{k-1} \mathbb{E}\left[\int_{t_i}^{t_{i+1}} \delta(X_{t_i}, X_s)\langle dX_s | dX_s\rangle | \mathcal{F}_{t_j}\right].$$

Définissons

$$\begin{aligned} M^\sigma =& CC' \frac{C^p_{p,\alpha}}{c^p_{p,\alpha}} \sum_{i=0}^{k-1} \int_{t_i}^{t_{i+1}} \delta(X_{t_i}, X_s)\langle dX_s | dX_s\rangle \\ =& CC' \frac{C^p_{p,\alpha}}{c^p_{p,\alpha}} \int_0^1 \sum_{i=0}^{k-1} 1_{\{s\in[t_i,t_{i+1}[\}} \delta(X_{t_i}, X_s)\langle dX_s | dX_s\rangle, \end{aligned}$$

et $M^\sigma_i = \mathbb{E}[M^\sigma | \mathcal{F}_{t_i}]$. Lorsque $|\sigma|$ tend vers zéro, par convergence dominée, la variable aléatoire M^σ converge vers zéro dans L^2, ce qui implique par l'inégalité de Doob que $\sup_i M^\sigma_i$ converge vers zéro dans L^2. Or $\delta(X_{t_j}, X^\sigma_{t_j}) \le M^\sigma_j$, donc $\sup_j \delta(X_{t_j}, X^\sigma_{t_j})$ converge dans L^2 vers 0 lorsque $|\sigma|$ tend vers zéro. Pour t quelconque, on définit $\sigma(t)$ l'élément t_j de σ qui vérifie $t_j \le t < t_{j+1}$. On a alors $\delta(X_t, X^\sigma_t) \le \delta(X_{\sigma(t)}, X^\sigma_{\sigma(t)}) + \delta(X_{\sigma(t)}, X_t)$. Comme les trajectoires de X sont uniformément continues, $\delta(X_{\sigma(t)}, X_t)$ converge en probabilité vers zéro, uniformément en t. Ceci achève la démonstration sous la condition (a).

Si la condition (b) est réalisée, on a

$$\delta(X_{t_j}, X^\sigma_{t_j}) \le CC' \frac{C^p_{p,\alpha}}{c^p_{p,\alpha}} \sum_{i=j}^{k-1} \mathbb{E}\left[\mathbb{E}\left[\int_{t_i}^{t_{i+1}} \delta(X_{t_i}, X_s) da(s) | \mathcal{F}_{t_i}\right]^p | \mathcal{F}_{t_j}\right]^{\frac{1}{p}},$$

ce qui donne par l'inégalité de Jensen, en posant $D_p = CC' \frac{C^p_{p,\alpha}}{c^p_{p,\alpha}}$,

$$\delta(X_{t_j}, X^\sigma_{t_j}) \le D_p \sum_{i=j}^{k-1} \mathbb{E}\left[(a(t_{i+1}) - a(t_i))^{p-1} \int_{t_i}^{t_{i+1}} \delta^p(X_{t_i}, X_s) da(s) | \mathcal{F}_{t_j}\right]^{\frac{1}{p}},$$

ou encore

$$\delta(X_{t_j}, X^\sigma_{t_j}) \le D_p \sum_{i=j}^{k-1} (a(t_{i+1}) - a(t_i))^{\frac{p-1}{p}} \left(\int_{t_i}^{t_{i+1}} \mathbb{E}\left[\delta^p(X_{t_i}, X_s) | \mathcal{F}_{t_j}\right] da(s)\right)^{\frac{1}{p}}.$$

Quitte à changer les constantes, on peut supposer que $p \ge 3$, et on peut majorer $\mathbb{E}\left[\delta^p(X_{t_i}, X_s) | \mathcal{F}_{t_j}\right]$ par $C' \int_{t_i}^{s} \mathbb{E}\left[\delta^{p-2}(X_{t_i}, X_u) | \mathcal{F}_{t_j}\right] da(u)$, et en majorant l'espérance conditionnelle par une constante, on obtient

$$\mathbb{E}\left[\delta^p(X_{t_i}, X_s) | \mathcal{F}_{t_j}\right] \le C''(a(s) - a(t_i)),$$

et

$$\int_{t_i}^{t_{i+1}} \mathbb{E}\left[\delta^p(X_{t_i}, X_s)|\mathcal{F}_{t_j}\right] da(s) \leq \frac{C''}{2}(a(t_{i+1}) - a(t_i))^2.$$

Ceci donne, quitte à changer C'', une majoration de la forme

$$\delta(X_{t_j}, X_{t_j}^\sigma) \leq C'' \sum_{i=j}^{k-1} (a(t_{j+1}) - a(t_j))^{\frac{p+1}{p}},$$

qui implique

$$\sup_j \delta(X_{t_j}, X_{t_j}^\sigma) \leq C'' \sum_{i=0}^{k-1} (a(t_{j+1}) - a(t_j))^{\frac{p+1}{p}}.$$

Le terme de droite tend vers zéro lorsque $|\sigma|$ tend vers zéro. Pour passer à une majoration de $\sup_t \delta(X_t, X_t^\sigma)$, on procède comme sous la condition (a). □

RÉFÉRENCES.

[A] Arnaudon (M.). — *Espérances conditionnelles et C-martingales dans les variétés*, Séminaire de Probabilités XXVIII, Lecture Notes in Mathematics, Vol 1583, Springer, 1994.

[E] Emery (M.). — *Stochastic calculus in manifolds.* — Springer, 1989.

[E,M] Emery (M.), Mokobodzki (G.). — *Sur le barycentre d'une probabilité dans une variété*, Séminaire de Probabilités XXV, Lecture Notes in Mathematics, Vol 1485, Springer, 1991.

[E,Z] Emery (M.), Zheng (W.). — *Fonctions convexes et semi-martingales dans une variété*, Séminaire de Probabilités XVIII, Lecture Notes in Mathematics, Vol 1059, Springer, 1984.

[K1] Kendall (W.S.). — *Probability, convexity and harmonic maps with small image I : uniqueness and fine existence*, Proc. London Math. Soc. (3), t. **61**, 1990, p. 371–406.

[K2] Kendall (W.S.). — *Convexity and the hemisphere*, J. London Math. Soc. (2), t. **43**, 1991, p. 567–576.

[M] Milnor (J.W.). — *Topology from the differentiable viewpoint.* — The University Press of Virginia, 1969.

[P1] Picard (J.). — *Martingales on Riemannian manifolds with prescribed limit*, J. Functional Anal. 99, t. **2**, 1991, p. 223–261.

[P2] Picard (J). — *Barycentres et martingales sur une variété*, à paraître aux Annales de l'Institut Henri Poincaré, 1994.

ÉQUATIONS DIFFÉRENTIELLES STOCHASTIQUES MULTIVOQUES

Emmanuel CÉPA

Département de Mathématiques, Université d'Orléans, BP 6759, 45067 Orléans cedex 2, France

1 Introduction

On s'intéresse à la résolution de l'équation différentielle stochastique associée à un opérateur maximal monotone multivoque A de $\mathbb{R}^d$, $d \in \mathbb{N}^*$, c'est-à-dire, formellement, du type :

$$\begin{cases} dX_t + A(X_t)dt \ni b(X_t)dt + \sigma(X_t)dW_t \\ X_0 = x_0 \in \overline{D(A)}. \end{cases} \tag{1.1}$$

La formulation précise de (1.1) que l'on utilisera est équivalente à celle donnée par A. Bensoussan et J.L. Lions dans [1] pour les processus de diffusion réfléchis où le problème est posé sous forme d'inéquations variationnelles stochastiques. Dans [6], P. Krée obtient l'existence d'une solution forte lorsque la composante brownienne et la composante multivoque agissent dans des directions séparées. Un cas très fréquent d'opérateur maximal monotone multivoque est constitué par le sous-différentiel d'une fonction convexe s.c.i. propre φ : en particulier quand cette fonction vaut 0 dans un convexe fermé et $+\infty$ en dehors, il s'agit simplement d'un problème de réflexion au bord d'un domaine convexe et le cas stochastique a déjà été abondamment traité, notamment par H. Tanaka [11] et J.L. Menaldi [9]. Dans le cas unidimensionnel ($d = 1$), on sait que tout opérateur maximal monotone multivoque A provient nécessairement de la sous-dérivée d'une fonction convexe s.c.i. propre φ et le problème d'existence et d'unicité a été résolu dans sa plus grande généralité par D. Lépingle-C. Marois [8] qui montrent que pour b, σ lipschitziens et tout opérateur maximal monotone multivoque $A = \partial\varphi$ de $\mathbb{R}$, le problème (1.1) admet une unique solution forte. Plus récemment, A. Bensoussan-A. Rascanu [2] ont montré l'existence et l'unicité d'une solution forte pour (1.1) dans le cas particulier où $A = \partial\varphi$, avec φ fonction convexe s.c.i. propre de $\mathbb{R}^d$ ($d \in \mathbb{N}^*$), et b, σ sont lipschitziens.

On se propose ici de montrer l'existence et l'unicité d'une solution forte pour l'équation différentielle stochastique associée à un opérateur maximal monotone multivoque *quelconque* A de $\mathbb{R}^d$ *pour tout* $d \in \mathbb{N}^*$. L'unicité trajectorielle est obtenue sensiblement comme en dimension 1 en utilisant uniquement la monotonie de A. L'existence d'une solution forte est démontrée en considérant un problème approché (pénalisation) puis, en montrant un résultat de compacité au sens de la convergence en loi pour la pseudo-topologie de Meyer-Zheng [10], en prouvant ensuite que la convergence est en fait uniforme (sur les compacts) grâce à une technique de changement de temps inspirée de T. Kurtz [7], et enfin en passant à la limite dans le problème approché. On obtient ainsi une solution faible du problème et, finalement, on conclut qu'il y a existence forte grâce à l'unicité trajectorielle et au théorème de Yamada-Watanabe (étendu au cas multivoque). Cette démonstration vise à construire la loi de la solution de (1.1) : elle consiste à voir une équation différentielle stochastique multivoque comme une perturbation multivoque d'une équation différentielle stochastique classique (cas $A = 0$) et donc cherche à régulariser (en espace) le terme multivoque A. Une technique de changement de temps est nécessaire pour "ralentir" les approximants de la solution car le passage à la limite "voit les singularités" de A, ce qui explique la relative complexité de la méthode.

La principale motivation de cette étude est l'obtention d'un résultat d'existence et d'unicité fortes pour des équations différentielles stochastiques réfléchies possédant un coefficient de diffusion satisfaisant aux conditions habituelles de type Lipschitz (sans hypothèse supplémentaire de non-dégénérescence) et, surtout, un coefficient de dérive très singulier (ne vérifiant donc pas les hypothèses habituelles de type Lipschitz). En particulier, le résultat essentiel démontré dans ce document permet d'obtenir des diffusions réfléchies dans un domaine D de $\mathbb{R}^d$ qui possèdent un coefficient de diffusion dégénéré et un coefficient de dérive discontinu, explosant au bord du domaine D.

La formulation du problème telle qu'elle a été faite dans (1.1) est purement formelle et n'a pour seul but que de donner une idée du problème posé au lecteur : une formulation précise sera donnée en même temps que les hypothèses et les résultats obtenus. Tout d'abord, on rappelle quelques résultats généraux sur les opérateurs maximaux monotones multivoques.

2 Opérateurs maximaux monotones multivoques

Après avoir rappelé la définition d'un opérateur maximal monotone multivoque, on donne quelques exemples et les propriétés élémentaires d'un tel opérateur, puis on construit une suite d'approximants univoques. Les résultats énoncés dans cette partie sont tout à fait classiques : le lecteur pourra trouver dans l'ouvrage de H. Brézis [3] à la fois un développement plus complet sur ce sujet et aussi les démonstrations que l'on a omises ici. Dans toute cette section, d est un entier naturel non nul fixé.

2.1 Définitions

Définition 2.1 *On appelle opérateur multivoque de $\mathbb{R}^d$ toute application A de $\mathbb{R}^d$ dans $\mathcal{P}(\mathbb{R}^d)$, où $\mathcal{P}(\mathbb{R}^d)$ désigne l'ensemble des parties de $\mathbb{R}^d$. Le domaine de A est l'ensemble*

$$D(A) = \{x \in \mathbb{R}^d : A(x) \neq \emptyset\}, \tag{2.2}$$

et l'image de A l'ensemble :

$$\operatorname{Im}(A) = \bigcup_{x\in\mathbb{R}^d} A(x). \tag{2.3}$$

Si pour tout $x \in \mathbb{R}^d$, $A(x)$ contient au plus un élément, on dira que A est un opérateur univoque de $\mathbb{R}^d$ (on retombe alors sur la théorie classique des opérateurs). Un opérateur multivoque de $\mathbb{R}^d$ est complètement déterminé par la donnée de son graphe défini par :

$$\operatorname{Gr}(A) = \{(x,y) \in \mathbb{R}^{2d} : x \in \mathbb{R}^d, y \in A(x)\}. \tag{2.4}$$

L'opérateur inverse de A, noté A^{-1}, est l'opérateur dont le graphe est symétrique de celui de A, c'est-à-dire :

$$y \in A^{-1}(x) \Leftrightarrow x \in A(y). \tag{2.5}$$

Définition 2.2 *Un opérateur multivoque A de $\mathbb{R}^d$ est dit monotone si et seulement si*

$$\langle y_1 - y_2, x_1 - x_2\rangle \geqslant 0\,, \qquad \forall (x_1,y_1),(x_2,y_2) \in \operatorname{Gr}(A), \tag{2.6}$$

où $\langle .,. \rangle$ est le produit scalaire usuel de $\mathbb{R}^d$.

Proposition 2.3 *Soit A un opérateur multivoque de $\mathbb{R}^d$. Alors A est monotone si et seulement si :*

$$|x_1 - x_2| \leqslant |(x_1 - x_2) + \lambda(y_1 - y_2)|, \qquad \forall (x_1, y_1), (x_2, y_2) \in \mathrm{Gr}(A), \forall \lambda > 0, \quad (2.7)$$

avec $|.|$ *la norme euclidienne associée à* $\langle .,. \rangle$.

Par conséquent, la condition de monotonie exprime que pour tout $\lambda > 0$, l'opérateur $(I + \lambda A)^{-1}$ est une contraction univoque de $\mathrm{Im}(I + \lambda A)$ dans $\mathbb{R}^d$. Autrement dit, pour tout $z \in \mathbb{R}^d$, l'équation $z \in x + \lambda A(x)$ admet au plus une solution, et si x_1, x_2 sont les solutions correspondant à z_1, z_2, on a : $|x_1 - x_2| \leqslant |z_1 - z_2|$. Les opérateurs que nous allons considérer dans la suite sont ceux pour lesquels l'équation $z \in x + \lambda A(x)$ possède exactement une solution pour tout $z \in \mathbb{R}^d$ et tout $\lambda > 0$.

L'ensemble des opérateurs multivoques de $\mathbb{R}^d$ est ordonné pour l'inclusion des graphes :

$$A \subset B \Leftrightarrow \{D(A) \subset D(B) \text{ et, pour tout } x \in D(A),\ A(x) \subset B(x)\}, \qquad (2.8)$$

et il est inductif pour cette relation d'ordre, d'où la

Définition 2.4 *Un opérateur monotone multivoque A de $\mathbb{R}^d$ est dit maximal monotone s'il est maximal dans l'ensemble des opérateurs monotones mutivoques, c'est-à-dire A est maximal monotone si et seulement si*

$$(x, y) \in \mathrm{Gr}(A) \quad \Leftrightarrow \quad \Big\{ \langle y - v, x - u \rangle \geqslant 0\,, \quad \forall (u, v) \in \mathrm{Gr}(A) \Big\}. \qquad (2.9)$$

Remarque. Insistons sur le fait que A est maximal dans l'ensemble des graphes monotones. Un opérateur qui est seulement maximal dans l'ensemble des opérateurs univoques monotones n'est pas nécessairement maximal monotone au sens de la définition 2.4.

La caractérisation suivante est fondamentale dans l'étude des opérateurs maximaux monotones multivoques.

Proposition 2.5 *Soit A opérateur multivoque de $\mathbb{R}^d$. Il y a équivalence entre les assertions suivantes :*

(i) A est maximal monotone ;

(ii) A est monotone et $\mathrm{Im}(I + A) = \mathbb{R}^d$;

(iii) pour tout $\lambda > 0$, $(I + \lambda A)^{-1}$ est une contraction univoque définie sur $\mathbb{R}^d$ tout entier.

2.2 Exemple fondamental d'opérateur maximal monotone multivoque : $\partial\varphi$

Définition 2.6 *Soit φ une fonction convexe de $\mathbb{R}^d$, c'est-à-dire une application φ : $\mathbb{R}^d \to]-\infty; +\infty]$ telle que :*

$$\varphi(tx + (1-t)y) \leqslant t\varphi(x) + (1-t)\varphi(y), \qquad \forall x, y \in \mathbb{R}^d, \forall t \in [0; 1]. \quad (2.10)$$

On appelle domaine effectif de φ l'ensemble $\mathrm{dom}(\varphi) = \{x \in \mathbb{R}^d : \varphi(x) < +\infty\}$ *et on dit que φ est propre si* $\mathrm{dom}(\varphi)$ *est non vide.*

Définition 2.7 *Soit φ fonction convexe propre de $\mathbb{R}^d$. On appelle sous-différentiel de φ, noté $\partial\varphi$, l'opérateur multivoque de $\mathbb{R}^d$ défini par son graphe de la façon suivante*

$$(x,y) \in \mathrm{Gr}(\partial\varphi) \Leftrightarrow \varphi(x) \leqslant \varphi(z) + \langle y, x-z\rangle, \qquad \forall z \in \mathbb{R}^d. \tag{2.11}$$

Proposition 2.8 *Le sous-différentiel d'une fonction convexe, semi-continue inférieurement (s.c.i. en abrégé), propre de $\mathbb{R}^d$ est un opérateur maximal monotone multivoque de $\mathbb{R}^d$.*

Cas particulier : sous-différentiel de la fonction indicatrice d'un convexe fermé. Soit D un convexe fermé non vide de $\mathbb{R}^d$. On appelle fonction indicatrice de D, notée I_D, la fonction définie par :

$$I_D(x) = \begin{cases} 0 & \text{si } x \in D \\ +\infty & \text{si } x \notin D. \end{cases} \tag{2.12}$$

On vérifie facilement que I_D est une fonction convexe, s.c.i., propre avec dom(I_D)=D. On a alors pour tout x dans D :

$$\partial I_D(x) = \{y \in \mathbb{R}^d : \langle y, x-z\rangle \geqslant 0, \forall z \in D\}, \tag{2.13}$$

c'est-à-dire :

$$\partial I_D(x) = \begin{cases} \emptyset & \text{si } x \notin D \\ \{0\} & \text{si } x \in \mathrm{Int}(D) \\ \Pi_x & \text{si } x \in \partial D, \end{cases} \tag{2.14}$$

où l'on a noté Int(D) l'intérieur de D et Π_x le cône normal extérieur à D en x.

Remarques.

(i) En dimension $d = 1$, tous les opérateurs maximaux monotones multivoques A sont de la forme décrite précédemment c'est-à-dire $A = \partial\varphi$ avec φ fonction convexe, s.c.i., propre de $\mathbb{R}$. Par contre dans le cas multidimensionnel ($d \geqslant 2$), il existe des opérateurs maximaux monotones multivoques qui ne sont pas des sous-différentiels de fonction convexe.

(ii) Soit A opérateur linéaire, univoque, monotone de $\mathbb{R}^d$. Alors A est maximal monotone multivoque si et seulement si $D(A)$ est dense dans $\mathbb{R}^d$ et A maximal dans l'ensemble des opérateurs univoques linéaires monotones.

(iii) Soit A opérateur univoque, monotone de $\mathbb{R}^d$ tel que $D(A) = \mathbb{R}^d$. Si A est hémicontinu (i.e. pour tous $x, y \in \mathbb{R}^d$, $A((1-t)x + ty) \to A(x)$ lorsque $t \to 0$) alors A est maximal monotone multivoque.

2.3 Propriétés élémentaires

Proposition 2.9 *Soit A opérateur maximal monotone multivoque de $\mathbb{R}^d$. Alors on a les propriétés :*

(i) Int $(D(A))$ *et* $\overline{D(A)}$ *(où* $\overline{D(A)}$ *désigne l'adhérence de* $D(A)$*) sont des convexes de* $\mathbb{R}^d$*, de plus* $\mathrm{Int}(D(A)) = \mathrm{Int}(\overline{D(A)})$;

(ii) pour tout $x \in \mathbb{R}^d$, $A(x)$ est un convexe fermé de $\mathbb{R}^d$.

Notations. La proposition 2.9 (ii) permet de définir l'opérateur univoque :

$$A^\circ(x) = \mathrm{proj}_{A(x)}(0), \qquad \forall x \in \mathbb{R}^d, \tag{2.15}$$

où pour tout convexe fermé D de $\mathbb{R}^d$, proj_D désigne la projection sur D et :

$$\mathrm{proj}_\emptyset(0) = \delta, \tag{2.16}$$

avec δ un point à l'infini de $\mathbb{R}^d$, fixé, et pour lequel on prendra $|\delta| = +\infty$. Par conséquent, pour tout $x \in \mathbb{R}^d$, $A^\circ(x)$ est l'élément de norme minimale de $A(x)$; on aura donc l'équivalence :

$$x \in D(A) \quad \Leftrightarrow \quad |A^\circ(x)| < +\infty. \tag{2.17}$$

Définition 2.10 *On appelle section principale d'un opérateur multivoque A de $\mathbb{R}^d$ tout opérateur univoque A' de $\mathbb{R}^d$ avec $D(A) \subset D(A')$ et tel que pour tout $(x,y) \in \overline{D(A)} \times \mathbb{R}^d$, l'inégalité*

$$\langle A'(z) - y, z - x\rangle \geqslant 0, \qquad \forall z \in D(A), \tag{2.18}$$

implique $y \in A(x)$.

Proposition 2.11 *Pour tout opérateur maximal monotone multivoque A de $\mathbb{R}^d$, l'opérateur A° est une section principale de A.*

Les opérateurs maximaux monotones multivoques de $\mathbb{R}^d$ sont localement bornés sur l'intérieur de leur domaine :

Proposition 2.12 *Soit A opérateur maximal monotone multivoque de $\mathbb{R}^d$. Alors A est borné au voisinage de tout point intérieur à $D(A)$, c'est-à-dire : pour tout x dans $\mathrm{Int}(D(A))$, il existe un voisinage V de x tel que*

$$\bigcup_{x' \in V} A(x') \tag{2.19}$$

soit une partie bornée de $\mathbb{R}^d$. En particulier, A est borné sur tout compact inclus dans $\mathrm{Int}\,(D(A))$.

2.4 Approximation Yosida "multivoque"

Proposition 2.13 *Soit A opérateur maximal monotone multivoque de $\mathbb{R}^d$. On pose*

$$J_n = \left(I + \frac{1}{n}A\right)^{-1}, \qquad n \in \mathbb{N}^*. \tag{2.20}$$

Alors J_n est une contraction univoque définie sur $\mathbb{R}^d$ tout entier, à valeurs dans $D(A)$ pour tout $n \in \mathbb{N}^*$. De plus :

$$J_n(x) \xrightarrow[n\to\infty]{} \mathrm{proj}_{\overline{D(A)}}(x), \qquad \forall x \in \mathbb{R}^d. \tag{2.21}$$

L'application J_n est appelée n^{ieme} résolvante de A.

Proposition 2.14 *Soit A opérateur maximal monotone multivoque de $\mathbb{R}^d$. On pose*

$$A_n = n(I - J_n), \qquad n \in \mathbb{N}^*. \tag{2.22}$$

Alors :

(i) A_n opérateur maximal monotone, univoque, défini sur $\mathbb{R}^d$ tout entier, lipschitzien de rapport n, pour tout $n \in \mathbb{N}^*$;

(ii) pour tout $x \in D(A)$,

$$A_n(x) \underset{n\to\infty}{\longrightarrow} A^\circ(x), \tag{2.23}$$

avec $|A_n(x)| \uparrow |A^\circ(x)|$ quand $n \uparrow \infty$ et

$$|A_n(x) - A^\circ(x)|^2 \leqslant |A^\circ(x)|^2 - |A_n(x)|^2 ; \tag{2.24}$$

(iii) pour tout $x \notin D(A)$,

$$|A_n(x)| \uparrow +\infty \quad quand \quad n \uparrow \infty ; \tag{2.25}$$

(iv) pour tout $x \in \mathbb{R}^d$, $A_n(x) \in A(J_n(x))$.

L'application A_n est appelée n^{ieme} approximée Yosida de A.

Remarque. On vérifie facilement que lorsque $A = \partial I_D$ avec D convexe fermé non vide de $\mathbb{R}^d$, on a pour tous $x \in \mathbb{R}^d$, $n \in \mathbb{N}^*$:

$$J_n(x) = \text{proj}_D(x) \quad \text{et} \quad A_n(x) = n(x - \text{proj}_D(x)). \tag{2.26}$$

3 Position du problème

3.1 Données

- $d \in \mathbb{N}^*$;
- A opérateur maximal monotone multivoque de $\mathbb{R}^d$ tel que :

$$\text{Int}(D(A)) \neq \emptyset ; \tag{3.27}$$

- $(\Omega, \mathcal{F}, \{\mathcal{F}_t\}, \mathbb{P})$ espace probabilisé filtré vérifiant les conditions habituelles ;
- $W = \{W_t, \mathcal{F}_t; 0 \leqslant t < \infty\}$ mouvement brownien d-dimensionnel standard défini sur $(\Omega, \mathcal{F}, \mathbb{P})$ avec $W_0 = 0$;
- $b : \mathbb{R}^d \rightarrow \mathbb{R}^d$ et $\sigma : \mathbb{R}^d \rightarrow \mathbb{R}^d \otimes \mathbb{R}^d$ applications mesurables ;
- $x_0 \in \overline{D(A)}$.

3.2 Énoncé du problème

On cherche un couple de processus (X, K) vérifiant :

$\boxed{1}$ $X = \{X_t, \mathcal{F}_t; 0 \leqslant t < \infty\}$ processus continu, adapté, défini sur $(\Omega, \mathcal{F}, \mathbb{P})$, à valeurs dans $\overline{D(A)}$ avec $X_0 = x_0$ $\mathbb{P}$-p.s. ;

$\boxed{2}$ $K = \{K_t, \mathcal{F}_t; 0 \leqslant t < \infty\}$ processus continu, adapté, à variation finie, défini sur $(\Omega, \mathcal{F}, \mathbb{P})$, à valeurs dans $\mathbb{R}^d$, $K_0 = 0$ $\mathbb{P}$-p.s. ;

$\boxed{3}$ $dX_t = b(X_t)dt + \sigma(X_t)dW_t - dK_t;$ $0 \leqslant t < \infty$, $\mathbb{P}$-p.s. ;

$\boxed{4}$ pour tout couple de processus (α, β), $\alpha = \{\alpha_t, \mathcal{F}_t; 0 \leqslant t < \infty\}$ et $\beta = \{\beta_t, \mathcal{F}_t; 0 \leqslant t < \infty\}$ continus, adaptés, définis sur $(\Omega, \mathcal{F}, \mathbb{P})$, à valeurs dans $\mathbb{R}^d$, vérifiant :

$$(\alpha_u, \beta_u) \in \mathrm{Gr}(A), \qquad \forall u \in [0; +\infty[, \tag{3.28}$$

la mesure

$$\langle X_u - \alpha_u, dK_u - \beta_u du \rangle \tag{3.29}$$

est $\mathbb{P}$-p.s. positive sur $\mathbb{R}^+$.

Le problème précédent $\left(\boxed{1}; \boxed{2}; \boxed{3}; \boxed{4}\right)$ sera noté $EDSM^d(A; b; \sigma; x_0)$ dans la suite.

Remarque. Lorsque $A = \partial I_D$ avec D convexe fermé non vide de $\mathbb{R}^d$, le problème $EDSM^d(A; b; \sigma; x_0)$ est équivalent au problème de diffusion (de coefficient de dérive b et de coefficient de diffusion σ) réfléchie (normalement) dans D.

3.3 Résultats

On sera amené à utiliser l'hypothèse suivante : il existe $0 < C < \infty$ telle que pour tous x, $y \in \mathbb{R}^d$,

$$|b(x) - b(y)| + |\sigma(x) - \sigma(y)| \leqslant C|x - y|. \tag{3.30}$$

On peut maintenant énoncer le résultat fondamental d'existence et d'unicité pour l'équation différentielle stochastique associée à un opérateur maximal monotone multivoque de $\mathbb{R}^d$. Pour tout $d \in \mathbb{N}^*$, on a :

Théorème 3.1 *($EDSM^d$) : Pour tout opérateur maximal monotone multivoque A de $\mathbb{R}^d$ vérifiant (3.27), tout $x_0 \in \overline{D(A)}$, toutes applications b, σ satisfaisant à (3.30), il existe une unique solution du problème $EDSM^d(A; b; \sigma; x_0)$.*

Remarques.

(*i*) On obtient le même résultat d'existence et d'unicité pour une condition initiale aléatoire η $\mathcal{F}_0$-mesurable, à valeurs dans $\overline{D(A)}$ et des coefficients $b(t, x)$ et $\sigma(t, x)$ dépendant du temps vérifiant alors au lieu de (3.30) : pour tout $T > 0$, il existe $0 < C(T) < \infty$ telle que pour tous $0 \leqslant t \leqslant T$ et x, $y \in \mathbb{R}^d$,

$$|b(t, x) - b(t, y)| + |\sigma(t, x) - \sigma(t, y)| \leqslant C(T)|x - y|, \tag{3.31}$$

$$|b(t, x)| + |\sigma(t, x)| \leqslant C(T)(1 + |x|). \tag{3.32}$$

(*ii*) Il est intéressant de noter que puisque l'on peut considérer en particulier A sous la forme $\partial\varphi$ avec φ seulement convexe, s.c.i., propre, le théorème 3.1 donne un résultat d'existence et d'unicité fortes pour des équations différentielles stochastiques réfléchies possédant un coefficient de dérive éventuellement très singulier. Le théorème 3.1 permet en particulier d'obtenir des diffusions réfléchies dans un domaine D de $\mathbb{R}^d$ qui possèdent un coefficient de diffusion dégénéré et un coefficient de dérive discontinu, explosant au bord du domaine D.

Conventions. On adopte la notation classique f^{-1} pour désigner l'application réciproque d'une bijection f. Dans les calculs, toutes les constantes, dont la valeur n'a pas lieu d'être précisée, seront notées indifféremment C.

4 Démonstration de l'unicité

On se propose de montrer dans cette section qu'il y a unicité trajectorielle pour l'équation différentielle stochastique associée à un opérateur maximal monotone multivoque A quelconque et des coefficients b, σ vérifiant les hypothèses Lipschitz habituelles.

Le lemme suivant étend un résultat analogue de D. Lépingle-C. Marois [8] en dimension 1. Ces derniers utilisent dans leur démonstration les propriétés particulières des convexes de $\mathbb{R}$: dans le cas multidimensionnel, il faut généraliser quelque peu leur preuve en faisant appel à l'approximation Yosida des opérateurs maximaux monotones multivoques.

Lemme 4.1 *(Lemme de monotonie) On suppose donné A opérateur maximal monotone multivoque de $\mathbb{R}^d$. Soient (x, k), (x', k') des fonctions continues sur $\mathbb{R}^+$, à valeurs dans $\mathbb{R}^d$, telles que :*

(*i*) *k, k' à variation finie ;*

(*ii*) *x, x' à valeurs dans $\overline{D(A)}$;*

(*iii*) *les mesures :*

$$\langle x(u) - \alpha(u), dk(u) - \beta(u)du \rangle, \tag{4.33}$$

$$\langle x'(u) - \alpha(u), dk'(u) - \beta(u)du \rangle, \tag{4.34}$$

sont positives sur $\mathbb{R}^+$ pour tout couple de fonctions (α, β) continues vérifiant :

$$(\alpha(u), \beta(u)) \in \mathrm{Gr}\,(A), \qquad \forall u \in \mathbb{R}^+. \tag{4.35}$$

Alors la mesure

$$\langle x(u) - x'(u),\ dk(u) - dk'(u) \rangle \tag{4.36}$$

est positive sur $\mathbb{R}^+$.

Démonstration. Soient (x, k), (x', k') vérifiant les hypothèses du lemme 4.1. Pour $n \in \mathbb{N}^*$, prenons :

$$\alpha^{(n)}(u) = J_n\Big(\frac{x(u) + x'(u)}{2}\Big), \tag{4.37}$$

$$\beta^{(n)}(u) = A_n\Big(\frac{x(u)+x'(u)}{2}\Big), \tag{4.38}$$

alors $\alpha^{(n)}$, $\beta^{(n)}$ continues et $(\alpha^{(n)}(u), \beta^{(n)}(u)) \in \mathrm{Gr}\,(A)$ car $A_n(x) \in A(J_n(x))$ pour tout x dans $\mathbb{R}^d$ d'après la proposition 2.14 *(iv)*.
En utilisant la positivité des mesures (4.33) et (4.34) pour $(\alpha^{(n)}, \beta^{(n)})$, $n \in \mathbb{N}^*$, on peut écrire :

$$0 \leqslant \Big\langle x(u) - J_n\Big(\frac{x(u)+x'(u)}{2}\Big), dk(u) - A_n\Big(\frac{x(u)+x'(u)}{2}\Big)du\Big\rangle, \tag{4.39}$$

$$0 \leqslant \Big\langle x'(u) - J_n\Big(\frac{x(u)+x'(u)}{2}\Big), dk'(u) - A_n\Big(\frac{x(u)+x'(u)}{2}\Big)du\Big\rangle, \tag{4.40}$$

d'où en ajoutant et retranchant $(x(u)+x'(u))/2$ dans le premier membre de chacun des produits scalaires :

$$0 \leqslant \Big\langle \frac{x(u)-x'(u)}{2} + \frac{1}{n}A_n\Big(\frac{x(u)+x'(u)}{2}\Big), dk(u) - A_n\Big(\frac{x(u)+x'(u)}{2}\Big)du\Big\rangle, \tag{4.41}$$

$$0 \leqslant \Big\langle \frac{x'(u)-x(u)}{2} + \frac{1}{n}A_n\Big(\frac{x(u)+x'(u)}{2}\Big), dk'(u) - A_n\Big(\frac{x(u)+x'(u)}{2}\Big)du\Big\rangle, \tag{4.42}$$

puis en sommant les équations (4.41) et (4.42) ainsi obtenues :

$$\begin{aligned} 0 \ \leqslant\ & \frac{1}{2}\Big\langle x(u)-x'(u)\,,\ dk(u)-dk'(u)\Big\rangle \ - \ \frac{2}{n}\Big|A_n\Big(\frac{x(u)+x'(u)}{2}\Big)\Big|^2 du \\ & + \Big\langle \frac{x(u)+x'(u)}{2} - J_n\Big(\frac{x(u)+x'(u)}{2}\Big),\ dk(u)+dk'(u)\Big\rangle, \end{aligned}$$

par conséquent :

$$\langle x(u)-x'(u),\ dk(u)-dk'(u)\rangle \geqslant -2\Big\langle \frac{x(u)+x'(u)}{2} - J_n\Big(\frac{x(u)+x'(u)}{2}\Big), dk(u)+dk'(u)\Big\rangle,$$

puis en passant à la limite quand $n \to \infty$ d'après la convergence

$$J_n(x) \xrightarrow[n\to\infty]{} \mathrm{proj}_{\overline{D(A)}}(x) \tag{4.43}$$

rappelée dans la proposition 2.13, et, grâce à la convexité de $\overline{D(A)}$ donnée par la proposition 2.9 :

$$\frac{x(u)+x'(u)}{2} \in \overline{D(A)}, \tag{4.44}$$

on obtient le résultat voulu. ∎

En vertu du lemme de monotonie précédent, la démonstration de l'unicité trajectorielle pour l'équation (1.1) est rapidement ramenée au cas des équations différentielles stochastiques classiques et ne pose donc aucun problème.

Théorème 4.2 *(Théorème d'unicité) : Pour tout opérateur maximal monotone multivoque A de $\mathbb{R}^d$, tout $x_0 \in \overline{D(A)}$, toutes applications b, σ satisfaisant à (3.30), l'équation différentielle stochastique multivoque (1.1) possède la propriété d'unicité trajectorielle.*

Démonstration. Soient (X,K) et (X',K') deux solutions faibles de l'équation différentielle stochastique multivoque (1.1) sur le même espace probabilisé filtré $(\Omega, \mathcal{F}, \{\mathcal{F}_t\}, \mathbb{P})$, par rapport au même mouvement brownien $W = \{W_t, \mathcal{F}_t; 0 \leqslant t < \infty\}$ et de même condition initiale. On pose :

$$S_p = \inf\{t \geqslant 0 : |X_t| + |X'_t| \geqslant p\}, \qquad p \in \mathbb{N}^*. \tag{4.45}$$

Alors S_p est un $\{\mathcal{F}_t\}$-temps d'arrêt pour tout $p \in \mathbb{N}^*$ et $S_p \uparrow \infty$ quand $p \uparrow \infty$ $\mathbb{P}$-p.s.. D'après la formule d'Itô, on a :

$$\begin{aligned} |X_{t\wedge S_p} - X'_{t\wedge S_p}|^2 \;=\; & 2\int_0^{t\wedge S_p} \langle b(X_s) - b(X'_s), X_s - X'_s \rangle ds \\ & + 2\int_0^{t\wedge S_p} \langle \sigma(X_s) - \sigma(X'_s), X_s - X'_s \rangle dW_s \\ & - 2\int_0^{t\wedge S_p} \langle X_s - X'_s, dK_s - dK'_s \rangle \\ & + \int_0^{t\wedge S_p} \operatorname{tr}\left[\{\sigma(X_s) - \sigma(X'_s)\}\{\sigma(X_s) - \sigma(X'_s)\}^*\right] ds . \end{aligned}$$

Grâce au lemme de monotonie (lemme 4.1), la mesure :

$$< X_u - X'_u,\, dK_u - dK'_u > \tag{4.46}$$

est $\mathbb{P}$-p.s. positive sur $\mathbb{R}^+$, d'où :

$$\begin{aligned} |X_{t\wedge S_p} - X'_{t\wedge S_p}|^2 \;\leqslant\; & 2\int_0^{t\wedge S_p} \langle b(X_s) - b(X'_s), X_s - X'_s \rangle ds \\ & + 2\int_0^{t\wedge S_p} \langle \sigma(X_s) - \sigma(X'_s), X_s - X'_s \rangle dW_s \\ & + \int_0^{t\wedge S_p} \operatorname{tr}\left[\{\sigma(X_s) - \sigma(X'_s)\}\{\sigma(X_s) - \sigma(X'_s)\}^*\right] ds , \end{aligned}$$

puis, en utilisant l'hypothèse (3.30) :

$$\begin{aligned} \mathbb{E}|X_{t\wedge S_p} - X'_{t\wedge S_p}|^2 &\leqslant C\,\mathbb{E}\int_0^{t\wedge S_p} |X_s - X'_s|^2 ds \\ &\leqslant C\int_0^t \mathbb{E}|X_{s\wedge S_p} - X'_{s\wedge S_p}|^2 ds , \end{aligned}$$

d'où $\mathbb{E}|X_{t\wedge S_p} - X'_{t\wedge S_p}|^2 = 0$ par le lemme de Gronwall puis, par le lemme de Fatou, $\mathbb{E}|X_t - X'_t|^2 = 0$ donc $X = X'$, $K = K'$. ■

5 Démonstration de l'existence

On se propose de démontrer l'existence d'une solution forte pour (1.1) en construisant la loi de la solution comme limite au sens de la convergence étroite d'une suite de lois approchées.

5.1 Topologie de Meyer-Zheng et changement de temps

5.1.1 Topologie de Meyer-Zheng

Dans [10], P.A. Meyer et W.A. Zheng munissent l'espace des trajectoires càdlàg de la topologie de la convergence en mesure (beaucoup plus faible que la topologie de Skorohod) afin d'obtenir des critères de compacité en loi pour les processus plus commodes que les critères usuels (Aldous, ...). On utilisera ici les résultats de P.A. Meyer et W.A. Zheng uniquement dans le cas particulier des processus croissants. Pour les résultats donnés sans démonstration, on renvoie systématiquement le lecteur à [10].

Notation. On note $\mathcal{D}$ l'ensemble des trajectoires càdlàg sur $[0;+\infty[$ à valeurs dans $\mathbb{R}$.

Définition 5.1 *On appelle pseudo-topologie sur $\mathcal{D}$ la topologie de la convergence en mesure pour la mesure de Lebesgue sur $\mathbb{R}^+$.*

Convention. On supposera systématiquement dans toute la suite que l'espace $\mathcal{D}$ est muni de la pseudo-topologie.

Le théorème suivant donne un critère de compacité en loi pour une suite de processus càdlàg croissants lorsque l'on munit l'espace des trajectoires càdlàg de la pseudo-topologie.

Théorème 5.2 *Soit $(\theta^{(n)})_{n\in\mathbb{N}^*}$ une suite de processus croissants, càdlàg, $\theta_0^{(n)} = 0$, définis sur un espace probabilisé $(\Omega, \mathcal{F}, \mathbb{P})$. On suppose que pour tout $t \in [0;+\infty[$ il existe une constante $C(t)$ finie telle que*

$$\sup_{n\in\mathbb{N}^*} \mathbb{E}\left(\theta_t^{(n)}\right) \leqslant C(t). \tag{5.47}$$

Alors $(\theta^{(n)})_{n\in\mathbb{N}^*}$ est tendue sur $\mathcal{D}$.

5.1.2 Convergence uniforme et changement de temps : un résultat déterministe

Dans [7], T. Kurtz montre que pour les processus, la convergence en loi pour la topologie de Meyer-Zheng implique à un changement de temps près la convergence en loi pour la topologie uniforme. De plus, T. Kurtz donne des conditions suffisantes pour que la convergence au sens Meyer-Zheng implique la convergence uniforme. On utilisera plus loin cette idée de changement de temps pour "tranformer" une convergence au sens Meyer-Zheng en une convergence uniforme (voir le changement de temps effectué à la section suivante) ; pour l'instant, on reprend de façon déterministe l'une des conditions suffisantes de T. Kurtz mentionnées précédemment. Plus précisément, on donne une condition (nécessaire et) suffisante pour avoir la convergence uniforme d'une suite de fonctions continues d-dimensionnelles $(x_n)_{n\in\mathbb{N}^*}$ dans le cas où il existe une suite de changements de temps $(\theta_n)_{n\in\mathbb{N}^*}$ convergeant p.p. (pour la mesure de Lebesgue) telle que la suite des "changées de temps" $(y_n(.) = x_n(\theta_n(.)))_{n\in\mathbb{N}^*}$ converge uniformément.

Théorème 5.3 *Soit $d \in \mathbb{N}^*$. On se donne $(x_n)_{n\in\mathbb{N}^*}$ dans $C([0;+\infty[;\mathbb{R}^d)$ et $(\theta_n)_{n\in\mathbb{N}^*}$ dans $C([0;+\infty[;\mathbb{R})$ avec θ_n strictement croissant, $\theta_n(0) = 0$, $\theta_n(\infty) = \infty$ pour tout*

$n \in \mathbb{N}^*$. On pose pour tout $n \in \mathbb{N}^*$ $\tau_n = (\theta_n)^{-1}$ et $y_n(t) = x_n(\tau_n(t))$. On suppose qu'il existe $y \in C([0;+\infty[;\mathbb{R}^d)$, $\tau \in C([0;+\infty[;\mathbb{R})$ et $\theta \in \mathcal{D}$ telles que :

$$(y_n, \tau_n, \theta_n) \xrightarrow[n\to\infty]{} (y, \tau, \theta) \qquad \text{dans } C([0;+\infty[;\mathbb{R}^{d+1}) \times \mathcal{D}. \tag{5.48}$$

On peut alors définir $x(t) = y(\theta(t))$. Une condition (nécessaire et) suffisante pour avoir la continuité de x et la convergence de $(x_n)_{n\in\mathbb{N}^*}$ vers x dans $C([0;+\infty[;\mathbb{R}^d)$ est :

$$\{(s,t) \in [0;+\infty[\times[0;+\infty[: \tau(s) = \tau(t)\} \subset \{(s,t) \in [0;+\infty[\times[0;+\infty[: y(s) = y(t)\}. \tag{5.49}$$

5.2 Approximation

Pour tout $n \in \mathbb{N}^*$, on considère l'équation différentielle stochastique classique (univoque) :

$$\begin{cases} dX_t^{(n)} = b(X_t^{(n)})dt - A_n(X_t^{(n)})dt + \sigma(X_t^{(n)})dW_t \\ X_0^{(n)} = x_0 \end{cases} \tag{5.50}$$

qui possède une unique solution continue adaptée $X^{(n)}$ sur $(\Omega, \mathcal{F}, \{\mathcal{F}_t\}, \mathbb{P})$ d'après le caractère Lipschitz des A_n (voir proposition 2.14) et l'hypothèse (3.30) faite sur b et σ.

Notations. Pour $n \in \mathbb{N}^*$, on définit les processus suivants sur $(\Omega, \mathcal{F}, \mathbb{P})$:

$$\theta_t^{(n)} = \int_0^t |A_n(X_u^{(n)})|du + t \; ; \; \tau^{(n)} = (\theta^{(n)})^{-1} \; ; \tag{5.51}$$

$$K_t^{(n)} = \int_0^t A_n(X_u^{(n)})du \; ; \; M_t^{(n)} = \int_0^t \sigma(X_u^{(n)})dW_u \; ; \tag{5.52}$$

$$Y_t^{(n)} = X_{\tau_t^{(n)}}^{(n)} \; ; \; H_t^{(n)} = K_{\tau_t^{(n)}}^{(n)} . \tag{5.53}$$

5.3 Tension

Dans le cas où A est le sous-différentiel de la fonction indicatrice d'un convexe fermé d'intérieur non vide D, J.L. Menaldi [9] montre qu'il existe $a \in \mathbb{R}^d$, $\gamma > 0$ (ne dépendant que de D) tels que pour tous $n \in \mathbb{N}^*$, $x \in \mathbb{R}^d$:

$$\langle(\partial I_D)_n(x), x-a\rangle \geqslant \gamma|(\partial I_D)_n(x)| . \tag{5.54}$$

On va maintenant montrer une inégalité qui généralise celle obtenue par J.L. Menaldi dans le cas de la fonction indicatrice d'un convexe fermé et qui sera fondamentale pour obtenir la tension au sens Meyer-Zheng de la suite des processus $X^{(n)}$ associés aux A_n.

Lemme 5.4 *Il existe $a \in \mathbb{R}^d$, $\gamma > 0$, $\mu \geqslant 0$ (ne dépendant que de A) tels que pour tous $n \in \mathbb{N}^*$, $x \in \mathbb{R}^d$:*

$$\langle A_n(x), x-a\rangle \geqslant \gamma|A_n(x)| - \mu|x-a| - \gamma\mu . \tag{5.55}$$

Démonstration. D'après l'hypothèse 3.27, il existe $a \in \text{Int}(D(A))$. Soit $\gamma > 0$ tel que $\bar{B}(a,\gamma) \subset \text{Int}(D(A))$. On pose :

$$\mu = \max\{|y| : y \in A(x),\, x \in \bar{B}(a,\gamma)\}. \tag{5.56}$$

D'après la proposition 2.12, on a $0 \leqslant \mu < \infty$. Soient $n \in \mathbb{N}^*$, $x \in \mathbb{R}^d$. L'inégalité (5.55) est triviale si $A_n(x) = 0$; supposons donc $A_n(x) \neq 0$. D'après la monotonie de A_n, on a pour tout $z \in \mathbb{R}^d$:

$$\langle A_n(x) - A_n(z),\, x - z\rangle \geqslant 0. \tag{5.57}$$

En appliquant (5.57) avec $z = a + \gamma \dfrac{A_n(x)}{|A_n(x)|}$, on obtient en utilisant la proposition 2.14 et (5.56):

$$\begin{aligned} \langle A_n(x),\, x - a\rangle &\geqslant \gamma|A_n(x)| + \langle A_n(z),\, x - z\rangle \\ &\geqslant \gamma|A_n(x)| - \mu|x - z| \\ &\geqslant \gamma|A_n(x)| - \mu|x - a| - \mu\gamma. \end{aligned}$$

■

Afin d'obtenir la convergence en loi de la suite $(X^{(n)})_{n\in\mathbb{N}^*}$, on va maintenant montrer un résultat de tension.

Proposition 5.5 *La suite* $(\tau^{(n)}, H^{(n)}, M^{(n)}, W, Y^{(n)}, \theta^{(n)})_{n\in\mathbb{N}^*}$ *est tendue sur l'espace* $C([0;+\infty[;\mathbb{R}^{4d+1}) \times \mathcal{D}$.

Démonstration. La tension de $(\tau^{(n)}, H^{(n)}, M^{(n)}, W, Y^{(n)})$ s'obtient facilement en utilisant le critère d'Aldous (voir [5]) et en remarquant que pour $0 \leqslant s \leqslant t < \infty$, $n \in \mathbb{N}^*$, l'on a :

$$|\tau_t^{(n)} - \tau_s^{(n)}| \leqslant |t - s|, \tag{5.58}$$

et (en effectuant le changement de variable $u = \tau_v^{(n)}$)

$$\begin{aligned} |H_t^{(n)} - H_s^{(n)}| &= \left| \int_{\tau_s^{(n)}}^{\tau_t^{(n)}} A_n(X_u^{(n)})\, du \right| \\ &= \left| \int_s^t \frac{A_n(Y_v^{(n)})}{|A_n(Y_v^{(n)})| + 1}\, dv \right| \\ &\leqslant \int_{s\wedge t}^{s\vee t} \frac{|A_n(Y_v^{(n)})|}{|A_n(Y_v^{(n)})| + 1}\, dv \\ &\leqslant |t - s|, \end{aligned}$$

c'est-à-dire :

$$|H_t^{(n)} - H_s^{(n)}| \leqslant |t - s|. \tag{5.59}$$

Montrons la tension de $(\theta^{(n)})$ grâce au critère du théorème 5.2. On pose :

$$S_p^{(n)} = \inf\{t \geqslant 0 : |X_t^{(n)}| \geqslant p\}; \qquad p \in \mathbb{N}^*, n \in \mathbb{N}^*. \tag{5.60}$$

Alors $S_p^{(n)}$ est un $\{\mathcal{F}_t\}$−temps d'arrêt pour tous $p \in \mathbb{N}^*, n \in \mathbb{N}^*$ et $S_p^{(n)} \uparrow \infty$ quand $p \uparrow \infty$. D'après la formule d'Itô appliquée à $X_{t\wedge S_p^{(n)}}^{(n)} - a$ (où a est donné par le lemme

5.4) et $x \to \frac{1}{2}|x|^2$, on a pour tous $n \in \mathbb{N}^*$, $0 \leqslant s \leqslant t < \infty$:

$$\begin{aligned}\tfrac{1}{2}|X^{(n)}_{s\wedge S_p^{(n)}} - a|^2 = \; & \tfrac{1}{2}|x_0 - a|^2 \\ & + \int_0^{s\wedge S_p^{(n)}} \langle b(X_u^{(n)}), X_u^{(n)} - a\rangle du \\ & + \int_0^{s\wedge S_p^{(n)}} \langle \sigma(X_u^{(n)}), X_u^{(n)} - a\rangle dW_u \\ & - \int_0^{s\wedge S_p^{(n)}} \langle A_n(X_u^{(n)}), X_u^{(n)} - a\rangle du \\ & + \tfrac{1}{2}\int_0^{s\wedge S_p^{(n)}} \mathrm{tr}(\sigma\sigma^*(X_u^{(n)}))du,\end{aligned}$$

puis en utilisant l'inégalité élémentaire

$$x.y \leqslant \frac{1}{2}x^2 + \frac{1}{2}y^2, \qquad \forall x, y \in \mathbb{R}, \tag{5.61}$$

l'hypothèse (3.30) et le lemme 5.4 :

$$\begin{aligned}\tfrac{1}{2}|X^{(n)}_{s\wedge S_p^{(n)}} - a|^2 \leqslant \; & \tfrac{1}{2}|x_0 - a|^2 + Ct + C\int_0^{s\wedge S_p^{(n)}} |X_u^{(n)} - a|^2 du \\ & + \int_0^{s\wedge S_p^{(n)}} \langle \sigma(X_u^{(n)}), X_u^{(n)} - a\rangle dW_u + \mu\gamma t \\ & - \gamma \int_0^{s\wedge S_p^{(n)}} |A_n(X_u^{(n)})| du + \mu \int_0^{s\wedge S_p^{(n)}} |X_u^{(n)} - a| du\,.\end{aligned}$$

En utilisant l'hypothèse (3.30), l'inégalité de Davis et l'inégalité (5.61), on peut écrire

$$\begin{aligned}\mathbb{E}\left[\sup_{0\leqslant s\leqslant t} \int_0^{s\wedge S_p^{(n)}} \langle \sigma(X_u^{(n)}), X_u^{(n)} - a\rangle dW_u\right] & \leqslant C\mathbb{E}\left[\left(\int_0^{t\wedge S_p^{(n)}} |X_u^{(n)} - a|^4 du\right)^{\frac{1}{2}}\right] + Ct^{\frac{1}{2}} \\ & \leqslant \tfrac{1}{4}\mathbb{E}\left[\sup_{0\leqslant s\leqslant t} |X^{(n)}_{s\wedge S_p^{(n)}} - a|^2\right] + Ct^{\frac{1}{2}} \\ & \quad + C\mathbb{E}\int_0^{t\wedge S_p^{(n)}} |X_u^{(n)} - a|^2 du\,,\end{aligned}$$

d'où, en revenant à l'équation précédente :

$$\mathbb{E}\left[\sup_{0\leqslant s\leqslant t} |X^{(n)}_{s\wedge S_p^{(n)}} - a|^2\right] \leqslant C(t+1) + C\int_0^t \mathbb{E}\left(\sup_{0\leqslant v\leqslant u} |X^{(n)}_{v\wedge S_p^{(n)}} - a|^2\right) du\,. \tag{5.62}$$

En particulier par le lemme de Gronwall :

$$\mathbb{E}\left[\sup_{0\leqslant s\leqslant t} |X^{(n)}_{s\wedge S_p^{(n)}} - a|^2\right] \leqslant C(t+1)e^{Ct}, \tag{5.63}$$

puis par le lemme de Beppo-Levi :

$$\mathbb{E}\left[\sup_{0\leqslant s\leqslant t} |X_s^{(n)} - a|^2\right] \leqslant C, \tag{5.64}$$

et en revenant à l'inégalité de départ, utilisant (5.63) :

$$\mathbb{E}\int_0^{t\wedge S_p^{(n)}} |A_n(X_u^{(n)})|du \leqslant C(t)\,, \tag{5.65}$$

puis appliquant de nouveau le lemme de Beppo-Levi :

$$\mathbb{E}\int_0^t |A_n(X_u^{(n)})|du \leqslant C(t)\,, \tag{5.66}$$

et a fortiori :

$$\sup_{n\in\mathbb{N}^*} \mathbb{E}(\theta_t^{(n)}) \leqslant C(t) < \infty, \qquad 0\leqslant t < \infty. \tag{5.67}$$

L'estimation uniforme (5.67) et le théorème 5.2 permettent d'affirmer que la suite $(\theta^{(n)})_{n\in\mathbb{N}^*}$ est tendue sur $\mathcal{D}$, où $\mathcal{D}$ est muni de la pseudo-topologie. ■

5.4 Convergence et réalisation

On note $\mathbb{L}_n, n \in \mathbb{N}^*$, la loi de $(\tau^{(n)}, H^{(n)}, M^{(n)}, W, Y^{(n)}, \theta^{(n)})$ sur $C([0;+\infty[;\mathbb{R}^{4d+1}) \times \mathcal{D}$. Puisque $C([0;+\infty[;\mathbb{R}^{4d+1}) \times \mathcal{D}$ est un espace métrique séparable, d'après le théorème de Prokhorov et la proposition 5.5, il existe une sous-suite (n_k) et une probabilité $\mathbb{L}$ sur l'espace $C([0;+\infty[;\mathbb{R}^{4d+1}) \times \mathcal{D}$ telles que :

$$\mathbb{L}_{n_k} \xrightarrow[k\to\infty]{} \mathbb{L}\,. \tag{5.68}$$

Pour alléger la présentation, on supposera que toute la suite $(\mathbb{L}_n)_{n\in\mathbb{N}^*}$ converge vers $\mathbb{L}$ quand $n \uparrow \infty$. D'après le théorème de réalisation de Skorohod, il existe un espace probabilisé $(\hat\Omega, \hat{\mathcal{F}}, \hat{\mathbb{P}})$ sur lequel sont définies des variables aléatoires $(\hat\tau^{(n)}, \hat H^{(n)}, \hat M^{(n)}, \hat W^{(n)}, \hat Y^{(n)}, \hat\theta^{(n)})_{n\in\mathbb{N}^*}$, $(\hat\tau, \hat H, \hat M, \hat W, \hat Y, \hat\theta)$ à valeurs dans $C([0;+\infty[;\mathbb{R}^{4d+1}) \times \mathcal{D}$ telles que pour tout $n \in \mathbb{N}^*$

$$(\hat\tau^{(n)}, \hat H^{(n)}, \hat M^{(n)}, \hat W^{(n)}, \hat Y^{(n)}, \hat\theta^{(n)}) \sim (\tau^{(n)}, H^{(n)}, M^{(n)}, W, Y^{(n)}, \theta^{(n)})\,, \tag{5.69}$$

$$\hat{\mathbb{P}}^{(\hat\tau,\hat H,\hat M,\hat W,\hat Y,\hat\theta)} = \mathbb{L}\,, \tag{5.70}$$

et on a la convergence $\hat{\mathbb{P}}$–p.s. sur $C([0;+\infty[;\mathbb{R}^{4d+1}) \times \mathcal{D}$:

$$(\hat\tau^{(n)}, \hat H^{(n)}, \hat M^{(n)}, \hat W^{(n)}, \hat Y^{(n)}, \hat\theta^{(n)}) \xrightarrow[n\to\infty]{} (\hat\tau, \hat H, \hat M, \hat W, \hat Y, \hat\theta)\,. \tag{5.71}$$

Notations. Pour $n \in \mathbb{N}^*$, on définit les processus suivants sur $(\hat\Omega, \hat{\mathcal{F}}, \hat{\mathbb{P}})$:

$$\hat X_t^{(n)} = \hat Y^{(n)}_{\hat\theta_t^{(n)}}\ ;\ \hat X_t = \hat Y_{\hat\theta_t}\ ;\ \hat K_t^{(n)} = \hat H^{(n)}_{\hat\theta_t^{(n)}}\ ;\ \hat K_t = \hat H_{\hat\theta_t}\,. \tag{5.72}$$

5.5 Convergence uniforme de la suite des approximants

On va appliquer les résultats du théorème 5.3 pour obtenir la convergence de la suite $(\hat X^{(n)})_{n\in\mathbb{N}^*}$ vers $\hat X$ $\hat{\mathbb{P}}$-p.s. dans $C([0;+\infty[;\mathbb{R}^d)$; on devra donc montrer le résultat suivant

Théorème 5.6 *Il existe $\hat\Omega_0 \in \hat{\mathcal{F}}$, $\hat{\mathbb{P}}(\hat\Omega_0) = 1$ tel que pour tout $\hat\omega \in \hat\Omega_0$, s'il existe $0\leqslant s\leqslant t < \infty$ vérifiant $\hat\tau_s(\hat\omega) = \hat\tau_t(\hat\omega)$ alors $\hat Y_s(\hat\omega) = \hat Y_t(\hat\omega)$.*

Démonstration. Ce théorème va être démontré en plusieurs propositions.

Proposition 5.7 $\hat{\mathbb{P}}\left(\hat{X}_t \in \overline{D(A)}; 0 \leqslant t < \infty\right) = 1.$

Démonstration. Puisque $\hat{X}$ est un processus càdlàg, il suffit de prouver que $\hat{\mathbb{P}}\left(\hat{X}_t \in \overline{D(A)}\right) = 1$ pour tout $0 \leqslant t < \infty$. Supposons au contraire qu'il existe $0 < t_0 < \infty$, $\hat{B}_0 \in \hat{\mathcal{F}}$ tels que $\hat{\mathbb{P}}(\hat{B}_0) > 0$ et $\hat{X}_{t_0}(\hat{\omega}) \notin \overline{D(A)}$ pour tout $\hat{\omega} \in \hat{B}_0$.
Comme $\hat{X}$ est continu à droite, il existe $\delta > 0$, $\hat{B}_1 \in \hat{\mathcal{F}}$ tels que $\hat{\mathbb{P}}(\hat{B}_1) > 0$ et $\hat{X}_t(\hat{\omega}) \notin \overline{D(A)}$ pour tout $\hat{\omega} \in \hat{B}_1$, $t \in [t_0; t_0 + \delta]$. D'après (5.66) et l'égalité en loi (5.69), on a pour tout $n \in \mathbb{N}^*$:

$$\mathbb{E}\int_{t_0}^{t_0+\delta} |A_n(\hat{X}_u^{(n)})| du \leqslant C, \tag{5.73}$$

et, a fortiori par le lemme de Fatou :

$$\int_{\hat{B}_1}\int_{t_0}^{t_0+\delta} \underline{\lim} |A_n(\hat{X}_u^{(n)})| du d\hat{\mathbb{P}} \leqslant C, \tag{5.74}$$

ce qui est impossible car $\underline{\lim}|A_n(\hat{X}_u^{(n)})| = +\infty$ sur $\hat{B}_1 \times [t_0; t_0+\delta]$ d'après la proposition 2.14. ■

On aura besoin dans la suite du lemme déterministe suivant :

Lemme 5.8 *Soient :*
• $(k^{(n)})_n$ *une suite de fonctions* $k^{(n)} : [0; +\infty[\longrightarrow \mathbb{R}^d$, $d \in \mathbb{N}^*$, *continues, à variation finie, telles que :*
(i) $\sup_n |k^{(n)}|_t \leqslant c(t) < \infty, \quad 0 \leqslant t < \infty$;
(ii) $k^{(n)} \xrightarrow[n\to\infty]{} k$ *dans* $C([0; +\infty[; \mathbb{R}^d)$;
• $(f^{(n)})_n$ *une suite de fonctions* $f^{(n)} : [0; +\infty[\longrightarrow \mathbb{R}^d$ *continues telles que :*
(iii) $f^{(n)} \xrightarrow[n\to\infty]{} f$ *dans* $C([0; +\infty[; \mathbb{R}^d)$;
Alors k *à variation finie et, pour tous* $0 \leqslant s \leqslant t < \infty$, *on a :*

$$\int_s^t \langle f^{(n)}(u), dk^{(n)}(u)\rangle \xrightarrow[n\to\infty]{} \int_s^t \langle f(u), dk(u)\rangle. \tag{5.75}$$

Proposition 5.9 *Il existe* $\hat{\Omega}_1 \in \hat{\mathcal{F}}$, $\hat{\mathbb{P}}(\hat{\Omega}_1) = 1$ *tel que pour tout* $\hat{\omega} \in \hat{\Omega}_1$, *s'il existe* $0 \leqslant s \leqslant t < \infty$ *tels que* $\hat{\tau}_s = \hat{\tau}_t$ *alors pour tout* $x \in \overline{D(A)}$:

$$\int_s^t \langle \hat{Y}_u - x, d\hat{H}_u\rangle \geqslant 0. \tag{5.76}$$

Démonstration. Soit $\hat{\Omega}_1$ le sous-ensemble de $\hat{\Omega}$ de $\hat{\mathbb{P}}$-probabilité 1 sur lequel a lieu la convergence p.s. (5.71). Il suffit de prouver (5.76) pour $x \in D(A)$. D'après la monotonie de A_n, on peut écrire pour tout $n \in \mathbb{N}^*$:

$$\int_{\hat{\tau}_s^{(n)}}^{\hat{\tau}_t^{(n)}} \langle \hat{X}_u^{(n)} - x, d\hat{K}_u^{(n)} - A_n(x)du\rangle \geqslant 0, \tag{5.77}$$

puis en faisant le changement de variable $u = \hat{\tau}_v^{(n)}$, en utilisant (5.72) et $\hat{\theta}^{(n)} = (\hat{\tau}^{(n)})^{-1}$

$$\int_s^t \langle \hat{Y}_u^{(n)} - x, d\hat{H}_u^{(n)}\rangle \geqslant \int_s^t \langle \hat{Y}_u^{(n)} - x, A_n(x)\rangle d\hat{\tau}_u^{(n)}. \tag{5.78}$$

Il suffit alors de passer à la limite quand n tend vers l'infini dans (5.78) en utilisant la convergence (5.71), le lemme 5.8, l'estimation (5.59) pour le membre de gauche et en ce qui concerne le membre de droite :

$$\begin{aligned}\left|\int_s^t \langle \hat{Y}_u^{(n)} - x,\ A_n(x)\rangle d\hat{\tau}_u^{(n)}\right| &\leqslant C.\Big(\sup_{0\leqslant u\leqslant t} |\hat{Y}_u^{(n)}| + |x|\Big).|A^\circ(x)|.|\hat{\tau}_t^{(n)} - \hat{\tau}_s^{(n)}| \\ &\leqslant C.|\hat{\tau}_t^{(n)} - \hat{\tau}_s^{(n)}| \underset{n\to\infty}{\longrightarrow} 0\end{aligned}$$

si $\hat{\tau}_s = \hat{\tau}_t$, grâce à la convergence (5.71). ■

Proposition 5.10 *Il existe $\hat{\Omega}_2 \in \hat{\mathcal{F}}$, $\hat{\mathbb{P}}(\hat{\Omega}_2) = 1$ tel que pour tout $\hat{\omega} \in \hat{\Omega}_2$, s'il existe $0\leqslant s\leqslant t < \infty$ vérifiant $\hat{\tau}_s(\hat{\omega}) = \hat{\tau}_t(\hat{\omega})$ alors $\hat{Y}_t(\hat{\omega}) - \hat{Y}_s(\hat{\omega}) = \hat{H}_s(\hat{\omega}) - \hat{H}_t(\hat{\omega})$.*

Démonstration. On note $\hat{\Omega}^{(n)}$ le sous-ensemble de $\hat{\Omega}$ de $\hat{\mathbb{P}}$-probabilité 1 sur lequel est vérifiée l'équation différentielle stochastique

$$d\hat{X}_t^{(n)} = b(\hat{X}_t^{(n)})dt + \sigma(\hat{X}_t^{(n)})d\hat{W}_t^{(n)} - d\hat{K}_t^{(n)}, \tag{5.79}$$

et on pose

$$\hat{\Omega}_2 = \hat{\Omega}_1 \cap \Big(\bigcap_{n\in\mathbb{N}^*} \hat{\Omega}^{(n)}\Big). \tag{5.80}$$

Alors $\hat{\mathbb{P}}(\hat{\Omega}_2) = 1$ et sur $\hat{\Omega}_2$, on a pour tous $n \in \mathbb{N}^*$, $0\leqslant s < t < \infty$:

$$\begin{aligned}\hat{Y}_t^{(n)} - \hat{Y}_s^{(n)} &= \int_{\hat{\tau}_s^{(n)}}^{\hat{\tau}_t^{(n)}} b(\hat{X}_u^{(n)})du + \int_{\hat{\tau}_s^{(n)}}^{\hat{\tau}_t^{(n)}} \sigma(\hat{X}_u^{(n)})d\hat{W}_u^{(n)} - (\hat{H}_t^{(n)} - \hat{H}_s^{(n)}) \\ &= \int_{\hat{\tau}_s^{(n)}}^{\hat{\tau}_t^{(n)}} b(\hat{X}_u^{(n)})du + (\hat{M}^{(n)}_{\hat{\tau}_t^{(n)}} - \hat{M}^{(n)}_{\hat{\tau}_s^{(n)}}) - (\hat{H}_t^{(n)} - \hat{H}_s^{(n)}).\end{aligned}$$

En utilisant la convergence (5.71) et l'hypothèse (3.30), on obtient

$$\begin{aligned}\left|\int_{\hat{\tau}_s^{(n)}}^{\hat{\tau}_t^{(n)}} b(\hat{X}_u^{(n)})du\right| &\leqslant \int_{\hat{\tau}_s^{(n)}}^{\hat{\tau}_t^{(n)}} |b(\hat{X}_u^{(n)})|du \\ &\leqslant C(\hat{\tau}_t^{(n)} - \hat{\tau}_s^{(n)}).\sup_{0\leqslant u\leqslant \hat{\tau}_t^{(n)}} (1 + |\hat{X}_u^{(n)}|) \\ &\leqslant C(\hat{\tau}_t^{(n)} - \hat{\tau}_s^{(n)}).\sup_{0\leqslant u\leqslant t} (1 + |\hat{Y}_u^{(n)}|) \\ &\leqslant C(\hat{\tau}_t^{(n)} - \hat{\tau}_s^{(n)}),\end{aligned}$$

donc si $\hat{\tau}_t = \hat{\tau}_s$ alors on a la convergence

$$\int_{\hat{\tau}_s^{(n)}}^{\hat{\tau}_t^{(n)}} b(\hat{X}_u^{(n)})du \underset{n\to\infty}{\longrightarrow} 0. \tag{5.81}$$

D'après la convergence (5.71), on a clairement

$$\hat{M}^{(n)}_{\hat{\tau}_t^{(n)}} - \hat{M}^{(n)}_{\hat{\tau}_s^{(n)}} \underset{n\to\infty}{\longrightarrow} \hat{M}_{\hat{\tau}_t} - \hat{M}_{\hat{\tau}_s}, \tag{5.82}$$

d'où, si $\hat{\tau}_s = \hat{\tau}_t$, alors

$$\hat{M}^{(n)}_{\hat{\tau}_t^{(n)}} - \hat{M}^{(n)}_{\hat{\tau}_s^{(n)}} \underset{n\to\infty}{\longrightarrow} 0. \tag{5.83}$$

Finalement, grâce aux convergences (5.71), (5.81), (5.83), on peut conclure en passant à la limite quand n tend vers l'infini dans l'équation vérifiée par $\hat{Y}^{(n)}$ que sur $\hat{\Omega}_2$ si l'on a $\hat{\tau}_t = \hat{\tau}_s$ alors nécessairement $\hat{Y}_t - \hat{Y}_s = \hat{H}_s - \hat{H}_t$. ■

Proposition 5.11 $\hat{\mathbb{P}}\left(\hat{Y}_t \in \overline{D(A)}\,; 0 \leqslant t < \infty\right) = 1$.

Démonstration. Puisque $\hat{Y}$ est un processus continu, il suffit de prouver $\hat{\mathbb{P}}(\hat{Y}_t \in \overline{D(A)}) = 1$ pour tout $0 \leqslant t < \infty$. Supposons au contraire qu'il existe $0 < t_0 < \infty$, $\hat{B} \in \hat{\mathcal{F}}$ tels que $\hat{\mathbb{P}}(\hat{B}) > 0$ et $\hat{Y}_{t_0}(\hat{\omega}) \notin \overline{D(A)}$ pour tout $\hat{\omega} \in \hat{B}$. On pose :

$$S = \sup\{0 \leqslant u \leqslant t_0 : \hat{Y}_u \in \overline{D(A)}\}. \tag{5.84}$$

Comme $\hat{Y}_0 = x_0 \in \overline{D(A)}$, et que $\hat{Y}$ est un processus continu, la variable aléatoire $\hat{Y}_S$ prend ses valeurs dans $\overline{D(A)}$ $\hat{\mathbb{P}}$-p.s.. D'après la proposition 5.7, on a nécessairement $\hat{\tau}_S = \hat{\tau}_{t_0}$ sur $\hat{B}$ puisque $\hat{Y}_s \notin \overline{D(A)}$ pour $s \in]S; t_0]$ et $\hat{\omega} \in \hat{B}$. Appliquant la proposition 5.9 sur $\hat{\Omega}_2 \cap \hat{B}$ (qui est de probabilité non nulle) entre S et t_0, on peut écrire :

$$\begin{aligned}
\tfrac{1}{2}|\hat{Y}_{t_0} - \hat{Y}_S|^2 &= \tfrac{1}{2}|\hat{H}_{t_0} - \hat{H}_S|^2 \\
&= \int_S^{t_0} \langle \hat{H}_u - \hat{H}_S, d\hat{H}_u \rangle \\
&= -\int_S^{t_0} \langle \hat{Y}_u - \hat{Y}_S, d\hat{H}_u \rangle \\
&\leqslant 0,
\end{aligned}$$

d'où $\hat{Y}_{t_0} = \hat{Y}_S$, ce qui est impossible sur $\hat{B}$. ■

Posons : $\hat{\Omega}_0 = \hat{\Omega}_1 \cap \hat{\Omega}_2 \cap \{\hat{Y}_t \in \overline{D(A)}; 0 \leqslant t < \infty\}$. On sait que $\hat{\mathbb{P}}(\hat{\Omega}_0) = 1$ d'après les étapes précédentes. De plus sur $\hat{\Omega}_0$, si $\hat{\tau}_s = \hat{\tau}_t$ alors :

$$\begin{aligned}
\tfrac{1}{2}|\hat{Y}_t - \hat{Y}_s|^2 &= \tfrac{1}{2}|\hat{H}_t - \hat{H}_s|^2 \\
&= \int_s^t \langle \hat{H}_u \quad \hat{H}_s, d\hat{H}_u \rangle \\
&= -\int_s^t \langle \hat{Y}_u - \hat{Y}_s, d\hat{H}_u \rangle \\
&\leqslant 0,
\end{aligned}$$

d'où $\hat{Y}_s = \hat{Y}_t$. Ceci termine la démonstration du théorème 5.6. ■

En appliquant $\hat{\mathbb{P}}$-p.s. le théorème 5.3 à $\hat{X}^{(n)}$, $\hat{\theta}^{(n)}$, $\hat{Y}^{(n)}$, $\hat{\tau}^{(n)}$, $\hat{X}$, $\hat{\theta}$, $\hat{Y}$, $\hat{\tau}$ d'une part, et à $\hat{K}^{(n)}$, $\hat{\theta}^{(n)}$, $\hat{H}^{(n)}$, $\hat{\tau}^{(n)}$, $\hat{K}$, $\hat{\theta}$, $\hat{H}$, $\hat{\tau}$ d'autre part, on obtient grâce au théorème 5.6, la convergence $\hat{\mathbb{P}}$-p.s. suivante

$$(\hat{X}^{(n)}, \hat{K}^{(n)}) \xrightarrow[n \to \infty]{} (\hat{X}, \hat{K}) \qquad \text{dans } C([0; +\infty[; \mathbb{R}^{2d}). \tag{5.85}$$

5.6 Passage à la limite dans $EDSM^d(A_n; b; \sigma; x_0)$

On va montrer dans la suite que $(\hat{X}, \hat{K})$ est solution de $EDSM^d(A; b; \sigma; x_0)$ sur l'espace probabilisé $(\hat{\Omega}, \hat{\mathcal{F}}, \hat{\mathbb{P}})$ et par rapport au mouvement brownien $\hat{W}$. La proposition suivante donne un premier résultat dans ce sens :

Proposition 5.12 *Le processus $\hat{X}$ est solution de l'équation différentielle stochastique*

$$d\hat{X}_t = b(\hat{X}_t)dt + \sigma(\hat{X}_t)d\hat{W}_t - d\hat{K}_t. \tag{5.86}$$

Démonstration. Il suffit de passer à la limite quand n tend vers l'infini dans

$$d\hat{X}_t^{(n)} = b(\hat{X}_t^{(n)})dt + \sigma(\hat{X}_t^{(n)})d\hat{W}_t^{(n)} - d\hat{K}_t^{(n)}, \tag{5.87}$$

grâce à la convergence (5.85) (pour l'intégrale stochastique, on pourra utiliser la technique classique d'approximation par des intégrales de Riemann-Stieltjes). ∎

Pour pouvoir affirmer que $(\hat{X}, \hat{K})$ est solution de $EDSM^d(A; b; \sigma; x_0)$, il reste à montrer la condition [4] du problème. On va pour cela passer à la limite quand n tend vers l'infini dans l'analogue de cette condition au rang n via le lemme 5.8.

Proposition 5.13 *Pour tout couple de processus $(\hat{\alpha}, \hat{\beta})$ continus, adaptés, définis sur $(\hat{\Omega}, \hat{\mathcal{F}}, \hat{\mathbb{P}})$, à valeurs dans $\mathbb{R}^d$, vérifiant :*

$$(\hat{\alpha}_u, \hat{\beta}_u) \in \mathrm{Gr}(A), \qquad \forall u \in [0; +\infty[, \tag{5.88}$$

la mesure

$$\langle \hat{X}_u - \hat{\alpha}_u, d\hat{K}_u - \hat{\beta}_u du \rangle \tag{5.89}$$

est positive sur $\mathbb{R}^+$.

Démonstration. Étape 1 : soient $\hat{\alpha}'$ processus continu, adapté, défini sur $(\hat{\Omega}, \hat{\mathcal{F}}, \hat{\mathbb{P}})$, à valeurs dans $D(A)$, et $0 \leqslant s \leqslant t$. On suppose que $\hat{\alpha}'$ vérifie :

$$\int_0^t |A^\circ(\hat{\alpha}'_u)| du < \infty. \tag{5.90}$$

On se propose de montrer :

$$\int_s^t \langle \hat{X}_u - \hat{\alpha}'_u, d\hat{K}_u - A^\circ(\hat{\alpha}'_u) du \rangle \geqslant 0. \tag{5.91}$$

On sait d'après la monotonie des A_n que pour tout $n \in \mathbb{N}^*$

$$\int_s^t \langle \hat{X}_u^{(n)} - \hat{\alpha}'_u, d\hat{K}_u^{(n)} - A_n(\hat{\alpha}'_u) du \rangle \geqslant 0. \tag{5.92}$$

De la convergence (5.71) et de la croissance des $\hat{\theta}^{(n)}$, on déduit

$$\hat{\theta}_t^{(n)}(\hat{\omega}) \leqslant C(t, \hat{\omega}). \tag{5.93}$$

D'après (5.59), (5.72) et (5.93), on peut écrire

$$|\hat{K}^{(n)}|_t \leqslant |\hat{H}^{(n)}|_{\hat{\theta}_t^{(n)}} \leqslant \hat{\theta}_t^{(n)} \leqslant C(t, \hat{\omega}), \tag{5.94}$$

d'où l'estimation uniforme

$$|\hat{K}^{(n)}(\hat{\omega})|_t \leqslant C(t, \hat{\omega}), \qquad \forall n \in \mathbb{N}^*. \tag{5.95}$$

D'après la convergence (5.85), l'estimation (5.95) et le lemme 5.8, on a la convergence

$$\int_s^t \langle \hat{X}_u^{(n)} - \hat{\alpha}'_u, d\hat{K}_u^{(n)} \rangle \xrightarrow[n \to \infty]{} \int_s^t \langle \hat{X}_u - \hat{\alpha}'_u, d\hat{K}_u \rangle. \tag{5.96}$$

D'autre part, en utilisant la proposition 2.14, la convergence (5.85), la majoration (5.90), et le théorème de convergence dominée de Lebesgue, on obtient :

$$\int_s^t \langle \hat{X}_u^{(n)} - \hat{\alpha}'_u, A_n(\hat{\alpha}'_u)du\rangle \xrightarrow[n\to\infty]{} \int_s^t \langle \hat{X}_u - \hat{\alpha}'_u, A^\circ(\hat{\alpha}'_u)du\rangle. \qquad (5.97)$$

On conclut donc que (5.91) est vérifiée en passant à la limite dans (5.92) via les convergences obtenues ci-dessus.

Étape 2 : on considère un couple $(\hat{\alpha}, \hat{\beta})$ de processus continus, adaptés, définis sur $(\hat{\Omega}, \hat{\mathcal{F}}, \hat{\mathbb{P}})$, à valeurs dans $\mathbb{R}^d$, vérifiant :

$$(\hat{\alpha}_u, \hat{\beta}_u) \in \mathrm{Gr}(A), \qquad \forall u \in [0; +\infty[. \qquad (5.98)$$

D'après l'étape 1, on sait que pour tout processus $\hat{\alpha}'$ continu, adapté, défini sur $(\hat{\Omega}, \hat{\mathcal{F}}, \hat{\mathbb{P}})$, à valeurs dans $D(A)$ et vérifiant (5.90) : on a :

$$\int_s^t \langle \hat{X}_u - \hat{\alpha}'_u, d\hat{K}_u - A^\circ(\hat{\alpha}'_u)du\rangle \geqslant 0, \qquad (5.99)$$

$$\int_s^t \langle \hat{\alpha}_u - \hat{\alpha}'_u, \hat{\beta}_u du - A^\circ(\hat{\alpha}'_u)du\rangle \geqslant 0. \qquad (5.100)$$

Pour $p \in \mathbb{N}^*$ prenons :

$$\hat{\alpha}'_u = J_p\Big(\frac{\hat{X}_u + \hat{\alpha}_u}{2}\Big). \qquad (5.101)$$

Alors α' est un processus continu, adapté, défini sur $(\hat{\Omega}, \hat{\mathcal{F}}, \hat{\mathbb{P}})$, à valeurs dans $D(A)$ et vérifiant (5.90) car :

$$\int_0^t |A^\circ(\hat{\alpha}'_u)|du \leqslant \int_0^t |A_p(\frac{\hat{X}_u + \hat{\alpha}_u}{2})|du < +\infty, \qquad (5.102)$$

d'après $A_p(x) \in A(J_p(x))$ pour tout x dans $\mathbb{R}^d$ (proposition 2.14 (iv)) et le fait que $A^\circ(\hat{\alpha}'_u)$ est par définition l'élément de norme minimale de $A(\hat{\alpha}'_u)$. En appliquant (5.99) et (5.100) avec $\hat{\alpha}'$, on peut écrire :

$$\int_s^t \Big\langle \hat{X}_u - J_p\Big(\frac{\hat{X}_u + \hat{\alpha}_u}{2}\Big), d\hat{K}_u - A^\circ(\hat{\alpha}'_u)du\Big\rangle \geqslant 0, \qquad (5.103)$$

$$\int_s^t \Big\langle \hat{\alpha}_u - J_p\Big(\frac{\hat{X}_u + \hat{\alpha}_u}{2}\Big), \hat{\beta}_u du - A^\circ(\hat{\alpha}'_u)du\Big\rangle \geqslant 0, \qquad (5.104)$$

d'où, en ajoutant et retranchant $(\hat{X}_u + \hat{\alpha}_u)/2$ dans le premier terme de chacun des produits scalaires :

$$\int_s^t \Big\langle \frac{\hat{X}_u - \hat{\alpha}_u}{2} + \frac{1}{p}A_p\Big(\frac{\hat{X}_u + \hat{\alpha}_u}{2}\Big), d\hat{K}_u - A^\circ(\hat{\alpha}'_u)du\Big\rangle \geqslant 0, \qquad (5.105)$$

$$\int_s^t \Big\langle \frac{\hat{\alpha}(u) - \hat{X}_u}{2} + \frac{1}{p}A_p\Big(\frac{\hat{X}_u + \hat{\alpha}_u}{2}\Big), \hat{\beta}_u du - A^\circ(\hat{\alpha}'_u)du\Big\rangle \geqslant 0, \qquad (5.106)$$

puis en sommant les équations ainsi obtenues :

$$\int_s^t \frac{1}{2}\langle \hat{X}_u - \hat{\alpha}_u \, , \, d\hat{K}_u - \hat{\beta}_u \, du\rangle \geqslant \frac{2}{p}\int_s^t \left\langle A_p\Big(\frac{\hat{X}_u + \hat{\alpha}_u}{2}\Big), A^\circ\Big(J_p\Big(\frac{\hat{X}_u + \hat{\alpha}_u}{2}\Big)\Big) du\right\rangle$$
$$- \int_s^t \left\langle \frac{\hat{X}_u + \hat{\alpha}_u}{2} - J_p\Big(\frac{\hat{X}_u + \hat{\alpha}_u}{2}\Big) \, , \, d\hat{K}_u + \hat{\beta}_u du\right\rangle,$$

or :

$$A_p\Big(\frac{\hat{X}_u + \hat{\alpha}_u}{2}\Big) \in A\Big(J_p\Big(\frac{\hat{X}_u + \hat{\alpha}_u}{2}\Big)\Big), \tag{5.107}$$

donc :

$$\int_s^t \left\langle A_p\Big(\frac{\hat{X}_u + \hat{\alpha}_u}{2}\Big) \, , \, A^\circ\Big(J_p\Big(\frac{\hat{X}_u + \hat{\alpha}_u}{2}\Big)\Big) du\right\rangle \geqslant 0 \tag{5.108}$$

par définition de A°, d'où :

$$\int_s^t \frac{1}{2}\langle \hat{X}_u - \hat{\alpha}_u \, , \, d\hat{K}_u - \hat{\beta}_u \, du\rangle \geqslant - \int_s^t \left\langle \frac{\hat{X}_u + \hat{\alpha}_u}{2} - J_p\Big(\frac{\hat{X}_u + \hat{\alpha}_u}{2}\Big) \, , \, d\hat{K}_u + \hat{\beta}_u du\right\rangle. \tag{5.109}$$

On passe ensuite à la limite quand $p \to +\infty$ dans l'équation précédente d'après la convergence

$$J_p(x) \xrightarrow[p\to\infty]{} \mathrm{proj}_{\overline{D(A)}}(x) \tag{5.110}$$

rappelée dans la proposition 2.13, et, grâce à la convexité de $\overline{D(A)}$ donnée par la proposition 2.9 :

$$\frac{\hat{X}_u + \hat{\alpha}_u}{2} \in \overline{D(A)}, \tag{5.111}$$

on obtient le résultat voulu après application du théorème de convergence dominée de Lebesgue via le caractère contractant de J_p. Ceci termine la démonstration de la proposition 5.13. ■

Globalement, on peut donc affirmer que $(\hat{X}, \hat{K})$ est une solution faible du problème $EDSM^d(A; b; \sigma; x_0)$. Par conséquent on a bien montré un résultat d'existence faible pour l'équation différentielle stochastique multivoque (1.1) d'où le théorème 3.1 d'existence et d'unicité fortes pour (1.1) d'après le théorème d'unicité trajectorielle pour (1.1) (voir théorème 4.2) et le théorème de Yamada-Watanabe.

⋆ ⋆ ⋆

Note : cet article est un résumé d'une partie des résultats obtenus par l'auteur dans sa thèse [4] : on renvoie le lecteur à [4] pour une rédaction plus détaillée et/ou pour d'éventuels compléments sur le sujet.

Références

[1] A. Bensoussan et J.L. Lions : *Applications des inégalités variationnelles en contrôle stochastique.* Dunod, Paris, 1978.

[2] A. Bensoussan et A. Rascanu : *d-Dimensional stochastic differential equation with a multivalued subdifferential operator in drift.* A paraitre, 1994.

[3] H. Brezis : *Opérateurs maximaux monotones et semi-groupes de contractions dans les espaces de Hilbert.* North Holland, Mathematics Studies, New York, 1973.

[4] E. Cépa : *Équations différentielles stochastiques multivoques.* Thèse, 1994.

[5] J. Jacod et A.N. Shiryaev : *Limit theorems for stochastic processes.* Springer Verlag, Berlin, 1987.

[6] P. Krée : *Diffusions equations for multivalued stochastic differential equations.* Journ. of Funct. Anal., 49, p. 73-90, 1982.

[7] T. Kurtz : *Random time change and convergencc in distribution under the Meyer-Zheng conditions.* The Annals of probability, vol.19, 3, 1010-1034, 1991.

[8] D.Lépingle et C. Marois : *Équations différentielles stochastiques multivoques unidimensionnelles.* Sém. Prob. 21, p. 520-533, 1987.

[9] J.L. Menaldi : *Stochastic variational inequality for reflected diffusion.* Indiana Univ. Math. Journ., 32, 5, p. 733-744, 1983.

[10] P.A. Meyer et W.A. Zheng : *Tightness criteria for laws of semimartingales.* Ann. Inst. H. Poincaré Proba. Stat., 20, 353-372, 1984.

[11] H. Tanaka : *Stochastic differential equations with reflecting boundary conditions in convex regions.* Hiroshima Math. Journ., 9, p. 163-177, 1979.

On the predictable representation property for superprocesses.

L. Overbeck*

Department of Statistics,
University of California, Berkeley,
367, Evans Hall
Berkeley, CA 94720,
U.S.A.†

Abstract

In this note a simple proof of the equivalence of the predictable representation property of a martingale with respect to a filtration associated with an orthogonal martingale measure and the extremality of the underlying probability measure P is given. The representation property enables us to characterize all measures which are locally absolutely continuous with respect to P. We apply this to superprocesses and remark on a related property of the excursion filtration of the Brownian motion.

Keywords: Predictable Representation, Orthogonal Martingale Measures, Superprocesses, Absolute Continuity.

1 Introduction

In this note we first extend the simple proof of the predictable representation property for superprocesses given in [EP1] to all orthogonal martingale measures provided the underlying probability measure P is extremal in the convex set of all solutions of the martingale problem which defines the martingale measure. The predictable representation says that every martingale of the underlying filtration can be written uniquely as a stochastic integral with respect to the orthogonal martingale measure. The proof follows easily from well-known techniques of Stochastic Calculus, cf. [JS, JY]. In the case of the historical process the predictable integrand is identified in [EP2] for a large class of martingales.

As our main new result we show in the Section 2.2 that every measure which is absolutely continuous with respect to P arises as a Girsanov transformation like in Dawson's lecture notes, [D].

*Supported by an EC-Fellowship under Contract No. ERBCHBICT930682 and a DFG-Fellowship.

†On leave from the Universität Bonn, Institut für Angewandte Mathematik, Wegelerstr. 6, 53115 Bonn, Germany.

Applied to superprocesses this means that every process which is absolutely continuous with respect to a superprocess is a superprocess with an additional (interacting) immigration, cf. Section 3.1.

The Fleming-Viot process is an example that the predictable representation does not hold if the martingale measure is not orthogonal, cf. Section 3.2.

Finally, in Section 3.3 we show that a related representation property of the excursion filtration of the Brownian motion (cf. [RW]) can (at least partially) deduced from the predictable representation property of a special superprocesses, namely of that with the trivial one-particle-motion.

2 Predictable Representation for orthogonal martingale measures

For the basic definition of martingale measures and their stochastic integrals we refer to Walsh [W]. We fix a worthy martingale measure $M(ds, dx)$ over a measurable space $(E, \mathcal{E})$ defined on the stochastic base $(\Omega, \mathcal{F}, (\mathcal{F}_t), P)$ where the starting σ-field $\mathcal{F}_0$ is P-trivial and $\mathcal{F} = \vee_{t\geq 0}\mathcal{F}_t$. The quadratic variation measure and its dominating measure are denoted by Q and K, resp. The set of $(M-)$integrable functions $\mathcal{P}_M$ equals the closure of simple predictable functions on $\Omega \times [0, \infty) \times E$ with respect to the norm $(\cdot, \cdot)_K^{1/2}$. The stocastic integral $\int_0^\cdot n(s, x)M(ds, dx)$ is denoted by $n.M$. We restrict to orthogonal martingale measures, i.e.

$$Q([0,t], A, B) \;=\; 0, \text{ if } A \cap B = \emptyset. \tag{2.1}$$

It follows that such a martingale measure is worthy with K = extension of Q, cf. [W].

2.1 Representation Theorem

We denote the set of square-integrable martingales over $(\Omega, \mathcal{F}, (\mathcal{F}_t), P)$ by $\mathbf{M}^2$.

Proposition 2.1 *Let N be in $\mathbf{M}^2$. Then there exists a unique function $n \in \mathcal{P}_M$ such that*

$$N_t \;=\; N_0 + \int_0^t \int_E n(s,x)M(ds,dx) + L_t, \tag{2.2}$$

where L is an L^2-martingale with $< L, \int_0^\cdot \int_E b(s,x)M(ds,dx) >= 0$ for every $b \in \mathcal{P}_M$.

Proof: We define

$$\mathbf{M}_0^2 := \{N \in \mathbf{M}^2 | N = \int_0^\cdot \int_E b(s,x)M(ds,dx), b \in \mathcal{P}_M\}.$$

By the orthogonality of Q we have

$$E[((b_k - b_l).M_\infty)^2] = (b_k - b_l, b_k - b_l)_K$$

for a sequence $(b_k)_{k\in\mathbb{N}} \subset \mathcal{P}_M$ and so this sequence is Cauchy in $\mathcal{P}_M$ iff the sequence $(b_k \cdot M_\infty)_{k\in\mathbb{N}}$ is Cauchy in $L^2(\Omega, P)$. Therefore $\mathbf{M}_0^2$ is a closed subspace of L^2 and the assertion follows. ◇

The next theorem gives conditions which are equivalent to the predictable representation property and well-known in the case of d-dimensional martingales, cf. [JS, Chapter 3]. But first, we extend the integration with respect to a martingale measure to the space of $\mathcal{P}_M$-valued measures.

Definition. Let $(E', \mathcal{E}')$ be a measurable space with generating field $\mathcal{A}'$. Let M' be a martingale measure over $\mathcal{A}'$. A function

$$m' : \Omega \times \mathbb{R}^+ \times E \times \mathcal{E}' \to \mathbb{R}$$

is called a $\mathcal{P}_M$-valued measure over $\mathcal{E}'$ iff it is finitely additive, σ-finite and continuous in $\emptyset$ as a function from $\mathcal{E}'$ to the Banach space $\mathcal{P}_M$. We then define a new martingale measure $m' \otimes M$ over $\mathcal{A}'$ by

$$m' \otimes M_t(A') \quad := \quad \int_0^t \int_E m'(s, x, A') M(ds, dx). \tag{2.3}$$

Theorem 2.2 *The following statements are equivalent:*

(i) The measure P is extremal in the convex set of all measures P^* on $(\Omega, \mathcal{F})$ such that M is a martingale measure with covariation Q under P^*.

(ii) Every local martingale N has a unique predictable representation

$$N_t \quad = \quad N_0 + \int_0^t \int_E n(s, x) M(ds, dx) \tag{2.4}$$

where $n \in \mathcal{P}_M^{loc}$. (Here "loc " means that there is a sequence of stopping times (S_n) such that $n(s \wedge T_n, x) \in \mathcal{P}_M$.)

(iii) Every martingale measure M' defined on $(\Omega, \mathcal{F})$ with respect to the filtration $(\mathcal{F}_t)$ over some measurable space $(E', \mathcal{E}')$ with generating field $\mathcal{A}'$ has a unique predictable representation

$$M' \quad = \quad m' \otimes M \tag{2.5}$$

where m' is a $\mathcal{P}_M$-valued measure over $\mathcal{E}'$.

Proof. (i) $\Rightarrow$ (ii): Follows by Proposition 2.1, cf. Theorem 38 in [Pr].

(ii) $\Rightarrow$ (iii): Let $m'(\omega, s, x, A')$ be the representing function of the martingale $(M'_t(A'))_{t \in \mathbb{R}^+}$. We just have to prove that the mapping $A' \to m'(\cdot, \cdot, \cdot, A')$ defines a $\mathcal{P}_M$-valued measure. The additivity of this mapping is obvious by the uniqueness of the predictable representation. Because

$$E[\int_0^\infty \int_E \int_E m'(s, x, A') m'(s, y, A') Q(ds, dx, dy)] = E[M'_\infty(A')^2],$$

the continuity of m' in $\emptyset$ follows by the continuity of M' in $\emptyset$.

(iii) $\Rightarrow$ (ii): Take as $(E', \mathcal{E}')$ a one point set.

(ii) ⇒ (i): Cf. [Pr, Theorem 37]. ◇

Remark. Assume condition (i) in Theorem 2.2. Then, according to [J],[JY], the set of all martingales coincides with $\mathcal{L}(M)$, the smallest closed subspace of martingales N with norm $E[\sup_{t\geq 0}|N_t|^2]^{1/2}$ which is closed under stopping and which contains all (ordinary) stochastic integrals $n.M(A)$ where n is $M(A)$ integrable and $A \in \mathcal{E}$. Hence we only have shown that $\{n.M | n \in \mathcal{P}_M\} = \mathcal{L}(M)$. This identification can however fail if M is not orthogonal, see Section 3.2.

2.2 Necassary condition of absolute continuity

In this section we want to show the converse of Dawson's Giransov transformation [D, ch.7]. In the present setting his result reads as follows:

For r be in $\mathcal{P}_M^{loc}$ we denote the corresponding exponential local martingale

$$\exp(r \cdot M_t - \frac{1}{2}\int_{[0,t)\times E\times E} r(s,x)r(s,y)Q(ds,dx,dy))$$

by $\mathcal{E}(r)$. If $\mathcal{E}(r)$ is a martingale, e.g. if $E[\exp(\frac{1}{2}\int_0^\cdot \int_E \int_E r(s,x)r(s,y)Q(ds,dx,dy))] < \infty$ or $\exp(r.M)$ is uniformly integrable, then we can define a new measure P^r on $(\Omega, \mathcal{F})$ by $\left.\frac{dP^r}{dP}\right|_{\mathcal{F}_t} := \mathcal{E}(r)_t$.

A modification of Dawson's argument shows that under P^r the process M^r defined by

$$M^r((0,t],A) \;:=\; M((0,t],A) + \int_0^t \int_E \int_E \mathbf{1}_A(x)r(s,y)Q(ds,dx,dy) \qquad (2.6)$$

is a martingale measure with covariation measure Q.

We shall show that every probability measure which is absolutely continuous with repect to P arises as a suitable P^r.

Theorem 2.3 *Let P be a measure on $(\Omega, \mathcal{F})$ such that M is a martingale measure with covariation goverened by Q which has the predictable representation property.*

Let $P' << P$.

- *There exits a predictable function $r \in \mathcal{P}_M^{loc.}$ such that the process $M(A)$ is a semimartingale with increasing process*

$$\int_0^t \int_{E\times E} \mathbf{1}_A(x)r(s,y)Q(ds,dx,dy). \qquad (2.7)$$

- *If we assume additionally that M is a continuous martingale measure, i.e. $h.M$ is continuous $\forall h \in \mathcal{P}_M$, then the density has the exponential form*

$$\left.\frac{dP'}{dP}\right|_{\mathcal{F}_t} = \exp(\int_0^t \int_E r(s,x)M(ds,dx) - \frac{1}{2}\int_0^t (\int_{E\times E} r(s,x)r(s,y)Q(ds,dx,dy))) \qquad (2.8)$$

Hence $P' = P^r$.

Proof. Let $Z_t = \frac{dP'}{dP}\Big|_{\mathcal{F}_t}$. The Girsanov transformation for one-dimensional martingales ([JS, Chapter 3]) implies that for every $A \in \mathcal{A}$ there exists a local P'-martingale $M'(A)$ such that

$$M_t(A) = M'_t(A) + \int_0^t \frac{1}{Z_s} d < Z, M(A) >s.$$

By the representation of martingales under P we have that

$$Z_t = 1 + \int_0^t \int_E z(s,x) M(ds,dx) \tag{2.9}$$

and hence

$$\int_0^t \frac{1}{Z_{s-}} d < Z, M(A) >_s = \int_0^t \int_{E\times E} \mathbf{1}_A(x) \frac{z(s,y)}{Z_{s-}} Q(ds,dx,dy).$$

The function r equals therefore $Z_s^{-1} z(s,x)$.

In order to prove the second assertion we notice that by (2.9) up to $T_n = \inf\{t | Z_t < \frac{1}{n}\}$ we have

$$\begin{aligned} Z_t &= 1 + \int_0^t Z_s \int_E r(s,x) M(ds,dx) \\ &= 1 + \int_0^t \int_E Z_s r(s,x) M(ds,dx). \end{aligned} \tag{2.10}$$

Hence by the exponential formula for martingales the assertion is proved for all (t,ω) such that $t \in [0, T_n(\omega)]$ for some $n \in I\!N$. Because the process

$$V := \int_0^\cdot \int_E \int_E r(s,x) r(s,y) Q(ds,dx,dy)$$

is continuous it "does not jump to infinity" in the terminology of [JS, Chapter 3.5a] and so the assertion is valid P-almost surely for all t. ◇

3 Examples

3.1 Superprocess.

The basic example motivating the present note is the (interacting) superprocess, cf. [D,P]. It is a process X defined on a filtered probability space $(\Omega, \mathcal{F}, (\mathcal{F}_t), P_{\mu_0}^{A,c})$ which takes values in the space $\mathcal{M}(E)$ of positive finite measures over a Polish space E. The basic data are a familiy of linear operators $A = (A(\omega, s))_{\omega\in\Omega, s\in[0,\infty)}$ with common domain $D \subset C_b([0,\infty) \times E)$, a positive bounded branching variation function c defined on $\Omega \times [0,\infty) \times E$ and a starting point $\mu_0 \in M(E)$. Then $(X, P_{\mu_0}^{A,c})$ satisfies by definition that

$$M_t[f] := X_t(f(t)) - \mu_0(f(0)) - \int_0^t X_s(A(s)f(s))ds \tag{3.1}$$

is a martingale under $P_{\mu_0}^{A,c}$ with quadratic variation

$$< M[f] >_t = \int_0^t X_s(c(s) f^2(s)) ds \tag{3.2}$$

for all $f \in D$ (where we use the notation $\mu(f) := \int_E f(x)\mu(dx), \mu \in \mathcal{M}(E), f \in C_b(E)$).

These linear martingales give rise to an orthogonal martingale measure M^A defined on $(\Omega, \mathcal{F}, (\mathcal{F}_t), P_{\mu_0}^{A,c})$ with covariation measure

$$Q^c(ds, dx, dy) = c(s,x)\delta_x(dy)X_s(dx)ds, \tag{3.3}$$

cf.[D]. Let us now assume that $P_{\mu_0}^{A,c}$ is extremal under all measures under which (3.1) is a martingale with quadratic variation (3.2). This is in particular the case if there is no interaction, i.e. $A(\omega, s) = A_0$ for one fixed operator A_0, which generates a Hunt process with state space E, and $c(\omega, s, x) = c(s, x)$. The unique solution of the martingale problem (3.1,3.2) is the superprocess over the one-partical-motion generated by (A_0, D). The martingale problem is also well-posed if $A(\omega, s)f(x) = A_0 f(x) + b(s, \omega, x)f(x)$ for some nice $b \in \mathcal{P}_{M^{A_0}}$, cf. [D]. The question of uniqueness of a general $P_{\mu_0}^{A,c}$ is intensively studied in [P].

Our results imply that every extremal $P_{\mu_0}^{A,c}$ has the predictable representation property and additionally that for every measure P' which is absolutely continuous with respect to $P_{\mu_0}^{A,c}$ there exists a $r \in \mathcal{P}_{M^A}$ such that $P' = P_{\mu_0}^{A^r,c}$, where $A^r(\omega, s)f(s, x) = A(\omega, s)f(s, x) + r(\omega, s, x)c(\omega, s, x)f(s, x)$, i.e. P' is a superprocess with additional immigration parameter rc.

If $P_{\mu_0}^{A,c}$ is a superprocess without interaction every process which is absolutely continuous with respect to $P_{\mu_0}^{A,c}$ is therefore a superprocess with immigration term

$c(s,x)r(\omega, s, x)$, i.e. the immigration term of a particle at place x depends on the history of the population ω up to time s.

3.2 Fleming-Viot-process

The Fleming-Viot-process X is a process on a filtered probability space $(\Omega, \mathcal{F}, (\mathcal{F}_t), P)$ taking values in the space $\mathcal{M}_1(E)$ of probability measures over a Polish space E. Its distribution is by definition the unique solution of the martingale problem characterize by the linear martingales (3.1) and their quadratic covariation $< M[f], M[g] >_t = \int_0^t X_s(fg) - X_s(f)X_s(g)ds$. Hence the associated martingale measure M is not orthogonal. Because the L^2-norm of a stochastic integral $a.M$ differs from the $\mathcal{P}_M$-norm of a the arguments in Section 2 do not work. Moreover, every predictable function $g(\omega, s)$, which does not depend on the space variable x has $g.M = 0$. Therefore,

$$F(\omega) := \int_0^T \int_E a(\omega, s, x)M(\omega, ds, dx) = \int_0^T \int_E (a(\omega, s, x) + g(\omega, s))M(\omega, ds, dx)$$

where we can choose the two representing functions a and $a + g$ different in $\mathcal{P}_M$, by assuming that

$$0 \neq E[\int_0^T g^2(s)ds] = (g, g)_K < \infty.$$

Hence already the uniqueness in Proposition 2.1 does not hold.

3.3 Excursion filtration

Rogers and Walsh consider in [RW] the following situation:
Let B_t be a Brownian motion on a complete probability space $(\Omega, \mathcal{F}, P)$ started in 0 and $L(t, x)$ its local time. For every $x \in I\!R$ define the increasing process $\tau(\cdot, x)$ by $\tau(t, x) = \inf\{u : \int_0^x L(u, y)dy > t\}$. τ is the inverse of the occupation time $A(u, x) := \int_0^x L(u, y)dy = \int_0^u \mathbf{1}_{B_s \le x} ds$. Define the σ-field $\mathcal{E}_x$ by the completion of

$$\sigma(B_{\tau(t,x)}, t \geq 0).$$

The family $(\mathcal{E}_x)_{x \in I\!R}$ is called the filtration of *excursion fields*. A function $\tilde{\phi}$ on $\Omega \times [0, \infty) \times I\!R$ is called $(\mathcal{E}_x)$-predictable if it is measurable with repect to the σ-algebra generated by all $(\mathcal{E}_x \times \mathcal{B}((0, \infty)))_{x \in I\!R}$-adapted processes which are left-continuous in t.

It is proved in [RW] that every $F \in L^2(\Omega, \mathcal{F}, P)$ can be written as

$$F = E[F] + \int_0^\infty \int_{I\!R} \phi(t, x) L(dt, dx) \tag{3.4}$$

with an *identifiable* ϕ satisfying $4E[\int_0^\infty \phi^2(t, B_t)dt] < \infty$. The property *identifiable* means that $\phi = \tilde{\phi} \circ \Gamma$ with a $(\mathcal{E}_x)$-predictable $\tilde{\phi}$ and $\Gamma(\omega, t, x) = (\omega, A(t, x, \omega), x)$. The integral above is defined by

$$\int\int \phi(t, x) L(dt, dx) := Z\,[L(T, b) - L(T, a) - L(S, b) + L(S, a)]$$

for a simple function $\phi(t, x) = Z \mathbf{1}_{(a,b]}(x) \mathbf{1}_{(S,T]}(t)$, where $Z \in b\mathcal{E}_a$ and $S = \tau(S^e, a), T = \tau(T^e, a)$ with $\mathcal{E}_a$-measurable times S^e and T^e and by a standard extension for all identifiable functions ϕ.

We will now point out how this result follows, at least partially, from the predictable representation property for superprocesses.

In their proof Rogers and Walsh use the following result from the Ray-Knight theory:
Suppose that $S < T < U < V$ are $\mathcal{E}_0$-identifiable times (, i.e. constructed as S and T above). Let $M_x := L(T, x) - L(S, x)$ and $N_x := L(V, x) - L(U, x)$. Then $(M_x)_{x \ge 0}$ is a continuous local $(\mathcal{E}_x)$-martingale with increasing process $4 \int_0^x M_y dy$. Moreover, M and N are orthogonal and $(M_x)_{x \ge 0}$ is an L^p-martingale iff $M_0 \in L^p$.

Here we easily notice the connection to a superprocess, namely that $(M_x)_{x \ge 0}$ is the superprocess with state space $\mathcal{M}([0, \infty))$ over the one-particle *motion* $Af = 0$, if we impose a time change of the Brownian motion B, i.e. a space transformation for '$L(\cdot, x)$' as a measure-valued process. Define the measure-valued process M by

$$M_x((a, b]) := L(\tau(b, 0), x) - L(\tau(a, 0), x).$$

By the covariation of M_x this can be extended to an orthogonal martingale measure with covariation $\delta_a(db) M_x(da)dx$. Hence every $F \in L^2(\Omega, \vee_{t \ge 0} \mathcal{E}_t, P)$ can be written as

$$F = E[F|\mathcal{E}_0] + \int_0^\infty \int_0^\infty \psi(a, x) M(dx, da).$$

By the definition of the different integrales $\psi.M$ and $\phi.L$ it is clear that $\psi.M = \phi.L$ iff $\phi(\omega, r, x) = \psi \circ \Gamma^0(\omega, r, x)$ with $\Gamma^0(\omega, r, x) = (\omega, A(\omega, r, 0), x)$. Hence we have to show that $\psi \circ \Gamma^0$ is identifiable if $\psi \in \mathcal{P}_M$. This follows for a simple function $\psi \in \mathcal{P}_M$ because in that case $\psi \circ \Gamma^0$ satisfies the conditions of Proposition 2.4 in [RW] and is therefore identifiable. For a general function $\psi \in \mathcal{P}_M$ the identifiability follows then by a monotone class argument.

Hence at least for $F \in L^2(\Omega, \vee_{x \geq 0} \mathcal{E}_x, P)$ the assertion (3.4) which is formula (2.1) in Theorem 2.1 of [RW] follows easily from the predictable representation property for superprocesses.

This remark should makes it plausible that in the case where we consider the reflecting Brownian motion $|B|$ instead of the Brownian motion the analog result of Rogers and Walsh follows completely from the predictable representation for orthogonal martingale measures, cf. also Remark 1.3b in [EP1].

Acknowledgement. I would like to thank J.F. Le Gall and M. Röckner for helpful dicussions about this note, especially J.F. Le Gall for telling me about the related results in [EP1].

References

[D] D. A. Dawson, Measure-valued Markov processes. In: P.L. Hennequin (ed.), Ecole d'Eté de Probabilité de Saint Flour XXI 1991, L.N.M. 1541. Springer, Berlin (1993).

[EP1] S.N.Evans, E.A. Perkins. Measure-valued branching diffusions with singular interactions. Canad. J. Math. 46 (1), 120-168 (1994).

[EP2] S.N.Evans, E.A. Perkins. Explicit stochastic integral representation for historical functionals. Preprint (1994).

[JS] J.Jacod, A.N.Shirayaev. Limit theorems for stochastic processes. Springer, Berlin (1987).

[JY] J. Jacod, M.Yor. Étude des solutions extrémales et représentation intégrale des solutions pour certains problàmes de martingales. Probab. Theory Relat. Fields 38, 83- 125 (1977).

[P] E.A. Perkins. On the martingale problem for interactive measure-valued branching diffusions. To appear in Mem. Amer. Math. Soc.

[Pr] P. Protter. Stochastic Integration and Differential Equation. Springer, Berlin (1990).

[RW] L.C.G. Rogers, J.B.Walsh. Local time and stochastic area integrals. The Annals of Probability 19, 457-482 (1991).

[W] J.B. Walsh. An introduction to stochastic partial differential equation. In: P.L. Hennequin (ed.), Ecole d'Eté de Probabilité de Saint Flour XIV 1984, L.N.M. 1180, 265-439. Springer, Berlin (1986).

Chaoticity on a stochastic interval $[0,T]$

A. Dermoune, Université du Maine,
Laboratoire de Statistique et Processus, B.P.535, 72017 Le Mans cédex, France.

Abstract

The chaotic representation property is given a meaning and established for a class of martingales X defined on some stochastic interval $[0,T]$ and having only finitely many jumps before $T-\varepsilon$.

1.Introduction

Let X be a martingale with predictable bracket $<X,X>_t = t$, $(\mathcal{F}_t)$ be its filtration and $\mathcal{F} = \bigcup_{t>0} \mathcal{F}_t$. We say that the martingale X has the chaotic representation property (C.R.P) or is chaotic, if for all $F \in L^2(\Omega, \mathcal{F})$, there exists a sequence (f_k) with $f_k \in L^2(\mathbb{R}_+^k, dt^{\otimes k})$, such that

$$F = \sum_{k=0}^{\infty} F_k,$$

where $F_0 = \mathbb{E}[F]$ and for $k > 0$

$$F_k = \int_{0<t_1<\ldots<t_k} f_k(t_1,\ldots,t_k)\, dX_{t_1}\ldots dX_{t_k}.$$

(For the definition of the latter multiple stochastic integral, see [7].)

The random variables $F_k, k \in N$, are such that

$$\mathbb{E}[F_k F_j] = \delta_j(k)\left(\int_{0<t_1<\ldots<t_k} f_k^2(t_1,\ldots,t_k)\, dt_1\ldots dt_k\right),$$

where $\delta_j(k) = 0$ if $k \neq j$ and $\delta_k(k) = 1$.

It is interesting to express the chaotic representation property as an isomorphism between $L^2(\Omega, \mathcal{F})$ and the symmetric Fock space over $H = L^2(\mathbb{R}_+, dt)$, defined by

$$Fock(H) = \oplus_{k=0}^{\infty} H^{\odot k}.$$

For, $k \in \mathbb{N}*$, the space $H^{\odot k} = L^2_{sym}(\mathbb{R}_+^k, dt_1\ldots dt_k)$ is the set of the class of square integrable functions with respect to $dt_1\ldots dt_k$, which are symmetric with respect to the k parameters $(t_1,\ldots,t_k)$. The scalar product over $H^{\odot k}$ is defined by

$$<f,g> = \int_{0<t_1<\ldots<t_k} f(t_1,\ldots,t_k) g(t_1,\ldots,t_k)\, dt_1\ldots dt_k,$$

and $H^{\odot 0} = \mathbb{R}$.

The well known examples of martingales having the chaotic representation property are the Brownian motion and the standard Poisson process [6].

Moreover, *He* and *Wang* [5] have characterized the Lévy processes which have the predictable representation property but until 1987 we did not know if these processes have the chaotic representation property.

In 1987, the author [2] proved that for the Lévy processes the chaotic representation property and the predictable representation property are equivalent.

In 1988, *Emery* [3] showed that a martingale earlier discovered by Azéma [1] has the chaotic representation property, introducing at the same time other examples which satisfy the "structure equation" of the form

$$d[X,X]_t = dt + \Phi(t)dX_t, \quad X_0 = x.$$

He later proved in [4] that if the predictable process $\Phi(t)$ is such that the integral $A_t = \int_0^t \Phi^{-2}(s)ds$ is a.s. finite for all t, then the predictable representation property implies the chaotic representation property. This applies to structure equations with Φ of the form

$$\Phi(t) = \phi_1(t)1_{]0,T_1]}(t) + \sum_{n\geq 2} \phi_n(t, T_{n-1}, ..., T_1)1_{]T_{n-1},T_n]}(t)$$

where ϕ_n are deterministic and the T_n's are the successive jumps of the solution X to the structure equation

$$d[X,X]_t = dt + \Phi(t)dX_t, \quad X_0 = x.$$

The hypothesis $A_t < \infty$ implies that there are only finitely many jumps on finite intervals since A_t is the predictable compensator of the number of jumps

$$C_t = \sum_{n\geq 1} 1_{[T_n,\infty[}(t).$$

The aim of this work is to study the following problem : Dropping the finiteness assumption for A_t and putting $T_\infty = \sup_n T_n$, we will allow T_∞ to be finite. The above formulas define (in law) the martingale X only on the interval $[0, T_\infty]$. We will prove that X still has the chaotic representation property, in the following sense : If M is a chaotic martingale independent of X(possibly defined on an enlargement of Ω), the martingale

$$Y_t = \begin{cases} X_t & \text{for } t \leq T_\infty \\ X_{T_\infty} + M_{t-T_\infty} - M_0 & \text{for } t \geq T_\infty \end{cases}$$

has the chaotic representation property (we will see in Lemma 2.2. that this does not depend on the choice of M).

2.Chaoticity before a stopping time

This section is devoted to giving a rigorous meaning to the chaotic representation property for a martingale defined only up to some stopping time.

Definition. *Let $(X_t)_{t\geq 0}$ be a martingale such that $< X, X >_t$ is equal to t, $(\mathcal{F}_t)$ be its filtration and T be a stopping time of $(\mathcal{F}_t)$. We say that X is chaotic on $[0,T]$ if $L^2(\mathcal{F}_T)$ is included in the chaotic space of X, i.e. if each $F \in L^2(\mathcal{F}_T)$ has an expansion $F = \sum_{k=0}^{\infty} F_k$ with $F_0 = \mathbb{E}[F]$ and for $k > 0$*

$$F_k = \int_{0<t_1<...<t_k} f_k(t_1, ..., t_k)dX_{t_1}...dX_{t_k}$$

with $(f_k)_{k\geq 0} \in Fock(H)$.

Lemma 2.1. *If the martingale $(X_t)_{t>0}$ is chaotic on $[0,T]$ and if a martingale $(Y_t)_{t>0}$ verifies $<Y,Y>_t = t$ and $X = Y$ on $[0,T]$, then T is a stopping time for the filtration generated by Y and Y is also chaotic on $[0,T]$.*

Proof. By proposition (1, ii) of [4], each element of $L^2(\mathcal{F}_T)$ is a sum of multiple integrals with respect to Y; so it only remains to prove that T is a stopping time for Y. For each $t \geq 0$, the indicator of the event $\{T \leq t\}$ is in both $L^2(\mathcal{F}_T)$ and $L^2(\mathcal{F}_t)$, so it is of the form

$$\mathbb{P}(T \leq t) + \sum_{k=1}^{\infty} \int_{0<t_1<\ldots<t_k<t} f_k(t_1, \ldots, t_k) dX_{t_1} \ldots dX_{t_k}.$$

By proposition (1, ii) of [4] again, it is also equal to

$$\mathbb{P}(T \leq t) + \sum_{k=1}^{\infty} \int_{0<t_1<\ldots<t_k<t} f_k(t_1, \ldots, t_k) dY_{t_1} \ldots dY_{t_k}$$

and this shows that T is a stopping time for Y.

Lemma 2.2. *Let T be a stopping time and X be a martingale defined on the interval $[0,T]$ only and verifying $<X,X>_t = t$ on this interval. The following conditions are equivalent.*

1) For some chaotic martingale M independent of X (and possibly defined on an enlargement of Ω), the martingale

$$Y_t = \begin{cases} X_t & \text{for } t \leq T \\ X_T + M_{t-T} - M_0 & \text{for } t \geq T \end{cases}$$

has the chaotic representation property.

2) Same statement as 1), with "for every M" instead of "for some M".

3) There exists a martingale $(X'_t)_{t>0}$ (possibly defined on an enlargement of Ω), verifying $<X',X'>_t = t$, chaotic on $[0,T]$, with restriction X to $[0,T]$.

4) Every martingale $(X'_t)_{t>0}$ (possibly defined on an enlargement of Ω), verifying $<X',X'>_t = t$, with restriction X to $[0,T]$, is chaotic on $[0,T]$.

Proof. The implications 2) $\Rightarrow$ 1) $\Rightarrow$ 3) are trivial and 3) is equivalent to 4) by Lemma 2.1. So it suffices to prove 3) $\Rightarrow$ 2). The proof is completely similar to the proof of Proposition (1, iii) of [4] and Corollary 2 of [4] except for one detail: With the notations of [4], X is no longer supposed to have the $C.R.P$ but only to be chaotic on $[0,T]$. So in the proof of $(1, iii)$, page 14, it is not obvious that there exists an element g in $Fock(H)$ such that

$$U = \int g(A) dX_A := \sum_{k=0}^{\infty} \int_{0<t_1<\ldots<t_k} g_k(t_1, \ldots, t_k) dX_{t_1} \ldots dX_{t_k}.$$

But we know that $U = \int_{A \subset [T,\infty[} f(A) dX_A$, so for almost every A, $\mathbb{E}[f^2(A) 1_{A \subset [T,\infty[}]$ is finite, and the chaoticity of X on $[0,T]$ implies that there exists $h(B,A)$ such that $\int h(B,A) dX_B$ is equal to $f(A) 1_{A \subset [T,\infty[}$. Since $f(A) 1_{A \subset [T,\infty[} \in L^2(\mathcal{F}_{\inf A})$, then $h(B,A)$ is null if $\sup B > \inf A$ and the existence of g is obtained by putting

$$g(\{t_1, \ldots, t_k\}) = \sum_{i=1}^{k+1} h(\{t_1, \ldots, t_{i-1}\}, \{t_i, \ldots, t_k\})$$

this proves the lemma.

Definition. *Let T be a stopping time and X be a martingale defined only on the interval $[0,T]$ and verifying $< X, X >_t = t$ on this interval. We say that X is chaotic on $[0,T]$ if the four conditions of Lemma 2.2 are met.*

Lemma 2.3. *Let $(T_n)_{n \in \mathbb{N}}$ be a non-decreasing sequence of stopping times and T_∞ its limit.*

1)If a martingale $(X_t)_{t \geq 0}$ verifying $< X, X >_t = t$ is chaotic on each interval $[0, T_n]$, it is also chaotic on $[0, T_\infty]$.

2)Let X be a martingale defined only on $[0, T_\infty]$ and verifying $< X, X >_t = t$. If for each n the restriction of X to $[0, T_n]$ is chaotic on $[0, T_n]$, then X is chaotic on $[0, T_\infty]$

Proof. 1) For each n, we know that $L^2(\mathcal{F}_{T_n})$ is included in the chaotic space of X. As this chaotic space is closed and as, by the martingale convergence theorem, $\bigcup_n L^2(\mathcal{F}_{T_n})$ is dense in $L^2(\mathcal{F}_{T_\infty})$, the latter is also included in the chaotic space of X.

2) Using Lemma 2.2, it suffices to apply 1) to the martingale

$$Y_t = \begin{cases} X_t & \text{for } t \leq T_\infty \\ X_{T_\infty} + B_{t-T_\infty} - B_0 & \text{for } t \geq T_\infty \end{cases}$$

where B is a Brownian motion independent of X.

3.Construction of the martingale

This section is devoted to constructing the martingale X announced in the introduction.

The set $\Omega = \mathbb{R}_+^{\mathbb{N}}$ is the set of the sequences $\omega = (S_n, n \in \mathbb{N})$ with S_0 equal to zero and $S_n \in \mathbb{R}_+$ for all $n \in \mathbb{N}$.

The sequence ω defines the following increasing sequence :

$$T_n = \sum_{i=0}^{n} S_i \quad \text{for} \quad n \in \mathbb{N}.$$

Let $T_\infty = \lim_{n \to \infty} T_n$.

For $i \in \mathbb{N}$, let ϕ_{i+1} be a measurable $\mathbb{R}_*$ valued function defined on $\mathbb{R}_+^{i+1}$. We define the point process p_t by

$$p_t = \begin{cases} 0 & \text{for } t \in [0, T_1[\\ \sum_{j=1}^{i} \phi_j(T_j, ..., T_1) & \text{for } t \in [T_i, T_{i+1}[. \end{cases}$$

The process (p_t) generates the increasing family of σ-fields $\mathcal{F}_t^0$ defined by

$$\mathcal{F}_t^0 = \sigma(p_s, s \leq t), \quad \mathcal{F}^0 = \sigma(p_s, s > 0).$$

We use the following notations:

$$\Phi_{i+1}(t) = \phi_{i+1}(t, T_i, ..., T_1) \quad \text{for} \quad i \geq 1,$$

$$\Phi(t) = \Phi_{i+1}(t) \quad \text{if} \quad t \in]T_i, T_{i+1}].$$

We suppose that, for all $i \in \mathbb{N}$, there exists a $\mathcal{F}_{T_i}^0$ measurable positive function $\tau_{i+1} > T_i$, such that

$$\int_{T_i}^{t} \Phi_{i+1}^{-2}(s) ds < +\infty \text{ for } t \in [T_i, \tau_{i+1}[\quad \text{and} \quad \int_{T_i}^{\tau_{i+1}} \Phi_{i+1}^{-2}(s) ds = \infty.$$

The probability measure $\mathbb{P}$ on $(\Omega, \mathcal{F}^0)$ is defined by the law of T_1, with density

$$\Phi_1^{-2}(t) exp\left\{-\int_0^t \Phi_1^{-2}(s)\, ds\right\} 1_{]0,\tau_1[}(t)dt$$

and the conditional law of T_{i+1}, with density

$$\Phi_{i+1}^{-2}(t) exp\left\{-\int_{T_i}^t \Phi_{i+1}^{-2}(s)\, ds\right\} 1_{]T_i,\tau_{i+1}[}(t)dt.$$

The σ-fields $\mathcal{F}_t^0$ are augmented with all subsets of $\mathbb{P}$-null sets of $\mathcal{F}^0$ and denoted by $\mathcal{F}_t$. For all $i \in \mathbb{N}$, T_i is a stopping time of $(\mathcal{F}_t)$.

Proposition 3.1.*Let $N(dt, dx)$ be the random measure on $\mathbb{R}_+ \times \mathbb{R}_*$ defined for $t > 0$ and A a measurable set of $\mathbb{R}_*$ by*

$$N(]0,t] \times A) = \sum_{T_n \leq t} 1_A(\Phi_n(T_n)).$$

The predictable projection of $N(dt, dx)$ with respect to the probability $\mathbb{P}$ is given by

$$\nu(dt, dx) = \Phi^{-2}(t) 1_{[0,T_\infty[}(t)\, dt\, \delta_{\Phi(t)}(dx).$$

Proof. Let $n \in \mathbb{N}_*$, f be a bounded measurable function on $\mathbb{R}_+^n$ and g be a bounded measurable function on $\mathbb{R}$.

Let us consider the predictable process

$$Z(t, x) = 1_{]T_n,T_{n+1}]}(t) f(T_1, ..., T_n) g(x).$$

We have to prove that

$$\mathbb{E}[\int_0^\infty \int_{\mathbb{R}*} Z(t,x) N(dt, dx)] = \mathbb{E}[\int_0^\infty Z(t, \Phi(t))\Phi^{-2}(t)dt].$$

From the equality

$$\int_0^\infty \int_{\mathbb{R}*} Z(t,x) N(dt, dx) = f(T_1, ..., T_n) g(\Phi_{n+1}(T_{n+1})),$$

and using the conditional law of T_{n+1}, with respect to $(T_1, ..., T_n)$, we obtain

$$\mathbb{E}[\int_0^\infty \int_{\mathbb{R}*} Z(t,x) N(dt, dx)]$$

$$= \mathbb{E}\left[f(T_1, ..., T_n) \int_{T_n}^{\tau_{n+1}} g(\Phi_{n+1}(t_{n+1}))\, \Phi_{n+1}^{-2}(t_{n+1})\right.$$

$$\left. exp\left(-\int_{T_n}^{t_{n+1}} \Phi_{n+1}^{-2}(s) ds\right) dt_{n+1}\right].$$

An integration by parts gives

$$\int_{T_n}^{T_{n+1}} \int_{T_n}^{t_{n+1}} g\left(\Phi_{n+1}(t)\right) \Phi_{n+1}^{-2}(t)\,dt$$
$$\left\{\Phi_{n+1}^{-2}(t_{n+1})\,exp\left(-\int_{T_n}^{t_{n+1}} \Phi_{n+1}^{-2}(s)ds\right)\right\} dt_{n+1}$$
$$= \int_{T_n}^{T_{n+1}} g\left(\Phi_{n+1}(t_{n+1})\right) \Phi_{n+1}^{-2}(t_{n+1})\,exp\left(-\int_{T_n}^{t_{n+1}} \Phi_{n+1}^{-2}(s)ds\right) dt_{n+1}.$$

Thus,

$$\mathbb{E}[f(T_1,...,T_n)\int_{T_n}^{T_{n+1}} g\left(\Phi_{n+1}(t)\right)) \Phi_{n+1}^{-2}(t)\,dt\,]$$
$$= \mathbb{E}[f(T_1,...,T_n)\,g\left(\Phi_{n+1}(T_{n+1})\right)],$$

which is exactly what was to be proved.

Proposition 3.2. *Let $m \in \mathbb{R}$.*

1) The process (X_t) defined on the predictable interval $[0, T_\infty[$ by

$$X_t = m + p_t - \int_0^t \Phi(s)^{-1} ds$$

is a $(\mathcal{F}_t)$ square integrable martingale with $< X, X >_t = t$, it verifies the structure equation

$$d[X,X]_t = dt + \Phi(t)dX_t, X_0 = m.$$

2) When the definition of X is extended to $[0, T_\infty]$ by $X_{T_\infty} = \lim_{n\to\infty} X_{T_n}$ on $T_\infty < \infty$, the martingale X has the chaotic representation property on $[0, T_\infty]$.

In the case when $T_\infty = \infty$ *a.s.*, the chaotic property 2) is a consequence of the Theorem 5 of [4].

Proof. 1) Since X_t is also equal to

$$X_t = m + \int_0^t \int_{\mathbb{R}_*} xp(dx,dt) - \int_0^t \int_{\mathbb{R}_*} x\nu(dx,dt)$$

by Proposition 3.1, X is a martingale with predictable bracket $< X, X >_{t\wedge T_\infty} = t \wedge T_\infty$ and satisfies the structure equation

$$d[X,X]_t = 1_{[t<T_\infty]}\,dt + \Phi(t)dX_t, \;\; X_0 = m.$$

2)By Lemma 2.3, it suffices to verify that, for each finite n, X is chaotic on $[0, T_n]$. Define a martingale X^n by the same construction as X, but with $\phi_i \equiv 1$ for $i > n$. The martingale M^n is identical in law to X on $[0, T_n]$ and is a compensated standard Poisson process after T_n. It has the chaotic representation property by Theorem 5 of [4]; this implies in particular that it is chaotic on $[0, T_n]$. So the restriction of X to $[0, T_n]$ is chaotic by Lemma 2.2, and X is chaotic on $[0, T_\infty]$ by Lemma 2.3.

4.Examples

Let $(\lambda_n, n \in \mathbb{N}*)$ be a sequence of strictly positive real numbers and $(T_n, n \in \mathbb{N}*)$ be the successive jumps such that the sojourn times $(T_{n+1} - T_n, n \in \mathbb{N})$ being independent exponentially distributed variables. The density of $T_n - T_{n-1}$ is

$$\lambda_n e^{-\lambda_n t}.$$

When

$$\sum_{n=1}^{\infty} \lambda_n^{-1} < \infty,$$

T_∞ is finite almost surely; or else, it is infinite almost surely. For $t \in [T_{n-1}, T_n[$ The predictable process Φ is given by
$\Phi(t) = \sqrt{\lambda_n^{-1}}$ and the martingale X by

$$X_t = -\sqrt{\lambda_n}(t - T_{n-1}) + \sum_{i=1}^{n-1} \left(\sqrt{\lambda_i^{-1}} - \sqrt{\lambda_i}(T_i - T_{i-1}) \right).$$

It is chaotic on $[0, T_\infty]$ by the preceding proposition.

Another example is given by the structure equation

$$d[X,X]_t = dt + f(X_{t-})dX_t, \quad X_0 = m$$

with m is such that $f(m) \neq 0$ and f is a deterministic continuous function. Let $T_\infty = \inf\{t > 0, X_t = 0\}$, for $t < T_\infty$, X_t can be constructed as follows: let $(T_n, n \in \mathbb{N}*)$ be the jump times of X_t, and suppose that the integral equation

$$x_t = f(X_{T_n} - \int_{T_n}^{t} x_s^{-1} ds), \, t > T_n,$$

has a unique solution $t \to \Phi_{n+1}(t, X_{T_n}, \tau_{n+1})$ on the widest interval $[T_n, \tau_{n+1}[$ of $[T_n, \infty[$ where x_t is defined.

If x_t is such that

$$\int_{T_n}^{t} x_s^{-2} ds < \infty, \quad \text{for } t \in [T_n, \tau_{n+1}[\quad \text{and} \quad \int_{T_n}^{\tau_{n+1}} x_s^{-2} ds = \infty,$$

then we can see that $x_{T_{n+1}} = \Delta X_{T_{n+1}}$ is the jump size at T_{n+1},

$$X_{T_{n+1}} = X_{T_n} + \Phi_{n+1}(T_{n+1}, X_{T_n}, \tau_{n+1}) - \int_{T_n}^{T_{n+1}} \Phi_{n+1}^{-1}(s, X_{T_n}, \tau_{n+1}) ds,$$

and for $t \in [T_n, T_{n+1}[$,

$$X_t = X_{T_n} - \int_{T_n}^{t} \Phi_{n+1}^{-1}(s, X_{T_n}, \tau_{n+1}) ds.$$

If we put $T_0 = 0$, then for all $n \in \mathbb{N}$ the law of T_{n+1}, with respect to $(T_0, ..., T_n)$, is supported by $]T_n, \tau_{n+1}[$ and has the density

$$\Phi_{n+1}^{-2}(t, X_{T_n}, \tau_{n+1}) exp \left\{ - \int_{T_n}^{t} \Phi_{n+1}^{-2}(s, X_{T_n}, \tau_{n+1}) \, ds \right\}.$$

By Proposition 3.2, X is chaotic on $[0, T_\infty]$.

If $f(x) = \beta x$ we find again the Azéma martingale with parameter $\beta \notin \{-1, 0\}$ on the interval $[0, T_\infty]$, where T_∞ is the first time when $X = 0$ (T_∞ is also the first accumulation point of jump times of X).

Remark.

The solution of the differential equation $x_t = f(a - \int_0^t x_s^{-1} ds)$ allows us to construct the martingale X on $[0, T_\infty]$; the existence and the uniqueness of the solution of this equation implies the existence and the uniqueness in law of X on $[0, T_\infty]$.

Acknowledgements

I would like to thank professor Jean-Pierre Lepeltier for his careful reading of the manuscript in several of its versions and Professor Michel Emery for his very useful help.

References

[1] J. Azéma: Sur les fermés aléatoires. Séminaire de probabilités XIX, Lect. Notes in Maths. 1123. Springer (1985).

[2] A. Dermoune: Distribution sur l'espace de Paul Lévy. Ann. Inst. Henri Poincaré, vol. 26, n° 1, 1990, p. 101-119.

[3] M. Emery: On the Azéma martingales. Séminaire de probabilités XXIII, Lect. Notes in Maths, 1372, Springer (1989).

[4] M. Emery: Quelques cas de représentation chaotique. Séminaire de probabilités XXV, Lect. Notes in Maths, 1485, Springer (1991).

[5] S. He, J. Wang: The total continuity of natural filtrations and the strong property of predictable representations for jump processes and processes with independent increments, Séminaire de probabilités XVI, Vol. 920, (1981).

[6] K. Ito: Spectral type of the shift transformation of differential processes with increments. Tr. Ann. Math. Soc, Vol. 81, (1956).

[7] P.A. Meyer: Un cours sur les intégrales stochastiques, Séminaire de probabilités X, Vol. 511, p.p 321-331, (1976).

On the rate of growth of subordinators with slowly varying Laplace exponent

Jean Bertoin(1) **and Ma.-Emilia Caballero**(2)

(1) *Laboratoire de Probabilités (CNRS), Université Paris VI, t. 56 4, Place Jussieu 75252 Paris, France*
(2) *Instituto de Matemáticas U.N.A.M., México 04510 D.F. Mexico*

ABSTRACT. Results of Fristedt and Pruitt [6, 7] on the lower functions of a subordinator are improved in the case when the Laplace exponent is slowly varying. This yields laws of the iterated logarithm for the local times of a class of Markov processes. In particular, this extends recent results of Marcus and Rosen [9] on certain Lévy processes close to a Cauchy process.

1 Introduction and main results

Consider a subordinator $\sigma = (\sigma_t, t \geq 0)$, that is σ is a right-continuous increasing process with stationary independent increments and $\sigma_0 = 0$. Denote its Laplace exponent by Φ,

$$\mathbf{E}(\exp -\lambda\sigma_t) = \exp -t\Phi(\lambda) \qquad (\lambda, t \geq 0)$$

and its inverse by S,

$$S_t = \sup\{s : \sigma_s \leq t\} \qquad (t \geq 0) .$$

Fristedt and Pruitt [6] proved the following law of the iterated logarithm for S. Introduce the inverse function φ of Φ and put

$$h(x) = \frac{\log\,|\log x|}{\varphi(x^{-1}\log\,|\log x|)} \qquad (x \in (0, 1/e) \cup (e, \infty)) .$$

The mapping $x \to h(x)$ increases both in the neighborhood of $0+$ and of ∞, and we denote its inverse by f. Then there exist two constants $c_0, c_\infty \in [1, 2]$ such that

$$\limsup_{t\to 0+} S_t/f(t) = c_0 \qquad \text{and} \qquad \limsup_{t\to\infty} S_t/f(t) = c_\infty \quad \text{a.s.} \tag{1}$$

Fristedt and Pruitt [7] also obtained precise estimates on the modulus of continuity of S. Specifically, denote by $\tilde{f}$ the inverse function of $\tilde{h}$, where

$$\tilde{h}(x) = \frac{|\log x|}{\varphi(x^{-1}\,|\log x|)} \qquad (0 < x < e)$$

There exist two constants $1 \leq \underline{c} \leq \overline{c} \leq 2$ such that

$$\liminf_{t\to 0+} \sup_{0\leq s\leq\sigma(1)} (S_{s+t} - S_s)/\tilde{f}(t) = \underline{c} \quad \text{a.s.} \tag{2}$$

$$\limsup_{t\to 0+} \sup_{0\leq s\leq\sigma(1)} (S_{s+t} - S_s)/\tilde{f}(t) = \overline{c} \quad \text{a.s.} \tag{3}$$

The constants c_0, c_∞, $\underline{c}$ and $\bar{c}$ do not seem to be known explicitly, except in the case when Φ is regularly varying with index $\alpha \in (0,1)$, see e.g. [1] and the references therein. The main results of this paper are that (1-3) can be made completely explicit when the Laplace exponent Φ is slowly varying. One can check that the argument of the proofs also applies when Φ is regularly varying with index 1; however this case has fewer applications than the preceding, and is left to the interested reader.

First, one has the following law of the iterated logarithm for S. Recall that $\Phi(\infty) < \infty$ only in the degenerate case when σ is a compound Poisson process.

Theorem 1 *Suppose that Φ is slowly varying at 0+ (respectively, at ∞ and $\Phi(\infty) = \infty$). Then*

$$\limsup \frac{S_t \, \Phi\left(t^{-1} \log |\log \Phi(1/t)|\right)}{\log |\log \Phi(1/t)|} = 1 \qquad a.s.$$

as $t \to \infty$ (respectively, as $t \to 0+$).

Remark. There is an analogue of Theorem 1 for increasing random walks. This can be deduced from Theorem 1 considering a subordinator with Lévy measure the step distribution of the random walk, and applying the law of large numbers.

Next, we specify the modulus of continuity of S.

Theorem 2 *Suppose that Φ is slowly varying at ∞ and $\Phi(\infty) = \infty$. Then*

$$\lim_{t\to 0+} \sup_{0\le s\le 1} \frac{(S_{t+s} - S_s)\, \Phi\left(t^{-1} \log \Phi(1/t)\right)}{\log \Phi(1/t)} = 1 \qquad a.s.$$

Presumably, Theorem 1 should follow (at least for large times) from a characterization due to Pruitt [11] of the lower functions of a general subordinator, but it does not seem straightforward. We will rather establish Theorems 1-2 using elementary lemmas in Fristedt and Pruitt [6, 7] and the observation that it is more fruitfull to work with the explicit functions

$$g(x) = \frac{\log |\log \Phi(1/x)|}{\Phi\left(x^{-1} \log |\log \Phi(1/x)|\right)} \text{ and } \tilde{g}(x) = \frac{\log \Phi(1/x)}{\Phi\left(x^{-1} \log \Phi(1/x)\right)} \tag{4}$$

than with the implicit functions f and $\tilde{f}$. The hint for this observation stems from a recent paper of Marcus and Rosen [9] where laws of the iterated logarithm are obtained for the local time of symmetric Lévy processes close to a Cauchy process. In section 3, we show that Theorem 1 can be applied both to give a short proof of the results of Marcus ans Rosen, and to extend them to a broader class of Lévy processes. (A related argument appears in [1] where the law of the iterated logarithm for a subordinator whose Laplace exponent is regularly varying with index $\alpha \in (0,1)$, is used to recover and extend earlier results of Marcus and Rosen [8]).

2 Proof of the theorems

To start with, we establish a simple lemma that holds even when Φ is not regularly varying. Recall that φ stands for the inverse function of Φ.

Lemma 1 *For every* $\gamma > 0$

$$\log\left|\log\Phi\left(\frac{\varphi(\gamma x^{-1}\log\,|\log x|)}{\log|\log x|}\right)\right| \sim \log|\log x|$$

both as $x \to 0+$ *and* $x \to \infty$.

Proof: First, we observe that since Φ increases

$$\begin{aligned}\Phi\left(\frac{\varphi(\gamma x^{-1}\log\,|\log x|)}{\log\,|\log x|}\right) &\leq \Phi\left(\varphi(\gamma x^{-1}\log\,|\log x|)\right)\\ &= \gamma x^{-1}\log\,|\log x|,\end{aligned}$$

provided that x being either small enough or large enough. On the other hand, recall that Φ is concave, so $\Phi(uv) \geq \Phi(u)v$ for all $u \geq 0$ and $v \in (0,1)$. As a consequence

$$\begin{aligned}\Phi\left(\frac{\varphi(\gamma x^{-1}\log\,|\log x|)}{\log\,|\log x|}\right) &\geq \Phi\left(\varphi(\gamma x^{-1}\log\,|\log x|)\right)/\log|\log x|\\ &= \gamma x^{-1},\end{aligned}$$

provided that x is either small enough or large enough. ■

The next lemmas give respectively the upper and lower bounds in Theorem 1. Recall that the function g is defined in (4).

Lemma 2 *Suppose that* Φ *is slowly varying at* $0+$ *(respectively, at* ∞ *and* $\Phi(\infty) = \infty$*). Then*

$$\limsup S_t/g(t) \leq 1 \quad a.s.$$

as $t \to \infty$ *(respectively, as* $t \to 0+$*).*

Proof: It follows readily from Lemma 4 in Fristedt and Pruitt [6] that for every $\gamma > 1$ and $\delta \in (0, \gamma - 1)$

$$\limsup S_t/f_{\gamma,\delta}(t) \leq 1 \quad \text{a.s.}$$

both as $t \to 0+$ and as $t \to \infty$, where $f_{\gamma,\delta}$ is the inverse function of

$$x \to \delta\frac{\log\,|\log x|}{\varphi\left(\gamma x^{-1}\log\,|\log x|\right)}.$$

So, all that we need is check that for every $\epsilon > 0$, there exists $\gamma > 1$ and $\delta \in (0, \gamma - 1)$ such that

$$(1-\epsilon)\, f_{\gamma,\delta}(x) \leq g(x) \tag{5}$$

for all x small enough (respectively, large enough). In this direction, we observe first that

$$g\left(\delta\frac{\log\,|\log x|}{\varphi(\gamma x^{-1}\log\,|\log x|)}\right) \sim g\left(\frac{\log\,|\log x|}{\varphi(\gamma x^{-1}\log\,|\log x|)}\right),$$

because g is slowly varying. Then a few lines of calculation based on Lemma 1 and the hypothesis that Φ is slowly varying show that the right-hand-side is equivalent to

$$\frac{\log\,|\log x|}{\Phi\left(\varphi(\gamma x^{-1}\log\,|\log x|)\right)} = x/\gamma\,.$$

We deduce that (5) holds provided that $\gamma < (1-\epsilon)^{-1}$. ∎

Lemma 3 *Suppose that Φ is slowly varying at $0+$ (respectively, at ∞ and $\Phi(\infty) = \infty$). Then*

$$\limsup S_t/g(t) \geq 1 \qquad a.s.$$

as $t \to \infty$ (respectively, as $t \to 0+$).

Proof: It follows now from Lemma 5 in Fristedt and Pruitt [6] that for every $\gamma < 1$ and $\delta > \gamma$

$$\limsup S_t/f_{\gamma,\delta}(t) \geq 1 \quad \text{a.s}$$

both as $t \to 0+$ and as $t \to \infty$, where $f_{\gamma,\delta}$ has been defined in the proof of Lemma 2. So, all we need is to check that for every $\epsilon > 0$, there exists $\gamma < 1$ and $\delta > \gamma$ such that

$$(1+\epsilon)\,f_{\gamma,\delta}(x) \geq g(x) \tag{6}$$

for all x small enough (respectively, large enough). But the argument in Lemma 2 shows that

$$g\left(\delta\frac{\log\,|\log x|}{\varphi(\gamma x^{-1}\log\,|\log x|}\right) \sim x/\gamma\,,$$

and hence (6) holds provided that $\gamma > (1+\epsilon)^{-1}$. ∎

The proof of Theorem 2 is similar. First, one checks readily the following analogue of Lemma 1. For every $\gamma > 0$

$$\log\Phi\left(\frac{\varphi(\gamma x^{-1}\,|\log x|)}{|\log x|}\right) \sim \log 1/x \qquad (x \to 0+) \tag{7}$$

(again, this holds even if Φ is not regularly varying). The upper-bound in Theorem 2 then follows from Lemma 5 of Fristedt and Pruitt [7] and (7) much in the same way as in Lemma 2. The lower-bound follows from Lemma 4 of [7] and (7) by an argument close to that in Lemma 3. We skip the details.

3 Applications to local times

We mentioned in the Introduction that the hint for Theorems 1-2 was the results of Marcus and Rosen [9] on the local time of certain symmetric Lévy processes. Conversely, it is interesting to discuss their results in our framework. In this direction, suppose that $X = (X_t, t \geq 0)$ is a standard Markov process started at a regular recurrent point, say 0. Then there exists a local time process at 0, $L = (L_t, t \geq 0)$, and the inverse local time $\sigma_\bullet = \inf\{s: L_s > \bullet\}$ is a subordinator. See Blumenthal and Getoor [2], section 5.3. The inverse S of σ obviously coincides with L, and thus Theorem 1 gives a law of the iterated logarithm for L as time goes to infinity, provided that

$$\text{the Laplace exponent } \Phi \text{ of } \sigma \text{ is slowly varying at } 0+. \tag{8}$$

Suppose now that the Markov process fulfills the duality conditions of chapter VI of Blumenthal and Getoor [2], and denote by $u^\lambda(x,y)$ the adequate version of the resolvent density. Then the local time L can be normalized such that

$$u^\lambda(0,0) = 1/\Phi(\lambda) \qquad (\lambda > 0),$$

and (8) holds if only if $u^\bullet(0,0)$ is slowly varying. When stronger dual conditions are fulfilled, namely when there exist semigroup densities $p_t(x,y)$ and $\widehat{p}_t(x,y)$ is duality, then

$$u^\lambda(0,0) = \int_0^\infty e^{-\lambda t} p_t(0,0)\, dt.$$

By a Tauberian theorem, we see that (8) holds if and only if the so called truncated Green function

$$G(t) = \int_0^t p_s(0,0)\, ds$$

is slowly varying at ∞, and in that case

$$G(t) \sim 1/\Phi(1/t) \qquad (t \to 0+).$$

Of course, the truncated Green function G is slowly varying at infinity whenever

$$p_\bullet(0,0) \quad \text{is regularly varying at } \infty \text{ with index } -1, \tag{9}$$

see e.g. Feller [5, Theorem VII. 9.1], but (9) is a strictly stronger requirement than (8).

Applying this to the case when X is a recurrent symmetric Lévy process having local time L for which (8) holds, we obtain the first part of Theorem 1.2 of Marcus and Rosen [9]. The second part, that is the law of the iterated logarithm for the difference $L - L^a$, where L^a is the local time at level $a \neq 0$, follows from the argument of section 4.2 in [1]. We point out that the result holds under the weaker assumption that the truncated Green function is slowly varying (this was conjectured by Marcus and Rosen) and that the symmetry condition can be dropped (actually, there are also some technical conditions in [9] which are now seen as unnecessary).

This reasoning also allows us to recover the law of the iterated logarithm for the local time process at level 1 for the two-dimensional Bessel process (see Meyre and

Werner [10], equation (1.c) on p. 51). More precisely, (9) holds when $X+1$ is a 2-dimensional Bessel process, and one then obtains

$$\limsup_{t\to\infty} \frac{L_t}{\log t \log_3 t} = 1 \text{ a.s.},$$

where $\log_3 = \log\log\log$.

Plainly, similar arguments apply when times tend to $0+$, and Theorems 1-2 provide relevant informations on the local rate or growth of the local time of certain Markov processes.

We conclude this section with simple conditions that guaranty that the semigroup density at 0, $p_t(0,0)$, of a real-valued Lévy process X, is regularly varying with index -1. Denote the characteristic exponent by ψ, i.e.

$$\mathbf{E}(\exp i\lambda X_t) = \exp -t\psi(\lambda)$$

for every $t \geq 0$ and $\lambda \in \mathbf{R}$.

Proposition 1 *Assume that the real part $\Re\psi$ of ψ is regularly varying at ∞ with index 1, and that the imaginary part $\Im\psi$ satisfies*

$$\lim_{\lambda\to\infty} \Im\psi(\lambda)/\Re\psi(\lambda) = c \in (-\infty,\infty).$$

Then there exists a continuous version of the semigroup density $x \to p_t(0,x)$, and

$$p_t(0,0) \sim \frac{1}{\pi(1+c^2)} r(1/t) \qquad (t \to 0+),$$

where r is an asymptotic inverse of $\Re\psi$. In particular, $p_\bullet(0,0)$ is regularly varying at $0+$ with index -1.

Proposition 2 *Assume that for some $t \geq 0$,*

$$\int_{-\infty}^{\infty} \exp\{-t\Re\psi(\lambda)\}\, d\lambda < \infty .$$

Then there exists a continuous version of the semigroup density $x \to p_t(0,x)$. Suppose moreover that $\Re\psi$ is regularly varying at $0+$ with index 1 and that

$$\lim_{\lambda\to 0+} \Im\psi(\lambda)/\Re\psi(\lambda) = c \in (-\infty,\infty).$$

Then

$$p_t(0,0) \sim \frac{1}{\pi(1+c^2)} r(1/t) \qquad (t \to \infty)$$

where r is an asymptotic inverse of $\Re\psi$. In particular, $p_\bullet(0,0)$ is regularly varying at ∞ with index -1.

The proofs of Propositions 1 and 2 are similar, we shall focus on the latter which is slightly more delicate than the former.

Proof of Proposition 2. The first assertion follows immediately from Fourier inversion, and more precisely, since the density must be real,

$$p_t(0,0) = \frac{1}{\pi}\int_0^\infty \exp\{-t\Re\psi(\lambda)\} \cos\{t\Im\psi(\lambda)\}\, d\lambda\,. \tag{10}$$

Then put $R(\lambda) = \min\{\Re\psi(\mu),\ 0 \le \mu \le \lambda\}$ and recall from Theorem 1.5.3 in [3] that $R \sim \Re\psi$. Denote by r the inverse of R, so that r is an asymptotic inverse of $\Re\psi$ and its regularly varying at 0+ with index 1, see Theorem 1.5.12 in [3]. On the one hand, we have by an Abelian theorem

$$\int_0^\infty \exp\{-tR(\lambda)\}\, d\lambda = \int_0^\infty \exp\{-\lambda t\}\, dr(\lambda) \sim r(1/t) \quad (t \to \infty). \tag{11}$$

On the other hand,

$$\int_0^\infty \exp\{-tR(\lambda)\}\, d\lambda = r(1/t)\int_0^\infty \exp\{-tR(\lambda r(1/t))\}\, d\lambda$$

and we know that $tR(\lambda r(1/t))$ converges pointwise to λ as $t \to \infty$. We deduce from (11) that

$$\lim_{t\to\infty}\int_0^\infty \exp\{-tR(\lambda r(1/t))\}\, d\lambda = 1 = \int_0^\infty e^{-\lambda}\, d\lambda$$

and this implies that the family of nonnegative functions

$$\lambda \to \exp\{-tR(\lambda r(1/t))\} \qquad (t \ge 1)$$

is uniformly integrable, see e.g. Theorem I.21 in Dellacherie-Meyer [4].

Then we re-express (10) as

$$\pi p_t(0,0)/r(1/t) = \int_0^\infty \exp\{-t\Re\psi(\lambda r(1/t))\} \cos\{t\Im\psi(\lambda r(1/t))\}\, d\lambda$$

By hypothesis, the integrand converges pointwise to $\exp\{-\lambda\}\cos\{c\lambda\}$ as $t \to \infty$ and its absolute value is bounded by $\exp\{-tR(\lambda r(1/t)\}$ which is uniformly integrable. Thus the integral converges to

$$\int_0^\infty e^{-\lambda}\cos(\lambda c)\, d\lambda = \frac{1}{1+c^2},$$

see e.g. Theorem I.21 in [4]. This proves our assertions. ■

Acknowledgment. This work was realized during a visit of the first author to the Instituto de Matemáticas (U.N.A.M.), whose support is gratefully acknowledged. The first author should like to thank Prof. M.B. Marcus for discussions on the article [9] which stimulated this work.

References

[1] J. Bertoin: *Some applications of subordinators to local times of Markov processes.* To appear in Forum Math.

[2] R. M. Blumental, R. K. Getoor: *Markov Processes and Potential Theory.* Academic Press, New York 1968.

[3] N. H. Bingham, C. M. Goldie, J. L. Teugels: *Regular Variation.* Cambridge University Press 1987.

[4] C. Dellacherie, P. A. Meyer: *Probabilités et Potentiel*, chapitres I à IV. Hermann 1975. Paris.

[5] W. Feller: *An Introduction to Probability Theory and its Application, vol 2.* Wiley 1971. New-York.

[6] B. E. Fristedt, W. E. Pruitt: *Lower functions for increasing random walks and subordinators.* Z. Wahrscheinlichkeitstheorie verw. Geb. 18, 167–182 (1971).

[7] B. E. Fristedt, W. E. Pruitt: *Uniform lower functions for subordinators.* Z. Wahrscheinlichkeitstheorie verw. Geb. 24, 63–70 (1972).

[8] M. B. Marcus, J. Rosen: *Laws of the iterated logarithm for the local times of symmetric Lévy processes and recurrent random walks.* Ann. Probab. 22, 626–658 (1994).

[9] M. B. Marcus, J. Rosen: *Laws of the iterated logarithm for the local times of recurrent random walks on Z^2 and of Lévy processes and random walks in the domain of attraction of Cauchy random variables.* Ann. Inst. Henri Poincaré 30-3, 467–499 (1994).

[10] T. Meyre, W. Werner: *Estimation asymptotique du rayon du plus grand disque recouvert par la saucisse de Wiener plane.* Stochastics and Stochastics Reports 48, 45–59 (1994).

[11] W. E. Pruitt: *An integral test for subordinators.* In: Random walks, Brownian motion and interacting particle systems (Eds: R. Durrett and H. Kesten), Prog. Probab. 28, 387-398, Birkhäuser (1991).

Une propriété de Markov pour les processus indexés par $\mathbb{R}$

S.Fourati

I.N.S.A.*de Rouen - 76130 Mont Saint Aignan.*

L'étude des processus de Markov indexés par toute la droite réelle a été remise à l'ordre du jour à l'occasion de travaux sur la mesure de Kuznetsov (cf. Kuznetsov [K], Fitzsimmons-Maisonneuve [F-M],[F],[DMM]...).

Lorsqu'on travaille avec des processus de Markov indexés par $\mathbb{R}$, les opérations de la théorie générale des processus (meurtre, translation, retournement du temps) prennent une forme agréable mais la propriété de Markov forte n'est pas conservée par ces opérations (par exemple, le retourné ,X_{-t}, d'un Markov fort n'est même pas en général modérément markovien). Ceci nous a amené à définir dans [F-L] une propriété de Markov générale, et le but de cet article est de montrer l'invariance de cette notion de processus markovien par les opérations de la théorie générale, et d'en donner des applications.

Voici l'origine de cette propriété :

Soient $\mathbb{E}$ un espace de Radon, $(\Omega,\mathcal{F}°)$ et X l'espace et le processus canoniques habituels associés. $\mathbb{P}$ une probabilité sur $(\Omega,\mathcal{F}°)$,$\mathcal{O}$ la tribu optionnelle associée à la filtration canonique $(\mathcal{F}_{t+})$, $\Pi^{\mathcal{O}}$ la projection optionnelle, et $\mathcal{G}$ la tribu sur $\mathbb{R}\times\Omega$ engendrée par les processus $\theta_s(X)$, $s > 0$ (c'est la tribu coprévisible d'Azéma [A]).

Sous $\mathbb{P}$, si le processus X vérifie la propriété de Markov forte et admet un semi-groupe de transition borélien et droit (P_t) alors :

Pour tout processus, $\mathcal{G}$-mesurable, positif Z, il existe une fonction mesurable f telle que :

$$\Pi^{\mathcal{O}}(Z) = f(X).$$

On peut définir une projection $\Pi^{\mathcal{G}}$ sur la tribu $\mathcal{G}$. Cette propriété est alors équivalente, dans le cas d'un processus transient, à la propriété : pour tout processus optionnel positif Y, il existe une fonction mesurable g telle que

$$\Pi^{\mathcal{G}}(Y) = g(X).$$

En général, la tribu $\check{\mathcal{G}}$, obtenue à partir de $\mathcal{G}$ par retournement du temps, n'est pas la tribu optionnelle (ni la tribu prévisible!) du processus retourné X_{-t}, ce qui fait que la propriété de Markov n'est pas invariante par retournement du temps.

Afin de pallier cet inconvénient, et de restaurer la symétrie passé-futur implicite dans la notion de processus de Markov ("le passé et le futur sont conditionnellement indépendants sachant le présent"), nous avons introduit dans [F-L] deux classes de tribus sur $\mathbb{R} \times \Omega$ les "tribus du passé" et les "tribus du futur" , qui jouent le rôle respectivement de $\mathcal{O}$ et $\mathcal{G}$, ces deux classes étant échangées par retournement du temps. Les tribus optionnelle et prévisible usuelles sont deux exemples de tribus du passé, la tribu coprévisible d'Azéma [A] est un exemple de tribu du futur.

Il existe sur de telles tribus des notions de projections qui généralisent les notions usuelles de projection (optionnelle, prévisible, etc.), qui ont été étudiées dans [F-L]. Si on se donne un couple $(\mathcal{H},\mathcal{G})$, formé d'une tribu du passé $\mathcal{H}$ et d'une tribu du futur $\mathcal{G}$, alors un processus sera dit $(\mathcal{H},\mathcal{G})$-markovien s'il vérifie : pour tout processus $\mathcal{G}$-mesurable positif Z, il existe une fonction mesurable f telle que $\Pi^{\mathcal{H}}(Z) = f(X)$. La définition de ces notions est rappelée au premier paragraphe de l'article.

Il résulte du théorème de commutation des projections de [F-L] que le retourné d'un processus $(\mathcal{H},\mathcal{G})$-markovien transient est un processus $(\hat{\mathcal{G}},\hat{\mathcal{H}})$-markovien.

De plus, en récupérant la symétrie passé-futur (au moins dans le cas transient), on récupère en même temps la stabilité de la propriété de Markov par translation par un temps aléatoire fini quelconque, ainsi que par des opérateurs d'oubli et de meurtre ; tout ceci fait l'objet du deuxième paragraphe.

C'est pour ces raisons que nous pensons que cette propriété de Markov est la bonne généralisation de la propriété de Markov forte des processus indexés par $\mathbb{R}_+$, qui en est un cas particulier, comme on le verra au paragraphe 3.

Un autre avantage de la définition proposée ici est que les résultats et démonstrations de la théorie usuelle s'y transposent mot à mot ce qui permet très facilement d'établir des résultats sur les mesures de Kuznetsov et les mesures de Palm associées. A titre d'exemple, on retrouve un résultat de Fitzsimmons [F] (qui généralisait lui-même un résultat de Dynkin et Getoor [D-G]), sur la représentation des mesures sur l'espace d'état d'un processus de Markov au moyen de fonctionnelles additives.

Le paragraphe 4 est consacré à cette représentation pour les processus markoviens de notre définition et dans le paragraphe 5 nous montrons pourquoi les mesures de Kuznetsov et leurs mesures de Palm vérifient notre définition.

I - Les tribus homogènes et la $(\mathcal{H},\mathcal{G})$-propriété de Markov.

1) Les tribus du passé :

On rappelle ici les principales définitions de [F-L].

Soit $(\Omega,\mathcal{F},P)$ un espace probabilisé ; si s est un réel et Z est une application de $\mathbb{R} \times \Omega$ dans un espace quelconque, on note $\theta_s(Z)$ l'application de $\mathbb{R} \times \Omega$ dans ce même espace définie par : $\theta_s(Z)_t(\omega) = Z_{t+s}(\omega)$.

Dorénavant, sauf mention explicite du contraire, le symbole Z désignera toujours une application mesurable de $\mathbb{R}\times\Omega$ dans $\mathbb{R}_+ \cup \{+\infty\}$.

Définitions.

On appelle *tribu du passé* une sous-tribu $\mathcal{H}$ de $\mathcal{B}(\mathbb{R})\times\mathcal{F}$ engendrée par une famille de processus réels à trajectoires continues à droite, limitées à gauche (càdlàg en abrègé) et vérifiant la propriété suivante: si Z est $\mathcal{H}$-mesurable et s un réel négatif, alors $\theta_s(Z)$ est encore $\mathcal{H}$-mesurable.

Un *temps aléatoire* ou plus brièvement *un temps* est une application de Ω dans $\mathbb{R} \cup \{-\infty, +\infty\}$.

Un *temps d'arrêt* de $\mathcal{H}$ est un temps tel que l'intervalle stochastique $[\![T,+\infty[\![$ appartienne à $\mathcal{H}$.

Le *temps de naissance* de la tribu $\mathcal{H}$, noté η est la borne inférieure essentielle (relativement à la probabilité P) des temps d'arrêt de $\mathcal{H}$ à valeurs dans $\mathbb{R} \cup \{+\infty\}$.

L'*intervalle de vie* que l'on notera ici $\{\eta,+\infty[\![$ (au lieu de $V(\mathcal{H})$ dans [F-L]) est l'intervalle stochastique $\{\eta,+\infty[\![\; = \;]\!]\eta,+\infty[\![\; \cup \; [\![\eta_a]\!]$ où η_a est la partie $\mathcal{H}$-accessible de η. [C'est-à-dire que η_a un temps d'arrêt de $\mathcal{H}$ tel que η_a = η sur $\eta_a<\infty$ et pour tout temps d'arrêt T de $\mathcal{H}$, $P(\eta = T < +\infty \text{ et } \eta_a = +\infty) = 0$].

Pour tout temps aléatoire T la *tribu des événements antérieurs à T* est la tribu $\mathcal{H}_T$ sur Ω engendrée par les variables aléatoires Z_T où Z est un processus $\mathcal{H}$-mesurable, admettant des limites en $-\infty$ et $+\infty$, (et que l'on prolonge en $\pm\infty$ par ces limites).

2) Projection sur une tribu du passé :

Proposition 0 :

Soit $\mathcal{H}$ une tribu du passé; pour tout processus Z, il existe un processus $\mathcal{H}$-mesurable $\Pi^{\mathcal{H}}(Z)$, unique à l'indistinguabilité près, nul hors de $\{\eta,+\infty[\![$, tel que :

Pour tout temps d'arrêt T de $\mathcal{H}$, on l'égalité

$$E[Z_T | \mathcal{H}_T]\, 1_{T\in\mathbb{R}} = \Pi^{\mathcal{H}}(Z)_T\, 1_{T\in\mathbb{R}} \text{ p.s.}$$

Le processus $\Pi^{\mathcal{H}}(Z)$ est appelé la *projection de Z sur $\mathcal{H}$.*

3) Les tribus du futur :

Symétriquement à ce qui précède, on appelle *tribu du futur* une sous-tribu $\mathcal{G}$ de $\mathcal{B}(\mathbb{R}) \times \Omega$, engendrée par des processus càglàd et vérifiant la propriété suivante : si Z est $\mathcal{G}$-mesurable et s est un réel positif, alors $\theta_s(Z)$ est encore $\mathcal{G}$-mesurable.

On remarque que $\mathcal{G}$ est une tribu du futur si et seulement si la tribu des processus $(t,\omega)\to Z_{-t}(\omega)$ où Z est $\mathcal{G}$-mesurable, est une tribu du passé.

Pour une tribu du futur $\mathcal{G}$, on définit de manière symétrique les notions de *co-temps d'arrêt, temps de mort* ξ, *intervalle de vie* $]\!]-\infty,\xi\}$, *tribu* $\mathcal{G}_T$ *des événements postérieurs à* T, *projection* $\Pi^{\mathcal{G}}$ *sur* $\mathcal{G}$.

4) La $(\mathcal{H},\mathcal{G})$-propriété de Markov :

Les définitions nous permettent d'introduire la notion suivante de processus markovien : considérons pour cela un espace de Radon $(\mathbb{E}^b_a,\mathcal{E}^b_a)$ dans lequel on a distingué deux points a et b tels que les singletons {a} et {b} appartiennent à $\mathcal{E}^b_a$.

Dans la suite, le symbôle f désigne une application mesurable de $\mathbb{E}^b_a$ dans $\mathbb{R}_+ \cup \{+\infty\}$, on dira que f est $\mathcal{E}^b$-mesurable (resp. $\mathcal{E}_a$- ou $\mathcal{E}$-mesurable) si on impose que f(a)=0 (resp. f(b)=0 ou f(a)=f(b)=0).

Définition I.1 :

Soient $\mathcal{H}$ ($\mathcal{G}$) une tribu du passé (du futur) sur $\mathbb{R}\times\Omega$, X une application de $\mathbb{R}\times\Omega$ dans $\mathbb{E}^b_a$.

On dira que X est un processus $(\mathcal{H},\mathcal{G})$-markovien si :

(a) X est $\mathcal{H}$-mesurable et $\mathcal{G}$-mesurable,

(b) $X_t(\omega) \neq a \Leftrightarrow (t,\omega) \in \{\eta,+\infty[\![$

(c) Tout processus $\mathcal{G}$-mesurable, Z admet pour projection sur $\mathcal{H}$ un processus de la forme f(X).

Commentaires :

a) La propriété (a) remplace les propriétés de régularité des trajectoires qu'on impose habituellement aux processus de Markov.

b) La propriété (b) impose, en particulier, qu'il existe une suite de temps d'arrêt de $\mathcal{H}$ ne prenant pas la valeur $-\infty$, qui décroit vers le temps de naissance α de X, $(\alpha=\inf\{t \mid X_t \neq a\})$. C'est une propriété de transience dans le passé, ou de "dissipativité".

c) La propriété de Markov ainsi énoncée est très générale :
On peut montrer que toutes les propriétés de Markov se trouvant dans la littérature (par exemple: Markov fort, modéré (Chung-Walsh [C-W] et Azéma [A]),processus canonique sous une mesure de Kuznetsov[1] ...) sont des propriétés de Markov relativement à un couple $(\mathcal{H},\mathcal{G})$ particulier.
Nous allons considérer plus loin (§III) en détails le cas de la propriété de Markov forte sur l'espace canonique et laisser au lecteur le soin de se convaincre dans les autres cas.

II - Translation, meurtre, retournement des processus de Markov.

1) Translation :

Soit T un temps à valeurs réelles finies. Pour tout Z on note $\theta_T(Z)$ le processus défini par $\theta_T(Z)_t(\omega) = Z_{t+T(\omega)}(\omega)$. $\mathcal{H}$ (resp $\mathcal{G}$) étant une tribu du passé (resp. du futur) *la translatée* $\theta_T(\mathcal{H})$ (resp. $\theta_T(\mathcal{G})$) est la tribu sur $\mathbb{R} \times \Omega$ engendrée par les processus $\theta_T(Z)$ où Z parcourt les processus $\mathcal{H}$-mesurables (resp. $\mathcal{G}$-mesurables).

Il est clair que $\theta_T(\mathcal{H})$ (resp. $\theta_T(\mathcal{G})$) reste une tribu du passé (resp.du futur);

[1] Il faut pour étudier cette mesure, étendre la définition au cas des mesures σ-finies, ce que nous ferons au paragraphe IV.

que le temps de naissance de $\theta_T(\mathcal{H})$ est η-T et que la projection sur $\theta_T(\mathcal{H})$ d'un processus Z s'obtient de la manière suivante :

$$\Pi^{\theta_T(\mathcal{H})}(Z)_t(\omega) = \Pi^{\mathcal{H}}(\theta_{-T}(Z))_{t+T}(\omega).$$

La proposition suivante est alors immédiate :

Proposition II.1 :

Si T un temps aléatoire fini alors :

$$X \text{ est } (\mathcal{H},\mathcal{G})\text{-markovien} \Leftrightarrow \theta_T(X) \text{ est } (\theta_T(\mathcal{H}),\theta_T(\mathcal{G}))\text{-markovien.}$$

2) Oubli :

T étant un temps non nécessairement fini, on définit le processus $a_T(X)$ et la tribu $a_T(\mathcal{H})$ de la manière suivante :

$$a_T(X)_t(\omega) = X_t(\omega) \text{ si } t \geq T(\omega)$$
$$= a \text{ sinon.}$$

$a_T(\mathcal{H})$ est la tribu engendrée par les processus de la forme $Z\,1_{[\![T,+\infty[\![}$ où Z parcourt $\mathcal{H}$.

Lemme : Si T est un temps d'arrêt de $\mathcal{H}$, alors $a_T(\mathcal{H})$ est une tribu du passé.

Démonstration : Il suffit pour cela d'exhiber une famille génératrice de $a_T(\mathcal{H})$ composée de processus càdlàg et stable par les opérateurs θ_s, $s < 0$.

Il suffit de prendre :

$$\mathcal{A} = \{Z\,1_{[\![T,+\infty[\![}\ Z \text{ càdlàg et } Z \in \mathcal{H}\}$$

$\mathcal{A}$ engendre $a_T(\mathcal{H})$ par définition. T étant un temps d'arrêt de $\mathcal{H}$, l'ensemble $[\![T,+\infty[\![$ est dans $\mathcal{H}$; d'où l'équivalence :

$$Z \in \mathcal{A} \Leftrightarrow Z \text{ càdlàg, } Z \in \mathcal{H} \text{ et } Z = Z1_{[\![T,+\infty[\![}.$$

Il est clair que $\mathcal{A}$ est stable par les opérateurs θ_s, $s < 0$.

La proposition et les faits suivants sont alors immédiats :

Le temps de naissance de $a_T(\mathcal{H})$ est $T\vee\alpha$ et la projection sur $a_T(\mathcal{H})$ d'un processus Z, notée $\Pi^{a_T(\mathcal{H})}(Z)$, est donnée par:

$$\Pi^{a_T(\mathcal{H})}(Z) = \Pi^{\mathcal{H}}(Z)1_{[\![T,+\infty[\![}$$

On note $a_T(\mathcal{G})$ la tribu du futur engendrée par les processus de la forme $Z1_{]\!]T-t,+\infty[\![}$ où Z parcourt $\mathcal{G}$ et t parcourt $\mathbb{R}_+^*$.

Proposition II.2 : Si T un temps d'arrêt de $\mathcal{H}$ et X un processus $(\mathcal{H},\mathcal{G})$ markovien, alors :

$$a_T(X) \text{ est } (a_T(\mathcal{H}),a_T(\mathcal{G}))\text{-markovien.}$$

Remarque : On peut remplacer l'hypothèse "T est un temps d'arrêt de $\mathcal{H}$" par "T est un temps aléatoire tel que l'intervalle stochastique $]\!]T,+\infty[\![$ est dans $\mathcal{H}$". Il faut remplacer l'intervalle stochastique $[\![T,+\infty[\![$ par $]\!]T,+\infty[\![$, le processus $a_T(X)$, les tribus $a_T(\mathcal{H})$ et $a_T(\mathcal{G})$ sont modifiées en conséquence et la proposition II.2 modifiée reste vraie.

3) Meurtre à un temps terminal :

Soit T un temps ; on définit le processus $b_T(X)$ et la tribu $b_T(\mathcal{G})$ de la manière suivante :

$$b_T(X)_t(\omega) = X_t(\omega) \text{ si } t<T(\omega)$$

$$= b \text{ sinon.}$$

$b_T(\mathcal{G})$ est la tribu engendrée par les processus de la forme $Z\, 1_{]\!]-\infty,T[\![}$, où Z parcourt $\mathcal{G}$.

On note $\mathcal{G}^-$ la tribu $\mathcal{G}^- = \bigcap_{t<0} \theta_t(\mathcal{G})$ (c'est la tribu co-progressive de [F-L]).

Lemme : Si T est le temps d'entrée dans un ensemble $\mathcal{G}^-$-mesurable, alors la tribu $b_T(\mathcal{G})$ est une tribu du futur.

Démonstration : Montrons d'abord : $1_{]\!]-\infty,T-s[\![} \in b_T(\mathcal{G})$; ce qui équivaut à :

(*) $$\forall s > 0,\ \exists Z \in \mathcal{G} \text{ tel que } 1_{]\!]-\infty,T-s[\![} = Z\, 1_{]\!]-\infty,T[\![}\,.$$

Soit $T(\omega) = D_A(\omega) = \inf\{t;(t,\omega) \in A\}$ (avec $A \in \mathcal{G}^-$).

On pose : $B = \bigcup_{n>0} \bigcap_{0<u<s+1/n} \theta_u({}^cA)$; on a l'égalité $1_{]\!]-\infty,T-s[\![} = 1_{]\!]-\infty,T[\![}\, 1_B$.

A partir de là on continue comme dans la démonstration du théorème I.6 de [F-L], pour établir successivement que :

· B est $\mathcal{G}$-analytique.

· Il existe un ensemble $\mathcal{G}$-mesurable tel que : $\{1_B \neq 1_C\}$ est $\lambda\otimes P$-négligeable ;

· Si on pose $Z_t(\omega)=\liminf_{s\to t} \operatorname{ess}_{s>t} 1_C(s,\omega)$; Z est $\mathcal{G}$-mesurable et : $Z1_{]\!]-\infty,T[\![}$ est indistinguable de $1_{]\!]-\infty,T-s[\![}$; et on a la propriété (*).

Ensuite, par un passage à la limite, on montre que $\forall s > 0$, $1_{]\!]-\infty,T-s]\!]} \in b_T(\mathcal{G})$.

On définit alors la famille de processus $\mathcal{B}$ par :

$$\mathcal{B} = \{Z\, 1_{]\!]-\infty,T-s]\!]} \ ;\ Z \ \ \mathcal{G}\text{-mesurable et càglàd, } s > 0\} \ ;$$

$\mathcal{B}$ est formée de processus càglàd, elle est stable par les opérateurs θ_s,s > 0 et d'après ce qui précéde, elle est incluse dans $b_T(\mathcal{G})$. Il est alors clair qu'elle engendre $b_T(\mathcal{G})$; $b_T(\mathcal{G})$ est donc une tribu du futur.

La proposition suivante est alors immédiate :

Proposition II.3 : Si T est un temps d'arrêt de $\mathcal{H}$ strictement supérieur à η, et un temps d'entrée dans un ensemble $\mathcal{G}^-$-mesurable, alors :

$$X \text{ est } (\mathcal{H},\mathcal{G})\text{-markovien} \Rightarrow b_T(X) \text{ est } (\mathcal{H},b_T(\mathcal{G}))\text{-markovien.}$$

Sur l'espace canonique décrit plus bas, les temps de cette proposition sont ceux égaux p.s. aux temps terminaux habituels.

4) *Retournement :*

On introduit la propriété $(\hat{b})$ symétrique de (b) : (cf. définition de (b) au paragraphe I.4).

$$(\hat{b}) \quad X_t(\omega)\neq b \Leftrightarrow (t,\omega) \in]\!]-\infty,\xi\}.$$

Si Z est un processus défini sur $\mathbb{R}\times\Omega$ à valeurs dans un espace quelconque, on note $\check{Z}$ le processus défini sur $\mathbb{R}\times\Omega$ à valeurs dans ce même espace défini par : $\check{Z}_t(\omega) = Z_{-t}(\omega)$.

Soit $\check{\mathcal{H}}$, (resp.$\check{\mathcal{G}}$) la tribu du futur (resp. du passé) engendrée par les processus $\check{Z}$ où Z est $\mathcal{H}$-mesurable (resp. $\mathcal{G}$-mesurable).

Théorème II.4 : On suppose que X vérifie les propriétés (b) et $(\hat{b})$, alors : X est un processus $(\mathcal{H},\mathcal{G})$-markovien $\Leftrightarrow$ $\check{X}$ est un processus $(\check{\mathcal{G}},\check{\mathcal{H}})$-markovien.

Démonstration : Quitte à retourner les processus et les tribus, il suffit de vérifier que si X est $(\mathcal{H},\mathcal{G})$-markovien, alors, pour tout Z $\mathcal{H}$-mesurable, $\Pi^{\mathcal{G}}(Z)$ est de la forme f(X).

Or, la $(\mathcal{H},\mathcal{G})$ propriété de Markov de X implique que, pour tout Z $\mathcal{G}$-mesurable, $\Pi^{\mathcal{H}}(Z)$ est de la forme f(X) et par conséquent est $\mathcal{G}$-mesurable. On en déduit que la tribu $\mathcal{H} \cap \mathcal{G}$ est indistinguable de $\sigma(X)$.

D'autre part, l'intervalle $\{\eta,\xi\} = \{X \in \mathbb{E}\}$ est $\mathcal{H} \cap \mathcal{G}$-mesurable, par conséquent on peut appliquer le théorème de commutation des projections (th. 2.2 de [F-L]) les projections $\Pi^{\mathcal{H}}$ et $\Pi^{\mathcal{G}}$ commutent donc, en particulier: Pour tout Z $\mathcal{H}$-mesurable, $\Pi^{\mathcal{G}}(Z)$ est $\mathcal{H}$-mesurable, donc $\mathcal{H} \cap \mathcal{G}$-mesurable et donc il existe une fonction mesurable f telle que $\Pi^{\mathcal{G}}(Z) = f(X)$.

Remarque : Pour les processus de Markov indéxés par $\mathbb{R}_+$ (identifiés ici aux processus valant "a" pour les temps négatifs),les opérations habituelles de translation par un temps d'arrêt, de meurtre à un temps de retour, de retournement à un temps de retour sont des compositions de translations, oublis et retournements au sens précédent.

III - Relation avec la propriété de Markov forte et les quasi-processus de Weil.

Soit (P_t) un semi-groupe sous-markovien droit sur $\mathbb{E}$, que l'on prolonge en un semi-groupe markovien au moyen du point cimetière b. Nous supposerons dans la suite que (P_t) est borélien, ce que l'on peut toujours faire quitte à utiliser la topologie de Ray (cf Sharpe [S]).

Nous dirons qu'un processus $(\mathcal{H},\mathcal{G})$-markovien admet (P_t) comme *semi-groupe de transition* si $\forall s \geq 0$, $\forall f$ $\mathcal{E}^b$-mesurable , $P_s f(X.) 1_{\{X \in \mathbb{E}^b\}}$ est une version de la projection sur $\mathcal{H}$ du processus $\mathcal{G}$-mesurable $(t,\omega)\to f(X_{s+t}(\omega))$.

On définit l'espace Ω des trajectoires de la manière suivante : un élément de Ω est une application de $\mathbb{R}$ dans $\mathbb{E}_a^b$ qui vérifie les propriétés suivantes :

· si $\omega^{-1}(\mathbb{R}) \neq \emptyset$ alors $\omega^{-1}(\mathbb{R})$ est un intervalle de $\mathbb{R}$ bornée ou non de la forme $[\alpha(\omega),\beta(\omega)[$ ou $]\alpha(\omega),\beta(\omega)[$, ω est continue à droite en tout point de cet intervalle, ω vaut a à gauche de $\alpha(\omega)$ et vaut b à droite de $\beta(\omega)$.

· si $\omega^{-1}(\mathbb{R}) = \emptyset$ alors ω est soit la fonction constante égale à a, soit la fonction constante égale à b. On pose dans le premier cas $\alpha(\omega) = \beta(\omega) = +\infty$ et dans le second $\alpha(\omega) = \beta(\omega) = -\infty$.

On désigne par X_t les applications coordonnées sur Ω : $X_t(\omega)=\omega(t)$, et par θ_s les opérateurs de translation sur Ω définis par : $X_t(\theta_s(\omega)) = X_{t+s}(\omega)$, $s,t \in \mathbb{R}$.

Ainsi, on a: $\theta_s(X)_t(\omega) = X_t(\theta_s(\omega)) = X_{t+s}(\omega)$ (voir la notation $\theta_s(X)$ dans le §I).

On désigne par $\mathcal{F}^\circ_{t+} = \bigcap_{\varepsilon>0} \sigma(X_s\ -\infty<s<t+\varepsilon)$ la filtration naturelle de X continue à droite, $\mathcal{F}^\circ = \mathcal{F}^\circ_{+\infty}$,et par $\hat{\mathcal{F}}^\circ_0 = \sigma(X_s,\ s\geq 0)$.

On considère une probabilité **P** sur $(\Omega,\mathcal{F}^\circ)$. La notation des complétées des tribus précédentes par rapport à la probabilité **P** sera obtenue en supprimant les signes "°".

On définit la tribu $\mathcal{G}$ sur $\mathbb{R}\times\Omega$, engendrée par les processus càd indistinguables d'un processus de la forme $(\mathfrak{z}\circ\theta_t(\omega))$ (noté $\mathfrak{z}\circ\theta.$ dans la suite) où $\mathfrak{z}$ est une variable $\hat{\mathcal{F}}^\circ_0$-mesurable. $\mathcal{G}$ est une tribu du futur (cf [F-L]).

Nous allons considérer deux tribus du passé : la *tribu du passé* $\mathcal{L}$, engendrée par les processus càdlàg <u>nuls hors de $\{X\in\mathbb{E}^b\}$</u> indistinguables d'un processus de la forme $(\mathfrak{z}\circ\theta.)$ avec $\mathfrak{z}$ $\mathcal{F}^\circ_{0+}$-mesurable.

On vérifie facilement qu'un temps aléatoire est un temps d'arrêt de $\mathcal{L}$ ssi son graphe est inclus dans $\{X\in\mathbb{E}^b\}$ et qu'il est p.s. égal à un temps intrinsèque T au sens de Weil [W][2]

[2] Notant $\mathcal{F}^*_{t+}$ la complétée universelle de $\mathcal{F}^\circ_{t+}$. T est un temps intrinsèque au sens de Weil si $\forall t, \{T \leq t\} \in \mathcal{F}^*_t$ et $T\circ\theta_t = T-t$.

Pour ces temps, la trace sur $\{T\in\mathbb{R}\}$ de la tribu $\mathcal{L}_T$, est celle de :

$\bigcap_{\varepsilon>0} \sigma(X_{T+s}\ ;\ -\infty<s<\varepsilon)$.

La *tribu optionnelle* $\mathcal{O}$, engendrée par les processus càdlàg nuls hors de $\{X\in\mathbb{E}^b\}$ et adaptés à la filtration $(\mathcal{F}_{t+})_{t\in\mathbb{R}}$.

Les temps d'arrêt de $\mathcal{O}$ sont les temps d'arrêt T de la filtration $(\mathcal{F}_{t+})$ [i.e. $\{T\le t\}\in\mathcal{F}_{t+}$, $\forall t\in\mathbb{R}$] dont le graphe est inclus dans $\{X\in\mathbb{E}^b\}$. Pour ces temps la trace sur $\{T\in\mathbb{R}\}$ de la tribu $\mathcal{O}_T$, est celle de la tribu habituelle $\mathcal{F}_{T+}$ $(A\in\mathcal{F}_{T+} \Leftrightarrow A\cap\{T\le t\}\in\mathcal{F}_{t+},\ \forall t\in\mathbb{R})$.

On a donc :

Proposition III.1 : Le processus X est $(\mathcal{O},\mathcal{G})$-markovien de semi-groupe de transition (P_t) si et seulement si : $\forall s\ge 0$, $\forall f$, $\forall T$ temps d'arrêt de $(\mathcal{F}_{t+})$

$$E[f(X_{s+T})|\ \mathcal{F}_{T+}]\ 1_{\{X_T\in\mathbb{E}^b,T\in\mathbb{R}\}} = P_s f(X_T)\ 1_{\{X_T\in\mathbb{E}^b,T\in\mathbb{R}\}}\ ;$$

ce qui est exactement *la propriété de Markov forte* usuelle, et on l'appellera ainsi par la suite.

Nous allons maintenant comparer la propriété de Markov forte à la $(\mathcal{L},\mathcal{G})$-propriété de Markov. Remarquons tout d'abord qu'en employant la technique utilisée dans [F-L] on peut généraliser le résultat classique de théorie générale des processus [DM p.198] "Tout processus Z càd adapté à $(\mathcal{F}_{t+})$ est optionnel", en "si X vérifie (b) alors X est $\mathcal{L}$-mesurable".

Proposition III.2 : Si X est un processus fortement markovien, et X vérifie l'hypothèse (b), alors X est $(\mathcal{L},\mathcal{G})$-markovien.

Démonstration : Cela découle immédiatement de la remarque ci-dessus et du fait que $\mathcal{L}\subseteq\mathcal{O}$.

La proposition suivante montre que dans le cadre des processus de Markov indexés par $\mathbb{R}_+$ (ou $\mathbb{R}_+^*$), les deux propriétés de Markov coïncident.

Proposition III.3 : Lorsque $\alpha = 0$ p.s., X est fortement markovien si et seulement si il est $(\mathcal{L},\mathcal{G})$-markovien.

Cette proposition découle du lemme général suivant:

Lemme : Lorsque $\alpha=0$ p.s. les tribus $\mathcal{O}$ et $\mathcal{L}$ sont égales.

Démonstration : On a d'abord $\forall t_0 \geq 0$, $\forall f$, $f(X_{t_0}) 1_{]\!]t_0,+\infty[\![}$ est indistinguable de $f(X_{\alpha+t_0})1_{]\!]\alpha+t_0,+\infty[\![} = \mathfrak{z}\circ\theta_.$ où $\mathfrak{z}$ est la variable $f(X_{\alpha+t_0})1_{\alpha+t_0<0} \in \mathcal{F}^\circ_{0-}$.

La famille de processus $\{f(X_{t_0}) 1_{]\!]t_0,+\infty[\![}$, f mesurable, $t_0 \in \mathbb{R}\}$ engendre la tribu prévisible habituelle associée à la filtration $(\mathcal{F}_{t+})_{t\in\mathbb{R}}$, donc, par un argument de classe monotone, on voit que si Z est prévisible, il existe une variable $\mathfrak{z}$ $\mathcal{F}^\circ_{0-}$-mesurable telle que $Z\,1_{]\!]0,+\infty[\![}$ et $\mathfrak{z}\circ\theta_.$ soient indistinguables.

Soit maintenant Y un processus càdlàg $\mathcal{O}$-mesurable et nul hors de $\{X\in\mathbb{E}^b\}$, Y^- est prévisible, il existe donc une variable $\mathfrak{z}$ $\mathcal{F}^\circ_{0-}$-mesurable telle que:

$Y^- 1_{]\!]0,+\infty[\![}$ et $\mathfrak{z}\circ\theta_.$ soient indistinguables.

En posant : $\mathfrak{z}^{(+)} = \liminf \mathfrak{z}\circ\theta_{1/n}$ on a : $\mathfrak{z}^{(+)} \in \mathcal{F}^\circ_{0+}$ et, Y et $\mathfrak{z}^{(+)}\circ\theta_.$ sont indistinguables sur $\{X \in \mathbb{E}^b\}$ et donc : $Y \in \mathcal{L}$.

D'où : $\mathcal{O} \subseteq \mathcal{L}$ et $\mathcal{O} = \mathcal{L}$.

Nous montrons maintenant que dans le cas général, la notion de processus $(\mathcal{L},\mathcal{G})$-markovien correspond à celle de quasi-processus au sens de Weil [W].

Théorème III.4 : Lorsque $\mathcal{L}$ vérifie l'hypothèse (b), les deux propositions suivantes sont équivalentes :

(i) X est $(\mathcal{L},\mathcal{G})$-markovien de semi-groupe de transition (P_t).

(ii) X est un quasi-processus markovien au sens de Weil [W][3], de semi-groupe de transition (P_t), et :

$\forall f$ $\mathcal{E}^b$-mesurable, $\forall s\geq 0$, $E[f(X_{\alpha+s})|\mathcal{L}_\alpha]\, 1_{\{-\infty<\alpha<+\infty, X_\alpha\in\mathbb{E}\}} = P_s f(X_\alpha)\, 1_{\{-\infty<\alpha<+\infty, X_\alpha\in\mathbb{E}\}}$.

Démonstration :

(i) ⇒ (ii) La propriété de Markov en α est vraie par définition. Ensuite, pour tout $n > 0$ et tout temps intrinsèque T, T+1/n est un temps intrinsèque dont le graphe est inclus dans $]\!]\alpha,+\infty[\![$ donc dans $\{X \in \mathbb{E}^b\}$: T+1/n

[2] Rappelons que cette dernière propriété signifie que pour tout temps intrinsèque T, le processus $(X_{T+t})_{t>0}$, défini sur $\Omega' = \{T\in\mathbb{R}\}$, est fortement markovien de semi-groupe de transition (P_t).

est un temps d'arrêt de $\mathcal{L}$ d'où $]\!]T,+\infty[\![= \cup\, [\![T+1/n,+\infty[\![\in \mathcal{L}$.

Le processus $X^{(T)}$ défini sur Ω par

$$X_t^{(T)}(\omega) = X_{t+T}(\omega) \text{ si } t>0 \text{ et } T(\omega) \in \mathbb{R}$$

$$= a \text{ sinon,}$$

s'obtient à partir de X par la composition d'une translation et d'un oubli, $X^{(T)}$ est donc $(\mathcal{L}^{(T)},\mathcal{G}^{(T)})$-markovien pour un couple de tribus $(\mathcal{L}^{(T)},\mathcal{G}^{(T)})$. Il est alors facile de vérifier que la réalisation canonique de $X^{(T)}$ est $(\mathcal{L},\mathcal{G})$ markovienne.

Ce processus ayant son temps de naissance en 0 (ou en $+\infty$ mais cela ne change rien!), la proposition précédente nous dit qu'il est fortement markovien; d'autre part, la translation et l'oubli ne modifiant pas le semi-groupe de transition, (P_t) reste celui de $X^{(T)}$.

(ii) $\Rightarrow$ (i) Il suffit de vérifier que pour toute fonction continue positive f de $\mathcal{E}^b$ et pour tout s positif, $\Pi^{\mathcal{L}}(f(X_{s+.}))$ est indistinguable de $P_sf(X.)$ sur $\{X \in E^b\}$. Cette propriété est vraie en α par hypothèse. Ensuite on prend une suite de temps intrinsèques à valeurs dans $\mathbb{R}\cup\{+\infty\}$ décroissant vers α, (T_n) (l'existence d'une telle suite est assurée par la propriété (b)).

D'après la propriété de Markov forte de $(X_{t+T_n})_{t>0}$, $P_sf(X_{.+T_n})$ est càd sur $]\!]0,+\infty[\![$, autrement dit, $P_sf(X.)$ est càd sur $]\!]T_n,+\infty[\![$ et par recollement sur $]\!]\alpha,+\infty[\![$.

D'autre part, $\Pi^{\mathcal{L}}(f(X_{s+.}))$ est càd sur $]\!]\alpha,+\infty[\![$ en tant que projection sur $\mathcal{L}$ d'un processus càd (cf [F-L]). Ces deux processus sont indistinguables sur $]\!]\alpha,+\infty[\![$ si et seulement si

Pour tout Z $\mathcal{L}$-mesurable :

$$E\left[\int_\alpha^{+\infty} Z_t\, \pi^{\mathcal{L}}(f(X_{.+s}))_t dt\right] = E\left[\int_\alpha^{+\infty} Z_t\, P_sf(X_t)dt\right].$$

Le membre de gauche de l'égalité vaut :

$E\left[\int_\alpha^{+\infty} Z_t\ f(X_{s+t})dt\right]$ (voir [F-L] pour cette propriété des projections sur une tribu du passé).

Il reste donc à vérifier l'égalité suivante pour tout Z $\mathcal{L}$-mesurable

$$(*) \qquad E\left[\int_\alpha^{+\infty} Z_t\, P_s f(X_t) dt\right] = E\left[\int_\alpha^{+\infty} Z_t\, f(X_{s+t}) dt\right]$$

Les processus X^n définis par :

$$X_t^n(\omega) = X_t(\omega) \text{ si } t > T_n(\omega)$$

$$= a \text{ sinon,}$$

s'obtiennent à partir de $(X_{t+T_n})_{t>0}$ par une translation . X^n est donc markovien relativement à un couple de tribus $(\mathcal{H}^n, \mathcal{G}^n)$.

D'après la propriété de projection sur $\mathcal{H}^n$, on a l'égalité pour tout Z $\mathcal{H}^n$-mesurable

$$E\left[\int_{T_n}^{+\infty} Z_t\, f(X_{s+t}) dt\right] = E\left[\int_{T_n}^{+\infty} Z_t\, \pi^{\mathcal{H}^n}(f(X_{.+s}))_t dt\right],$$

$$= E\left[\int_{T_n}^{+\infty} Z_t\, P_s f(X_t) dt\right] \text{ d'après la propriété de Markov de } X^n.$$

On a donc l'égalité (*) pour de tels Z.

Prenons maintenant Z de la forme:

$Z = \prod_{i=1}^{m} f_i(X_{.-s_i})$ (les s_i étant positifs). posons $Z^n = \prod_{i=1}^{m} f_i(X^n_{.-s_i})$;

Z^n est $\mathcal{H}^n$-mesurable et: $\forall\omega, \forall t \notin \{\alpha(\omega)+s_i ; 1<i<n\}$:$Z_t(\omega) = \lim Z_t^n(\omega)$; en particulier : $Z_t(\omega) = \lim Z_t^n(\omega)$ sauf pour un ensemble de (t,ω) $\lambda \otimes$ P-négligeable.

Et donc par passage à la limite, on obtient l'égalité (*) pour de tels Z. Par un argument de classe monotone, on étend ce résultat aux Z de la forme $\mathfrak{z}\circ\theta_.$, avec $\mathfrak{z} \in \mathcal{F}^\circ_{0-}$, puis aux Z càdlàg tels que Z_- soit de la forme précédente, (car $\{Z \neq Z_-\}$ est $\lambda \times P$ négligeable) ce qui est le cas pour tout Z càdlàg $\mathcal{L}$-mesurable.

Un nouvel argument de classe monotone donne l'égalité (*) pour tout Z $\in \mathcal{L}$.

IV - Représentation des mesures sur l'espace d'états par les fonctionnelles additives.

Définitions : On appelle *ensemble* P-*polaire* un borélien A de $\mathbb{E}$ tel que l'ensemble $\{\omega \in \Omega ; \exists t, X_t(\omega) \in A\}$ est P-négligeable.

On appelle *fonctionnelle additive droite* un noyau $A(\omega,ds)$ sur $\mathbb{R}$ vérifiant les propriétés suivantes :

(i) $\forall\omega \in \Omega,\ A(\omega, \{t,\ X_t(\omega) \notin \mathbb{E}\}) = 0$

(ii) $\forall\omega \in \Omega\ \forall a,b,t \in \mathbb{R}\ ,\ a \le b,$

$$A(\theta_t(\omega),\ [a,b[) = A(\omega,[a+t,b+t[) \text{ et :}$$

(iii) l'application : $\omega \to A(\omega,[a,b[)$ est mesurable pour la tribu $\sigma(X_s,\ a \le s < b)^*$ (complétée universelle de $\sigma(X_s,\ a \le s < b)$).

On se donne dans tout ce paragraphe une probabilité $\mathbf{P}$ sur $(\Omega,\mathcal{F}°)$ sous laquelle les propriétés (b),$(\hat{b})$ et la $(\mathcal{L},\mathcal{G})$-propriété de Markov de X sont vérifiées.

Théorème IV.1 : Si μ est une mesure sur $\mathbb{E}$ finie ne chargeant pas les ensembles P-polaires alors il existe une fonctionnelle additive droite unique à une l'indistinguabilité près, telle que :

Pour toute fonction f $\mathcal{E}$-mesurable, on a: $\mu(f) = E\left[\int_{-\infty}^{+\infty} f(X_s)A(\omega,ds)\right]$.

Avant de passer à la démonstration on rappelle (cf [F-L]) qu'une mesure aléatoire intégrable de $\mathcal{L}$ (resp. de $\mathcal{G}$) est une famille de mesures sur $\mathbb{R}$ $(A(\omega,ds))_{\omega\in\Omega}$ telle que le processus $A(\omega,]-\infty,t])$ est $\mathcal{L}$-mesurable (resp. $A(\omega,[t,+\infty[)$ est $\mathcal{G}$-mesurable) et tel que $E[A(\omega,\mathbb{R})]<+\infty$.

On démontrera en appendice le lemme :

Lemme : Une mesure aléatoire intégrable de $\mathcal{L}$ et de $\mathcal{G}$ est indistinguable d'une fonctionnelle additive droite.

Démonstration du théorème IV.1 :

Après avoir remplacé la tribu optionnelle par la tribu $\mathcal{L}$, on copie mot à mot la démonstration d'Azéma [A page 491], nous nous contentons ici de rappeler son idée essentielle:

On pose pour tout processus Z : $L(Z) = \mu(f)$ où f est choisie telle que f(X.) soit une version de $\Pi^{\mathcal{L}}\Pi^{\mathcal{G}}(Z)$, alors :

Il existe une mesure aléatoire intégrable unique $A(\omega,ds)$ telle que:

$$\forall Z \quad L(Z) = E\left[\int_{-\infty}^{+\infty} Z_s A(\omega,ds)\right].$$

Et A est à la fois une mesure aléatoire intégrable de $\mathcal{L}$ et de $\mathcal{G}$.
A partir de là on applique le lemme pour conclure.

<u>*Remarque*</u> : D'après la proposition III.3 ce résultat contient le résultat analogue pour les processus de Markov indexés par $\mathbb{R}_+$; en fait nous allons voir que c'est une forme équivalente du résultat de Dynkin et Getoor [D-G] complété par Fitzsimmons [F] faisant intervenir les mesures de Kuznetsov.

V – Complément : extension du travail précédent aux mesures de Palm associées aux mesures dissipatives.

1) *Extension à certaines mesures σ-finies* :

Sur un espace mesurable quelconque $(\Omega,\mathcal{F})$, $\mathcal{H}$ étant une tribu du passé on peut encore projeter sur $\mathcal{H}$, relativement à une mesure Q à condition que Q vérifie la propriété :

(c) pour tout temps d'arrêt T de $\mathcal{H}$ à valeurs dans $\mathbb{R}\cup\{+\infty\}$, la trace de Q sur $(\{\eta<+\infty\},\mathcal{H}_T)$ est σ-finie.

Et cette propriété est équivalente à la propriété suivante :

(c') Il existe une suite décroissante T_n de temps d'arrêt de $\mathcal{H}$ à valeurs dans $\mathbb{R}\cup\{+\infty\}$ telle que : $\bigcup_n \,]\!]T_n\,;+\infty[\![\,=\, \{\eta,+\infty[\![$ et pour tout n, la trace de Q sur $(\{\eta<+\infty\},\ \mathcal{H}_{T_n})$ est σ-finie.

Lorsque (c) est vérifiée, on peut étendre naturellement la notion d'espérance conditionnelle sur $\mathcal{H}_T\cap\{\eta<+\infty\}$, et à fortiori sur $\mathcal{H}_T\cap\{T\in\mathbb{R}\}$ (car $\{T\in\mathbb{R}\}\subseteq\{\eta<+\infty\}$).
En procédant par recollement à l'aide d'une suite de temps d'arrêt (T_n) choisie comme dans la propriété (c') on établit que si $\mathcal{H}$ vérifie (c) la proposition de projection O reste vraie.
On procède de même pour étendre tout ce qui a été fait dans [F-L] sous l'hypothèse que $\mathcal{H}$ vérifie (c) ; pour la commutation des projections, il faut supposer que la tribu du futur vérifie la propriété $(\hat{c})$ symétrique de (c).
A partir de là, quitte à adjoindre la propriété (c) à la propriété (b) et $(\hat{c})$ à $(\hat{b})$, tout ce qui a été fait ici reste vrai sans changement à l'exception du théorème III.4 : Pour étendre celui-ci, il faut supposer outre que $\mathcal{L}$ vérifie (b) et (c), que la réalisation canonique du processus défini sur $\{T\in\mathbb{R}\}$, $(X_{t+T})_{t>0}$ vérifie aussi (c) pour tout temps intrinsèque T [sans quoi la

quasi-propriété de Markov n'a pas de sens!][4]. Dans ce cas, (qui sera en vigueur par la suite) la $(\mathcal{L},\mathcal{G})$ propriété de Markov reste équivalente à la quasi-propriété de Markov au sens de Weil, plus la propriété de Markov forte en α.

2) Lien avec les mesures de Palm associées aux mesures dissipatives et application du théorème IV.1 à ces mesures :

On appelle ensemble invariant toute partie A de Ω $\mathcal{F}^*$-mesurable ($\mathcal{F}^*$ est la complètée universelle de $\mathcal{F}°$) telle que pour tout réel t on ait $\theta_t(A) = A$. On notera $\mathcal{I}^*$ la tribu formée des ensembles invariants.

Nous allons énoncer un résultat essentiellement dû à Weil [W] modernisé par Fitzsimmons [F] et [DMM] que nous prolongeons ici de façon maximale en α à la manière de Fitzsimmons.

Théorème V.1 : Soit (P_t) un semi-groupe sousmarkovien droit et borélien, ν une mesure dissipative pour (P_t), σ-finie :

Il existe une unique mesure sur $(\Omega,\mathcal{I}^*)$ notée Π_ν telle que :

(i) $\Pi_\nu(\alpha=+\infty) = \Pi_\nu(\beta=-\infty) = 0$

(ii) pour toute fonction mesurable positive f sur $\mathbb{F}$,

$$\nu(f) = \Pi_\nu\left(\int_{-\infty}^{+\infty} f(X_s)ds\right)$$

(iii) sous Π_ν, X vérifie la quasi-propriété de Markov de Weil.

(iv) si on note η la mesure sur $(\mathbb{E},\mathcal{E})$ définie par :

$$\eta(f) = \Pi_\nu(f(X_\alpha)\ ;\ \alpha > -\infty \text{ et } X_\alpha \in \mathbb{E})$$

alors $\nu - \eta U$ est harmonique (U est le noyau potentiel de (P_t)).

4 Posant:Pour tout temps intrinsèque T et tout réel t strictement positif: $\mathcal{F}^T_t = \sigma(X_{T+s}; 0<s<t)$; On peut voir facilement que la réalisation canonique de $(X_{t+T})_{t>0}$ vérifie (c) si et seulement si: $\forall t>0$, la trace de **Q** sur $((T<+\infty),\mathcal{F}^T_t)$ est σ-finie. C'est une hypothèse un peu plus faible que celle de Weil [W].

Il faudra remarquer les faits suivants :

1) La propriété (ii) implique que pour tout temps intrinsèque T la réalisation canonique du processus $(X_{T+t})_{t>0}$ vérifie aussi (c) et donc que la quasi-propriété de Markov a un sens.

2) De même les propriétés (i) et (ii) impliquent encore que sous Π_ν les propriétés (b),(c),$(\hat{b})$ et $(\hat{c})$ sont vérifiées (remarquer au préalable que ces propriétés ne concernent que les traces des mesures sur la tribu $\mathcal{I}^*$ des invariants).

3) Selon [DMM], il existe un temps stationnaire S (i.e. S est $\mathcal{F}^*$-mesurable et $\forall t$, $\forall \omega$ $S\circ\theta_t(\omega) = S(\omega)-t$) fini Π_ν-p.s. et on peut prolonger Π_ν sur $(\Omega,\mathcal{F}^*)$ par une mesure notée $\Pi_\nu^{(S)}$ en posant: $\forall A\in\mathcal{F}^*$, $\Pi_\nu^{(S)}(A) = \Pi_\nu(A\circ\theta_S)$.

On obtient donc, avec le théorème III.4 que X vérifie la $(\mathcal{L},\mathcal{G})$ propriété de Markov sous $\Pi_\nu^{(S)}$ et nous pouvons appliquer le théorème IV.1 pour obtenir :

Corollaire : Si μ est une mesure finie sur $\mathbb{E}$ ne chargeant pas les ensembles Π_ν-polaires alors il existe une fonctionnelle additive droite $A(\omega,ds)$ telle que pour toute fonction f $\mathcal{E}$-mesurable, $\mu(f) = \Pi_\nu\left[\int_{-\infty}^{+\infty} f(X_s)\, A(\omega,ds)\right]$.

Remarques et compléments :

1) le lecteur pourra vérifier qu'une mesure Q sur $(\Omega,\mathcal{F}^*)$ coïncide avec une mesure de la forme Π_ν sur la tribu $\mathcal{I}^*$ si et seulement si les trois propriétés suivantes sont vérifiées simultanément:

i) Q vérifie (b) (c) $(\hat{b})$ $(\hat{c})$

ii) sous Q, X est $(\mathcal{L},\mathcal{G})$-markovien de semi-groupe de transition (P_t)

iii) $-\infty < \alpha < +\infty$ et $X_\alpha \in \mathbb{E} \Leftrightarrow -\infty<\alpha<+\infty$ et la limite à droite de X_t en α existe dans $\mathbb{E}$ au sens de la topologie de Ray associée à (P_t).

2) On peut étendre le corollaire à une mesure Σ-finie.

3) [DMM] démontrent que si ν est une mesure dissipative σ-finie, Q_ν la mesure de Kuznetsov associée et $A(\omega,ds)$ une fonctionnelle additive droite, alors on a la relation :

$$\forall \mathfrak{z} \in \mathcal{F}^\circ, \mathfrak{z} \geq 0, \quad \Pi_\nu\left[\int_{-\infty}^{+\infty} \mathfrak{z}\circ\theta_s \, A(\omega,ds)\right] = Q_\nu\left[\int_{]0,1]} \mathfrak{z}\circ\theta_s \, A(\omega,ds)\right]$$

Comme les ensembles Π_ν et Q_ν-polaires coïncident, le corollaire précédent s'exprime à l'aide de Q_ν, ce qui redonne le résultat de Dynkin-Getoor [D-G], généralisé par Fitzsimmons [F].

De plus, comme le fait ce dernier, on peut généraliser ce résultat à une mesure excessive σ-finie quelconque en faisant un détour par le semi-groupe $P'_t = e^{-t}P_t$.

Appendice

On se place sur l'espace canonique $(\Omega,\mathcal{F}^\circ)$, **P** est une probabilité sur cet espace et on suppose que X vérifie la $(\mathcal{L},\mathcal{G})$ propriété de Markov sous **P**. On va démontrer ici le

Lemme : Toute mesure aléatoire intégrable $A(\omega,ds)$ de $\mathcal{L}$ et de $\mathcal{G}$ est indistinguable d'une fonctionnelle additive droite (F.A.D. en abrégé).

Lemme préliminaire : $\forall\varepsilon\geq 0$, si Z est un processus $\mathcal{L}\cap\theta_{-\varepsilon}(\mathcal{G})$ mesurable alors il est indistinguable d'un processus de la forme $\mathfrak{z}\circ\theta.$ où $\mathfrak{z}$ est une v.a. mesurable pour le tribu $\mathcal{F}^\circ_{[-\varepsilon,0]} = \sigma(X_s\ ;\ -\varepsilon \leq s \leq 0)$.

Démonstration : On projette d'abord sur $\mathcal{L}$ un processus de la forme : $Z=\mathfrak{z}_1\circ\theta.\times\mathfrak{z}_2\circ\theta$ avec $\mathfrak{z}_1 \in \mathcal{F}^\circ_{[-\varepsilon,0]}$ et $\mathfrak{z}_2 \in \mathcal{F}^\circ_0$, on obtient un processus de la forme : $\mathfrak{z}\circ\theta$ où $\mathfrak{z} \in \mathcal{F}^\circ_{[-\varepsilon,0]}$. La propriété $\Pi^{\mathcal{L}}(Z) = \mathfrak{z}\circ\theta$. Avec $\mathfrak{z} \in \mathcal{F}^\circ_{[-\varepsilon,0]}$ est étendue par un argument de classe monotone à tout processus $\theta_{-\varepsilon}(\mathcal{G})$-mesurable, et donc en particulier à un processus Z $\mathcal{L}\cap\theta_{-\varepsilon}(\mathcal{G})$-mesurable pour lequel $\Pi^{\mathcal{L}}(Z) = Z1_{\{X\in E^b\}}$.

De plus, pour un tel processus, il existe un réel k telque Z=k sur {X=a}, (c'est vrai pour tout processus $\mathcal{L}$-mesurable!), d'où :

$Z = k1_{\{X=a\}}+\Pi^{\mathcal{L}}(Z)$, et on en déduit une v.a. $\mathfrak{z} \in \mathcal{F}^\circ_{[-\varepsilon,0]}$ telle que $\mathfrak{z}\circ\theta.$ soit indistinguable de Z.

<u>*Démonstration du lemme*</u> : Soit $A(\omega,ds)$ une mesure aléatoire intégrable de $\mathcal{L}$ et de $\mathcal{G}$. Il est facile de voir que $A(\omega,.)$ est portée par l'ensemble $\{s, X_s(\omega)\in\mathbb{E}\}$;

Soit $A = B + C$ sa décomposition en somme d'une mesure ponctuelle et d'une mesure diffuse.

a) $\forall\omega \in \Omega$, $\forall\Gamma \in \mathcal{B}(\mathbb{R})$,

$$B(\omega,\Gamma) = \sum_{t\in\Gamma} A(\omega,\{t\})$$

Le processus $Z_t(\omega) = A(\omega,\{t\})$ est $\mathcal{L} \cap \mathcal{G}$-mesurable et nul hors de $\{X \in \mathbb{E}\}$; $B(\omega,ds)$ est donc une mesure aléatoire de $\mathcal{L}$ et de $\mathcal{G}$ et, Z est indistinguable d'un processus de la forme $f(X)$ où f est une fonction $\mathcal{E}$-mesurable. Posons $B°(\omega,\Gamma) = \sum_{t\in\Gamma} f(X_t(\omega))$. $B°(\omega,ds)$ est une F.A.D. indistinguable de $B(\omega,ds)$.

b) Reste à établir que C est indistinguable d'une F.A.D. :

C est une mesure aléatoire diffuse, intégrable et différence de deux mesures aléatoires de $\mathcal{L}$ et de $\mathcal{G}$, c'est donc aussi une mesure aléatoire de $\mathcal{L}$ et de $\mathcal{G}$.

Quitte à remplacer $C(\omega,ds)$ par : $1_{C(\omega,\mathbb{R})<+\infty}$ $C(\omega,ds)$, on peut supposer de plus que $C(\omega,\mathbb{R})<+\infty$ pour tout ω ; de plus $C(\omega,.)$ est portée par $\{s, X_s(\omega)\in\mathbb{E}\}$.

Pour tout $n\in\mathbb{N}$, notons Z^n le processus : $(t,\omega)\mapsto C(\omega,]t-1/n,t])\wedge 1$, Z^n est un processus $\mathcal{L} \cap \theta_{-1/n}(\mathcal{G})$-mesurable.

D'après le lemme précédent, il est indistinguable d'un processus (borné!) Z^{on} de la forme $\mathfrak{z}^n\circ\theta.$ où $\mathfrak{z}^n$ est une variable $\mathcal{F}_{[-1/n,0]}$-mesurable.

De plus, quitte à remplacer $\mathfrak{z}^n$ par $\mathfrak{z}^n 1_{\{X_0\neq a, X_{-1/n}\neq b\}}$, on peut choisir Z^{on} nul hors de $[\alpha(\omega),\beta(\omega)+1/n]$. D'autre part, pour tout ω, la mesure $C(\omega,.)$ est limite faible de la suite de mesures nZ^n_t dt.

Notons D l'événement $\mathbb{P}$-négligeable :

$$D = \{\omega \ ; \ \exists t \in \mathbb{R}, \exists n \in \mathbb{N} \ \text{ tels que } Z^{on}_t(\omega) \neq Z^n_t(\omega)\}.$$

On a : $\forall\omega \notin D$, $C(\omega,.)$ est la limite faible des mesures $nZ^{on}_t(\omega)dt$.

On emploie alors les méthodes développées dans l'appendice de [D-G] qu'on rappelle ici brièvement :

On se donne une limite médiale de Mokobodzki (cf [D-M]) .

I désigne dans la suite un intervalle borné de $\mathbb{R}$, on pose:

l'application $\omega \to A(\omega,]a,b[)$ est $\sigma(X_u ; a<u<b)^*$-mesurable.

Et donc, en prenant C° la partie diffuse de A°, on obtient une F.A.D.

b) d'autre part, pour $\omega \notin D$: $C(\omega,.) = A^\circ(\omega,.)$; $C(\omega,.)$ étant diffuse, on a aussi $C(\omega,.) = C^\circ(\omega,.)$.

Et donc C et C° sont indistinguables.

Références

[A] **Azéma** : Théorie générale des processus et retournement du temps. *Ann. sci. de l'ENS. 4ème s.t6, 1973.*

[C-W] **Chung et Walsh** : To reverse a Markov process. *Acta Math. 123, 1970, p.225-251.*

[D-G] **Dynkin et Getoor** : Additive functionals and entrance laws. *J. Fonct. Anal. 62. 1985, p.221-265.*

[D-M] **Dellacherie et Meyer** : Probabilités et potentiel.*T1.Hermann 1980.*

[DMM] **Dellacherie, Maisonneuve et Meyer** : Probabilités et potentiel. *T5. Hermann. 1992.*

[F] **Fitzsimmons** : Homogeneous random measures and a weak order for the excessive measures of a Markov processes. *Trans. Amer. Math. Soc.303 (1987), p. 431-478.*

[F-M] **Fitzsimmons et Maisonneuve** : Excessive measures and Markov processes with random birth and death. *Proba. th. rel. fields. Vol. 72, 1986.*

[K] **Kuznetsov** : Construction of Markov Processes with random times of birth and death. *Th. Prob. Appl. Vol.18 ; 1974.*

[L] **Lenglart** : Tribus de Meyer et théorie des processus. *Sem. Prob.XIV. LN in Math. n° 784, Springer-Verlag.*

[F-L] Lenglart et Fourati : Tribus homogènes et commutation des projections. *Sem. Prob.XXI.LN in Math.1247.Springer-Verlag 1987.*

[S] Sharpe : General Theory of Markov processes. *Académic Press,1988.*

[W] Weil : Quasi-processus et énergie. *Sem. Prob. V. LN in Math 191. Springer-Verlag 1971.*

Non-linear Wiener-Hopf theory, 1: an appetizer

by

David Williams

The first use of non-linear Wiener-Hopf theory occurred in Neveu's very important paper (Neveu, 1987) which was further developed in (Kaj & Salminen, 1993). See also (Jansons & Rogers, 1992) for further probabilistic insight.

In case the Reader thinks that Section 2 is just too bizarre, I emphasize that the method gives useful information on integral curves for dynamical systems of which (4.1) is the simplest possible case.

1. The classical problem

1(a) *Infinitesimal generator of a Markov chain.* Let

$$\mathbf{g}(t) = \{g_{ij}(t) : i, j \in J\}$$

be the transition matrix at time t of a Markov chain Y on a finite set J:

$$g_{ij}(t) := \mathbb{P}_i[Y(t) = j].$$

Then,

$$\mathbf{g}(u+t) = \mathbf{g}(u)\mathbf{g}(t), \quad \dot{\mathbf{g}}(t) = \frac{d\mathbf{g}}{dt} = G\mathbf{g}(t), \qquad \text{where } G := \dot{\mathbf{g}}(0).$$

Here, G is the infinitesimal generator, or Q-matrix, of $\{\mathbf{g}(t)\}$.

1(b) *Time transformation of a Markov chain.* Suppose now that $X = \{X(t)\} = \{X_t\}$ is a Markov chain, possibly of finite lifetime, with Q-matrix Q on a finite set E. Let $V : E \to \mathbb{R} \setminus \{0\}$. We shall also write V for the diagonal matrix $\operatorname{diag}(V(i))$. Set $E^{\pm} := \{i \in E : \pm V(i) > 0\}$ and partition $V^{-1}Q$ as

$$V^{-1}Q = \begin{array}{c} \\ E^+ \\ E^- \end{array} \!\!\begin{array}{c} \begin{array}{cc} E^+ & E^- \end{array} \\ \begin{pmatrix} A & B \\ -C & -D \end{pmatrix} \end{array}.$$

For $t \geq 0$, define

$$\varphi(t) := \int_0^t V(X_s)ds, \quad \tau_t^{\pm} := \inf\{u : \pm\varphi(u) > t\}, \quad Y_t^{\pm} := X(\tau_t^{\pm}).$$

Note that $Y_t^+ \in \mathbb{E}^+$ (or Y_t^+ is in the coffin state). It is clear that Y^+ is a Markov chain on E^+; let G^+ be its Q-matrix. We think of a $^+$Universe in which $^+$Observers use $S^+(t) := \sup\{\varphi(s) : s \leq t\}$ as a clock; such observers can see our particle only for part of the time when it is in E^+. Define

$$g_{jk}^+(t) := \mathbb{P}_j(Y_t^+ = k) \quad (j, k \in E^+),$$
$$h_{ik}^+(t) := \mathbb{P}_i(Y_t^+ = k) \quad (i \in E^-,\ k \in E^+),$$

$\mathbb{P}_i$ referring to the situation when $X_0 = i$. With H^+ as the $E^- \times E^+$ matrix with $H^+_{ik} := h^+_{ik}(0)$, we have

$$\dot{\mathbf{g}}^+(t) = G^+\mathbf{g}^+(t), \quad \mathbf{h}^+(t) = H^+\mathbf{g}^+(t),$$

the latter being probabilistically obvious.

Decompositions according to the time and nature of the first jump from $E^\pm$ to $E^\mp$ yield:

$$\mathbf{g}^+(t) = e^{tA} + \int_0^t e^{(t-s)A}B\mathbf{h}^+(s)ds,$$

whence

$$\dot{\mathbf{g}}^+(t) = A\mathbf{g}^+(t) + B\mathbf{h}^+(t),$$

and

$$\mathbf{h}^+(t) = \int_0^\infty e^{uD}C\mathbf{g}^+(u+t)dt = e^{-tD}\int_t^\infty e^{vD}C\mathbf{g}^+(v)dv.$$

whence

$$\dot{\mathbf{h}}^+(t) = -C\mathbf{g}^+(t) - D\mathbf{h}^+(t).$$

We therefore have

$$G^+ = A + BH^+, \quad H^+G^+ = -C - DH^+.$$

With the obvious notations G^- and H^-, we have

$$V^{-1}Q\begin{pmatrix} I^+ & H^- \\ H^+ & I^- \end{pmatrix} = \begin{pmatrix} I^+ & H^- \\ H^+ & I^- \end{pmatrix}\begin{pmatrix} G^+ & 0 \\ 0 & -G^- \end{pmatrix},$$

where $I^\pm$ is the identity matrix on $E^\pm$. Of course, this does not in itself tell us what $G^\pm$ and $H^\pm$ are.

For a survey of this type of problem and some continuous-state-space generalizations, see Williams (1991).

2. The non-linear version: the simplest case

2(a) *Infinitesimal-generator function of a continuous-parameter branching process.* Consider the following model. At time 0, there is one particle. Each particle dies at constant rate K, and at the moment of death gives birth to n particles ($n = 0, 1, 2\ldots$ or ∞) with probability p_n. The usual independence assumptions hold.

Let $N(t)$ be the number of particles alive at time $t \geq 0$, and set

$$g(t,\theta) := \mathbb{E}\theta^{N(t)} = \sum_{0\leq n\leq\infty} \theta^n \mathbb{P}(N(t) = n) \quad (0 \leq \theta < 1).$$

Then, by the branching property,

$$g(u+t,\theta) = \mathbb{E}\mathbb{E}\left(\theta^{N(t+u)} \,\middle|\, N(u)\right) = \mathbb{E}g(t,\theta)^{N(u)} = g(u, g(t,\theta)),$$

whence, on differentiating with respect to u and setting $u = 0$,

$$\dot{g}(t,\theta) = G\big(g(t,\theta)\big),$$

where G is the 'infinitesimal-generator function'

$$G(\theta) := \frac{\partial}{\partial u} g(u,\theta)\Big|_{u=0} = K \sum_{0 \le n \le \infty} p_n(\theta^n - \theta), \quad (0 \le \theta < 1).$$

As $\theta \uparrow 1$, $G(\theta) \uparrow G(1-) = -Kp_\infty \le 0$.

2(b) *Multi-time substitution.* In Our universe, we have a branching process of particles which (while alive) live on the real line. Each particle is either of type $+$ or of type $-$. A particle of type $+$ has constant velocity $b_+ > 0$, while a particle of type $-$ has constant velocity $-b_- < 0$. A particle of type $i = \pm$ dies at rate $K_i := \lambda_i + q_i + r_i$; at its moment of death it

either just dies with probability λ_i/K_i,
or gives birth to a particle of the opposite type with probability q_i/K_i,
or gives birth to two particles of its own type with probability r_i/K_i.

A particle's child is born at the same position as its parent. The usual independence assumptions hold. The λ, q, r parameters satisfy

$$\lambda_\pm \ge 0, \quad q_\pm > 0, \quad r_\pm > 0.$$

Particle k has lifespan denoted by $[\beta_k, \delta_k)$. We write $S_k^+(u)$ for the furthest right (the 'sup') which particle k *or any of its ancestors* has reached by time u.

Observers in the $^+$Universe can see only particles of type $+$, and they use our S_k^+ as a clock in which to observe particle k. This means that our particle k will feature as a $^+$Particle in the $^+$Universe if and only if $S_k^+(\delta_k) > S_k^+(\beta_k)$, and will then have $^+$Lifespan $[S_k^+(\beta_k), S_k^+(\delta_k))$; it will be a $^+$Child of the (almost certainly) unique ancestor in our system with death time $S_k^+(\beta_k)$. A few moments' thought will convince you that, if we ignore the positions of the $^+$Particles, then the evolution of the *number* of $^+$Particles in $^+$Time is a branching process just like that in Section 2(a).

For $t \ge 0$, define

$$\begin{aligned} N^+(t) &:= \text{Number of } {}^+\text{Particles alive at } {}^+\text{Time } t \\ &= \#\left\{k : t \in [S_k^+(\beta_k), S_k^+(\delta_k))\right\}. \end{aligned}$$

We wish to study, for $0 \le \theta < 1$,

$$g^+(t,\theta) := \mathbb{E}_+ \theta^{N^+(t)},$$
$$h^+(t,\theta) := \mathbb{E}_- \theta^{N^+(t)}.$$

Here, $\mathbb{E}_\pm$ is the expectation given that at time 0 in our universe, there is just one particle of type $\pm$ at position 0 on $\mathbb{R}$. We shall have

$$\dot{g}^+(t,\theta) := \frac{\partial}{\partial t} g(t,\theta) = G^+\big(g^+(t,\theta)\big),$$

where G^+ is the infinitesimal-generator function for N^+, and

$$h^+(t,\theta) = \mathbb{E}_-\left(\theta^{N^+(t)}\,\middle|\,N^+(0)\right) = \mathbb{E}_- g^+(t,\theta)^{N^+(0)} = H^+(g^+(t,\theta)).$$

It is therefore enough to find G^+ and H^+.

3. Calculations.

By decomposing the behaviour according to the first jump, we have

$$g^+(t,\theta) = \theta e^{-tK_+/b_+} + \int_0^{t/b_+} e^{-K_+ s}\left\{\lambda_+ + q_+ h^+(t - b_+ s,\theta) + r_+ g^+(t - b_+ s,\theta)^2\right\} ds,$$

equivalently,

$$e^{K_+ t/b_+} g^+(t,\theta) = \theta + b_+^{-1}\int_0^t e^{K_+ v/b_+}\left\{\lambda_+ + q_+ h^+(v,\theta) + r_+ g^+(v,\theta)^2\right\} dv,$$

whence

$$b_+ \dot{g}^+(t,\theta) = \lambda_+[1 - g^+(t,\theta)] + q_+[h^+(t,\theta) - g^+(t,\theta)] + r_+[g^+(t,\theta)^2 - g^+(t,\theta)].$$

Similarly,

$$h^+(t,\theta) = \int_0^\infty e^{-K_- t}\left\{\lambda_- + q_- g^+(t + b_- u,\theta) + r_- h^+(t + b_- u,\theta)^2\right\} du,$$

whence it follows easily that

$$-b_- \dot{h}^+(t,\theta) = \lambda_-[1 - h^+(t,\theta)] + q_-[g^+(t,\theta) - h^+(t,\theta)] + r_-[h^+(t,\theta)^2 - h^+(t,\theta)].$$

Now recall that $h^+(t,\theta) = H^+(g^+(t,\theta))$.

4. First conclusions. We have shown that $y = H^+(x)$ $(0 \le x < 1)$ *is an integral curve for the dynamical system*

$$\text{(4.1)(a)} \qquad +b_+ x'(t) = \lambda_+[1 - x(t)] + q_+[y(t) - x(t)] + r_+[x(t)^2 - x(t)],$$
$$\text{(4.1)(b)} \qquad -b_- y'(t) = \lambda_-[1 - y(t)] + q_-[x(t) - y(t)] + r_-[y(t)^2 - y(t)].$$

This curve connects the point $(0, H^+(0))$ on the (closed) left-hand edge of the unit square to the point $(1, H^+(1-))$ on the right-hand edge. Note that H^+ is a convex function on $[0,1)$: indeed, all its derivatives are non-negative.

The symmetry of (4.1) shows that, with obvious notation relating to the $^-$Universe, *the curve* $x = H^-(y)$ $(0 \le y < 1)$ *is also an integral curve for the system (4.1)*, and it links the top and bottom edges of the unit square. Any point at which the two 'probabilistic' curves $y = H^+(x)$ and $x = H^-(y)$ cross within the unit square must be an equilibrium point of the system (4.1). But, of course, more is true: if these curves do cross, they dominate the topology of the system (4.1).

This note *is* only an appetizer; and the only remaining thing for it to do is to provide some pictures of some integral curves for the system (4.1) for certain sets of parameters. Of course, there are lots of fascinating questions, even for this simplest case. The n-dimensional and infinite-dimensional generalizations are still more interesting. But that is work to share with research students, one of whom, Owen Lyne, has already done nice work on related travelling-wave problems. If we put $u(t,a) = 1 - x(a-ct)$ and $v(t,a) = 1 - y(a-ct)$ in (4.1), we obtain a system with first equation

$$\frac{\partial u}{\partial t} = -(c+b_+)\frac{\partial u}{\partial a} - (\lambda_+ + q_+)u + q_+ v + r_+ u[1-u],$$

containing a drift term, a death term, a mutation term, and a logistic term.

5. Pictures. In all the pictures,

$$b_+ = b_- = 1.$$

The unit square and the two probabilistic curves within it are shown via solid lines.

Figure (a). Here, $\lambda_+ = 1$, $\lambda_- = 1$, $q_+ = 1$, $q_- = 1$, $r_+ = 2$, $r_- = 2$. The point $(\frac{1}{2}, \frac{1}{2})$ is an equilibrium point, through which our probabilistic curves pass with slopes $2-\sqrt{3}$ and $2+\sqrt{3}$. But we see that the two probabilistic curves are here part of the same orbit which encloses the point $(1,1)$. As the computer picture suggests, for these parameters, the point $(1,1)$ is a *centre*: all orbits close to it are periodic. The Reader will discover a simple proof of this assertion for these particular parameters. We have $\mathbb{P}_-[N^+(0) = \infty] > 0$.

Figure (b). Here, $\lambda_+ = 0$, $\lambda_- = 0$, $q_+ = 1$, $q_- = 4$, $r_+ = 4$, $r_- = 4$. Our particles spend more time in state $+$ than in state $-$, but the birth rates are high enough to guarantee that $\mathbb{P}_+[N^-(0) = \infty] > 0$. Orbits started near $(1,1)$ spiral in towards $(1,1)$.

Figure (c). Here, $\lambda_+ = 0$, $\lambda_- = 0$, $q_+ = 1$, $q_- = 4$, $r_+ = 0.4$, $r_- = 0.4$. Large-deviation theory makes it *intuitively clear* that since $r_+ + r_- < (\sqrt{q_+} - \sqrt{q_-})^2$, we must have $\mathbb{P}_-[N^+(0) = \infty] = 0$. But it is *topologically obvious* that, since there are no equilibria in the interior of the unit square when $\lambda_+ = \lambda_- = 0$, and since $\mathbb{P}_+[N^-(0) = 0] > 0$, we must have $\mathbb{P}_-[N^+(0) = \infty] = 0$. In this case there are infinitely many integral curves connecting the top and bottom of the unit square, one going steeply (initial slope 5) up from $(0,0)$ and the others contained in the 'triangle' formed by the probabilistic curves.

Figure (d). Here, $\lambda_+ = 1$, $\lambda_- = 1$, $q_+ = 1$, $q_- = 1$, $r_+ = 1$, $r_- = 1$. This case is 'critical' in many ways.

Some integral curves for the system (4.1)

In each case, $b_+ = b_- = 1$

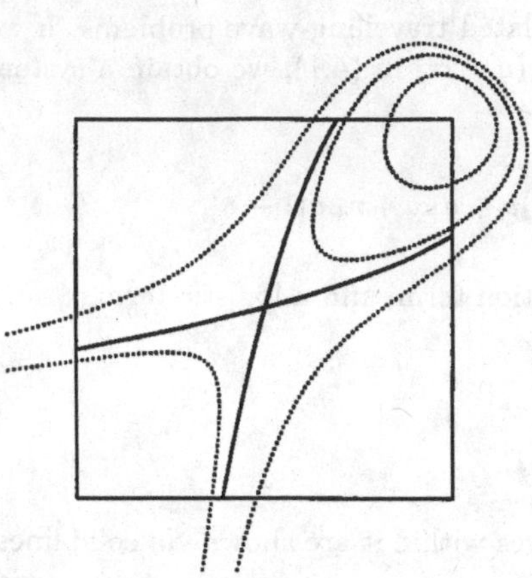

(a) $\lambda_+ = 1,\ \lambda_- = 1,\ q_+ = 1,\ q_- = 1.$
$r_+ = 2,\ r_- = 2.$

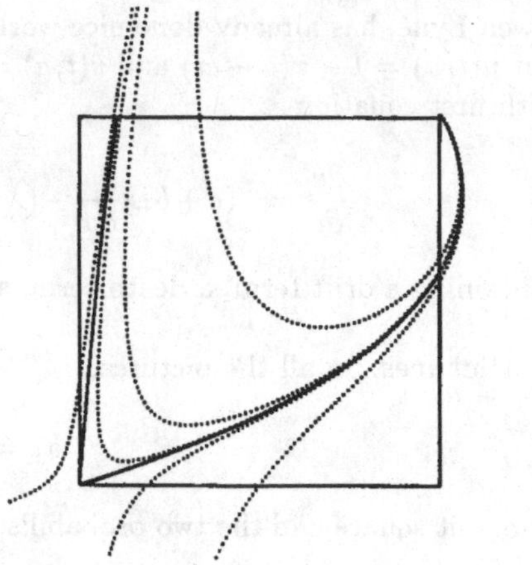

(b) $\lambda_+ = 0,\ \lambda_- = 0,\ q_+ = 1,\ q_- = 4.$
$r_+ = 4,\ r_- = 4.$

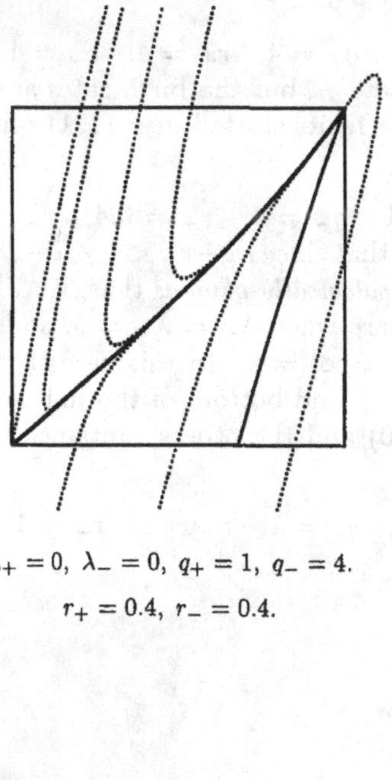

(c) $\lambda_+ = 0,\ \lambda_- = 0,\ q_+ = 1,\ q_- = 4.$
$r_+ = 0.4,\ r_- = 0.4.$

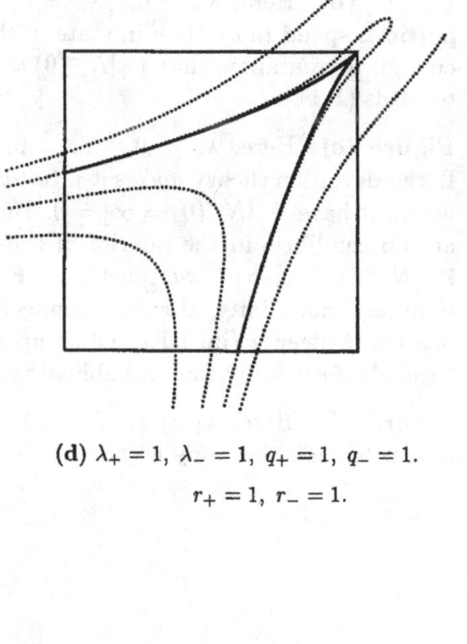

(d) $\lambda_+ = 1,\ \lambda_- = 1,\ q_+ = 1,\ q_- = 1.$
$r_+ = 1,\ r_- = 1.$

Acknowlegments. My thanks to Owen Lyne and to John Toland. Special thanks to Chris Rogers whose suggestions have resulted in a much cleaner exposition.

References

Jansons, K.M. & Rogers, L.C.G.R. 1992 Decomposing the branching Brownian path, Ann. Applied Prob. **2**, 973–986.

Neveu, J. 1987 Multiplicative martingales for spatial branching processes. *Seminar on Stochastic Processes* (ed. E. Çinlar, K.L. Chung and R.K. Getoor), Progress in Probability and Statistics **15**, pp. 223–241. Boston: Birkhauser.

Kaj, I. & Salminen, P. 1993 On a first passage problem for branching Brownian motions, *Ann. Applied Prob.* **3**, 173–185.

Williams, D. 1991 Some aspects of Wiener-Hopf factorization, *Phil. Trans. R. Soc. Lond. A* **335**, 593–608.

From an Example of Lévy's

Yukuang Chiu
Department of Mathematics, 0112
University of California, San Diego
La Jolla, CA 92093-0112, USA

The motive of this paper is to prove completely an assertion of P. Lévy [3], who claimed that for each positive integer n, there exists a polynomial F_n of degree n such that the Wiener integral with respect to Brownian motion $\{B(u); 0 \le u\}$

$$X(t) = \int_0^t F_n\left(\frac{u}{t}\right) dB(u)$$

is again a Brownian motion and

$$\mathbf{B}(X;t) \underset{\neq}{\subset} \mathbf{B}(B;t).$$

Here $\mathbf{B}(X;t)$ is the σ–field generated by $\{X(s); s \le t\}$.

Over the last few years, the non-canonical representation of Brownian motion of this kind related has been of interest to many authors, in particular Th. Jeulin & M. Yor [4,5,6,7] and M. Hitsuda [2].

Let $X_0(t) = B(t), (t \ge 0)$ be a standard Brownian motion. In this paper, our precise purpose is to construct a sequence of Brownian motion $\{X_n(t); t \ge 0\} (n \ge 0)$ such that $X_n(t)$ can be represented as a Wiener integral

$$X_n(t) = \int_0^t F_n\left(\frac{u}{t}\right) dB(u), \quad (n \ge 1).$$

Here $F_n(t)$ is a polynomial of degree n in t. And if $\mathbf{M}(X_n;t)$ is a linear span of $\{X_n(u); 0 \le u \le t\}$ in $L^2(\Omega, P)$, then for all $t > 0$,

(1) $$\mathbf{M}(X_{n+1};t) \underset{\neq}{\subset} \mathbf{M}(X_n;t), \ (n \ge 0),$$

and further, for all $t > 0$ and $n \ge 1$,

(2) $$\int_0^t F_n\left(\frac{u}{t}\right) u^k du = 0, \ k = 1, 2, \ldots, n.$$

Let $F_n(t) = a_n t^n + a_{n-1} t^{n-1} + \cdots + a_1 t + a_0$. We know by calculating the covariance that $X_n(t)$ is a Brownian motion if and only if the coefficients of $F_n(t)$ satisfy the following equations

(3) $$\begin{cases} \frac{a_0 a_0}{1} + \frac{a_0 a_1}{2} + \cdots + \frac{a_0 a_n}{n+1} = 1, \\ \frac{a_1 a_0}{2} + \frac{a_1 a_1}{3} + \cdots + \frac{a_1 a_n}{n+2} = 0, \\ \cdots\cdots\cdots\cdots\cdots\cdots \\ \frac{a_n a_0}{n+1} + \frac{a_n a_1}{n+2} + \cdots + \frac{a_n a_n}{2n+1} = 0. \end{cases}$$

In the simplest case

$$\begin{cases} a_n = \frac{2n+1}{n}, a_0 = -\frac{n+1}{n}, \\ a_1 = a_2 = \cdots = a_{n-1} = 0. \end{cases}$$

is a solution of equation (3).

Theorem 1 *If* $P_n(t) = \frac{2n+1}{n}t^n - \frac{n+1}{n}$ *and if* $F_n(t)$ *are defined by the following recursive formula*

$$\begin{cases} F_1(t) = P_1(t) \\ F_n(\frac{u}{t}) = F_{n-1}(\frac{u}{t}) - \int_u^t F_{n-1}(\frac{u}{\tau})\frac{\partial}{\partial\tau}P_n(\frac{\tau}{t})d\tau, \ (n \geq 2) \end{cases},$$

then $F_n(t)$ *satisfies (2), the coefficients of* $F_n(t)$ *are given by*

$$a_k = (-1)^{n+k}\binom{n}{k}\binom{n+1+k}{n}, \ k = 0, 1, \cdots, n$$

and

$$X_n(t) := \int_0^t F_n\left(\frac{u}{t}\right) dB(u), \ (n \geq 1)$$

are Brownian motions satisfying condition (1). Further, $X_n(t)$ *and* $X_{n+1}(t)$ *are related by*

$$X_{n+1}(t) = \int_0^t P_{n+1}\left(\frac{u}{t}\right) dX_n(u), \ (n \geq 0). \tag{4}$$

In order to prove the theorem, we prepare the following lemma.

Lemma 1 *If* $s < n$, *we have*

$$\sum_{k=0}^{n}(-1)^k\binom{n}{k}\binom{n+k}{s} = 0.$$

To prove this, we note

$$\begin{aligned} \frac{1}{s!}\left(\frac{d}{dx}\right)^s((1+x)^n x^n) &= \frac{1}{s!}\left(\frac{d}{dx}\right)^s\left(\sum_{k=0}^{n}\binom{n}{k}x^{n+k}\right) \\ &= \sum_{k=0}^{n}\binom{n}{k}\binom{n+k}{s}x^{n+k-s}. \end{aligned}$$

The result follows by letting $x = -1$.

The validity of the coefficients of $F_n(t)$ can be established by mathematical induction. The assertion is trivial for $n = 1$. Suppose the assertion holds for n. Now using lemma 1 and then noting

$$\binom{n}{k}\binom{n+1+k}{n}\left(1 - \frac{2(n+1)+1}{n+1-k}\right) = -\binom{n+1}{k}\binom{n+2+k}{n+1},$$

$$\binom{n}{k}\binom{n+1+k}{n}\frac{2(n+1)+1}{n+1-k} = \frac{2(n+1)+1}{n+1}\binom{n+1}{k}\binom{n+1+k}{n},$$

we see that

$$\begin{aligned} F_{n+1}\left(\frac{u}{t}\right) &= F_n\left(\frac{u}{t}\right) - \int_u^t F_n\left(\frac{u}{\tau}\right)\frac{\partial}{\partial\tau}P_{n+1}\left(\frac{\tau}{t}\right)d\tau \\ &= \sum_{k=0}^{n}(-1)^{n+k}\binom{n}{k}\binom{n+1+k}{n}\left(1 - \frac{2(n+1)+1}{n+1-k}\right)\left(\frac{u}{t}\right)^k \\ &\quad + \sum_{k=0}^{n}(-1)^{n+k}\binom{n}{k}\binom{n+1+k}{n}\frac{2(n+1)+1}{n+1-k}\left(\frac{u}{t}\right)^{n+1} \end{aligned}$$

$$= \sum_{k=0}^{n}(-1)^{n+1+k}\binom{n+1}{k}\binom{n+2+k}{n+1}\left(\frac{u}{t}\right)^k + \binom{2n+3}{n+1}\left(\frac{u}{t}\right)^{n+1}$$
$$= \sum_{k=0}^{n+1}(-1)^{n+1+k}\binom{n+1}{k}\binom{n+2+k}{n+1}\left(\frac{u}{t}\right)^k.$$

which shows the assertion holds for coefficients of $F_{n+1}(t)$.

We next show (2). By the recursive formula of F_n, we obtain

$$\begin{aligned}\int_0^t F_n\left(\frac{u}{t}\right)u^k du &= \int_0^t F_{n-1}\left(\frac{u}{t}\right)u^k du - \int_0^t\int_u^t F_{n-1}\left(\frac{u}{\tau}\right)\frac{\partial}{\partial\tau}P_n\left(\frac{\tau}{t}\right)u^k d\tau du\\ &= \int_0^t F_{n-1}\left(\frac{u}{t}\right)u^k du - \int_0^t d\tau\frac{\partial}{\partial\tau}P_n\left(\frac{\tau}{t}\right)\int_0^\tau F_{n-1}\left(\frac{u}{\tau}\right)u^k du\end{aligned}$$

This equals zero if $k < n$ by induction; and when $k = n$, this becomes

$$\int_0^t F_{n-1}\left(\frac{u}{t}\right)u^n du - \left[P_n\left(\frac{\tau}{t}\right)\int_0^\tau F_{n-1}\left(\frac{u}{\tau}\right)u^n du\Big|_{\tau=0}^t - \int_0^t P_n\left(\frac{u}{t}\right)u^n du\right]$$
$$= \int_0^t P_n\left(\frac{u}{t}\right)u^n du = 0.$$

which is what we needed to prove.

Again we easily verify, by mathematical induction, that

$$\int_0^1 F_n(u)du = \frac{(-1)^n}{n+1}.$$

Thus we have proved, in combination with the previous equation, that the coefficients of F_n are another solution to equation (3).

Now if we write

$$X_n(t) = \int_0^t F_n\left(\frac{u}{t}\right)dB(u),$$

then by the above argument, $X_n(t)$ is again a Brownian motion. The differential of $X_n(t)$, by Itô's formula [1], is seen to be

$$dX_n(u) = dB(u) + \int_0^u \frac{\partial}{\partial u}F_n\left(\frac{\tau}{u}\right)dB(\tau)du.$$

Therefore

$$\begin{aligned}&\int_0^t P_{n+1}\left(\frac{u}{t}\right)dX_n(u)\\ =\ &\int_0^t P_{n+1}\left(\frac{u}{t}\right)dB(u) + \int_0^t\left\{P_{n+1}\left(\frac{u}{t}\right)\int_0^u\frac{\partial}{\partial u}F_n\left(\frac{\tau}{u}\right)dB(\tau)\right\}du\\ =\ &\int_0^t P_{n+1}\left(\frac{u}{t}\right)dB(u) + \int_0^t dB(\tau)\int_\tau^t P_{n+1}\left(\frac{u}{t}\right)\frac{\partial}{\partial u}F_n\left(\frac{\tau}{u}\right)du\\ =\ &\int_0^t P_{n+1}\left(\frac{u}{t}\right)dB(u) +\\ &\qquad\int_0^t dB(\tau)\left\{F_n\left(\frac{\tau}{u}\right)P_{n+1}\left(\frac{u}{t}\right)\Big|_{u=\tau}^t - \int_\tau^t F_n\left(\frac{\tau}{u}\right)\frac{\partial}{\partial u}P_{n+1}\left(\frac{u}{t}\right)du\right\}\\ =\ &\int_0^t F_n\left(\frac{u}{t}\right)dB(u) - \int_0^t\left\{\int_u^t F_n\left(\frac{u}{\tau}\right)\frac{\partial}{\partial\tau}P_{n+1}\left(\frac{\tau}{t}\right)d\tau\right\}dB(u)\\ =\ &X_{n+1}(t)\end{aligned}$$

This establishes (4).

To show (1), let us fix $t_0 > 0$ and let $z = \int_0^{t_0} u^{n+1} dX_n(u)$. Now $z \in \mathbf{M}(X_n;\, t)$ and note that for all t such that $0 < t \leq t_0$,

$$E[X_{n+1}(t) \cdot z] = \int_0^t P_{n+1}\left(\frac{u}{t}\right) u^{n+1} du = 0.$$

This verifies (1). The proof of theorem is thus completed.

Remark 1 *This construction was suggested by P. Lévy in his book* [3] *and $F_1(t)$ was given there.*

Remark 2 *Although we have (1), for all $n > 0$, we notice,*

(5) $$\mathbf{M}(B; \infty) = \mathbf{M}(X_n; \infty).$$

This equation has the following interpretation. For each finite time t, as we have already seen, $\mathbf{B}(X_n; t)$ contains less information than $\mathbf{B}(B; t)$. Nevertheless, $\mathbf{B}(X_n; t)$ will "catch up" with $\mathbf{B}(B; t)$ by increasing time to infinity.

Acknowledgements

I sincerely thank M. Yor for his suggestions and preprints which have greatly enriched my knowledge on this topic.

References

[1] K. Itô, Introduction to Probability Theory. vol.3. (in Japanese) Iwanami-Shoten, (1978).

[2] M. Hitsuda, Gaussian innovations and integral operators. Proceedings of the Sixth USSR-Japan Symposium. Kiev, Aug. 5-10, 1991. World Scientific.

[3] P. Lévy, Processes Stochastiques et mouvement brownien. Gauthier-Villars. (1965), p. 299–304.

[4] Th. Jeulin, M. Yor, Filtration des ponts browniens et équations différentielles linéaires. Séminaire de Probabilités XXIV. Lect. Notes in Maths. 1426. Springer (1990), p. 227–265.

[5] Th. Jeulin, M. Yor, Une décomposition non-canonique du drap brownien. Séminaire de Probabilités XXVI. Lect. Notes in Maths. 1526. Springer (1992), p. 322–347.

[6] Th. Jeulin, M. Yor, Moyennes mobiles et semimartingales. Séminaire de Probabilités XXVII. Lect. Notes in Maths. 1557. Springer (1993), p. 53–77.

[7] Th. Jeulin, M. Yor, Sur certaines décompositions non-canoniques des diffusions réelles. Manuscrit non publié (Décembre 1990).

A HORIZONTAL LEVY PROCESS ON THE BUNDLE OF ORTHONORMAL FRAMES OVER A COMPLETE RIEMANNIAN MANIFOLD

by

David Applebaum

Department of Mathematics, Statistics and
Operational Research,
The Nottingham Trent University, Burton Street, Nottingham,
England, NG1 4BU.

ABSTRACT

We establish an existence theorem for a class of SDE's driven by Lévy processes on a manifold. As an application we consider an SDE driven by horizontal vector fields on the orthonormal frame bundle of a Riemannian manifold. The canonical projection of the solution of this equation on to the base is considered as a candidate for a "Lévy process on a Riemannian manifold".

1) Introduction

A Lévy process in $\mathbb{R}^n$ is essentially a stochastic process with independent and stationary increments. All the random variables comprising such a process are infinitely divisible. Conversely, as was shown by Itô, any infinitely divisible random variable can be embedded in a Lévy process (see [Itô], theorem 3.1). Hence Lévy processes can be characterised by the Lévy-Khintchine formula, through their characteristic functions. Alternatively, at the level of random variables we have the Lévy-Itô decomposition which exhibits every Lévy process as a combination of a Brownian motion, a Poisson point process (suitably renormalised) and a drift. Itô's formula then extends this decomposition to C^2 functions of the process (see e.g. [IkWa] for details). An obvious generalisation of the above would be to replace $\mathbb{R}^n$ by an arbitrary Lie group G. A major advance in this direction was the work of Hunt in 1956. He showed that there was a one to one correspondence beween convolution semigroups of probability measures p on G and a class of linear operators on $C^2(G)$, the correspondence being that each such operator generates a Markov semigroup with kernel p [Hun]. This can be shown to be equivalent to the Lévy-Khintchine formula when $G = \mathbb{R}^n$. More recently, H.Kunita and the present author have obtained an analogue of the Lévy-Itô decomposition for smooth functions of such processes. As in the abelian case the decomposition is obtained with the aid of a Brownian motion, a Poisson random measure and a drift ([ApKu]).

The aim of the present paper is to begin the work of generalising the above ideas to a Riemannian manifold M. The procedure we adopt herein is to imitate the well-known Eels-Elworthy construction for obtaining Brownian motion on a manifold M by canonical projection of a suitable process in the bundle of orthonormal frames O(M) (see e.g. [Elw], [IkWa]). An existence theorem for solutions of SDE's in compact manifolds has been established by Fujiwara however as O(M) is not compact we cannot use this result herein. §2 of this paper is then devoted to proving a general existence result for a class of SDE's driven by Lévy processes on not-necessarily compact manifolds. In §3 we specialise to the case of O(M) and construct a process which we call a horizontal Lévy process on O(M) which satisfies an SDE driven by a Lévy process taking values in the horizontal vector fields. We note that there is some similarity here with recent work by A.Estrade and M.Pontier who have constructed the horizontal lift of a manifold-valued càdlàg semimartingale [EsPo].We wish to go in the opposite direction and obtain a

Lévy process on the manifold as the canonical projection of the horizontal Lévy process.
We make two observations

(a) Intuitively a Lévy process on M is a combination of a drift, a Brownian motion on M and a Poisson point process which is constrained to jump along geodesics of arbitrary length. In order to ensure that there is a rich supply of the latter we will assume that M is geodesically complete.

(b) Brownian motion on a manifold is obtained by projection of the appropriate frame bundle-valued process on to the base manifold and is characterised by its generator (see e.g.[Eme] p. 62) which is of course the Laplace-Beltrami operator. In our case, the operator which is our natural candidate to be the generator of a "Lévy process on a manifold" exhibits a manifest time dependence which indicates that our process is not, in general, Markovian.

Note: - After writing this paper, it was brought to my attention that the existence and uniqueness result of §2 is in fact a special case of a more general construction given in [Coh]. I have however retained my original proof as I think there is some value in showing that the elegant method of [Elw] extends to the case of SDE's with jumps.

Notation: If M is a manifold, Diff(M) is the group of all diffeomorphisms of M with identity id. Every complete vector field Y on M generates a one-parameter subgroup of Diff(M) which we denote as $\{\mathrm{Exp}(tY), t \in \mathbb{R}\}$. If S is a topological space, $\mathcal{B}(S)$ will denote the Borel σ-algebra of S and $C_0(S)$ is the space of continuous functions on S which vanish at ∞. Einstein summation convention will be used throughout.

Acknowledgement: I would like to thank Anne Estrade, Michel Emery and Serge Cohen for helpful comments on an earlier version of this paper.

2. Existence of Solutions to an SDE on A Manifold Driven by a Lévy Process

Let $\mathcal{V}$ be a d-dimensional connected paracompact smooth manifold and let $Y_1, \ldots, Y_n$ be smooth complete vector fields on $\mathcal{V}$. We denote by $\mathcal{T}$ the linear span of $\{Y_1, \ldots, Y_n\}$ and make the assumption that every element of $\mathcal{T}$ is complete. We note that this condition is automatically satisfied if the Lie algebra $\mathcal{L}$ generated by $\mathcal{T}$ is finite-dimensional, but we do not make this latter assumption here. For each $x \in \mathbb{R}^n$, we denote by $\xi(x)$ the diffeomorphism of $\mathcal{V}$ defined by

$$\xi(x) = \operatorname{Exp}(x^j Y_j)$$

Let X be an n-dimensional Lévy process on some complete probability space $(\Omega, \mathfrak{F}, P)$ equipped with a filtration $(\mathfrak{F}_t, t \in \mathbb{R}^+)$. Hence there exists an m-dimensional Brownian motion $B = (B(t), t \in \mathbb{R}^+)$ where $m \leq n$ and a Poisson random measure N on $\mathbb{R}^+ \times (\mathbb{R}^n - \{0\})$ which is independent of B and has associated Lévy measure ν on $\mathbb{R}^n - \{0\}$ given by
$\mathbb{E}(N(t,G)) = t\,\nu(G)$ for all $t \in \mathbb{R}^+$, $G \in \mathcal{B}(\mathbb{R}^n - \{0\})$,
such that $X = (X^1, \ldots, X^n)$ has Lévy-Itö decomposition

$$X^j(t) = c^j t + \sigma^j_k B^k(t) + \int_0^{t+} \int_{|x| \geq 1} x^j N(dt, dx) + \int_0^{t+} \int_{|x| < 1} x^j \tilde{N}(dt, dx) \qquad \ldots(2.1)$$

$$\text{for } 1 \leq j \leq n, \; t \in \mathbb{R}^+$$

Here $c = (c^1, \ldots, c^n) \in \mathbb{R}^n$, $\sigma = (\sigma^j_k)$ is a real $(n \times m)$ matrix and $\tilde{N}$ is the compensated process $\tilde{N}(t,G) = N(t,G) - t\,\nu(G)$.

We introduce the $\mathcal{T}$-valued Lévy process $X_{\mathcal{T}} = (X_{\mathcal{T}}(t), t \in \mathbb{R}^+)$ given by

$$X_{\mathcal{J}}(t) \;=\; X^j(t)\,Y_j \qquad \text{for } t \in \mathbb{R}^+ \qquad \ldots(2.2)$$

We aim to study SDE's on V driven by $X_{\mathcal{J}}$.

To the extent that we are thus attempting to construct a "stochastic exponential" we might write such an SDE as

$$d\phi(t) \;=\; dX_{\mathcal{J}}(t)(\phi(t-)) \qquad \ldots(2.3)$$

$$\text{with} \qquad \phi(0) \;=\; p \quad \text{a.s.}$$

More precisely we are seeking a unique càdlàg adapted process $\phi = (\phi(t),\ 0 \le t \le \sigma)$ taking values in V with explosion time $\sigma \le \infty$ which satisfies the stochastic integro-differential equation

$$\begin{aligned} f(\phi(t)) \;=\; & f(p) + \int_0^t (Z_0 f)(\phi(s-))ds + \int_0^t (Z_k f)(\phi(s-))\circ dB^k(s) \\ & + \int_0^{t+}\int_{|x|\ge 1} [f(\xi(x)\phi(s-)) - f(\phi(s-))]\; N(ds,dx) \\ & + \int_0^{t+}\int_{|x|<1} [f(\xi(x)\phi(s-)) - f(\phi(s-))]\; \tilde{N}(ds,dx) \\ & + \int_0^{t+}\int_{|x|<1} [f(\xi(x)\phi(s-)) - f(\phi(s-)) - x^j\, Y_j f(\phi(s-))]\; \nu(dx)\, ds \end{aligned} \qquad \ldots(2.4)$$

for each $f \in C^\infty(V)$, $t \in \mathbb{R}^+$. Note that we have, for convenience, introduced the notation

$$Z_0 \;=\; c^j\, Y_j \quad , \quad Z_k \;=\; \sigma^j_k\, Y_j \qquad (1 \le k \le m)$$

For further examination of the relationship between (2.3) and (2.4), see pages 1105-6 of [ApKu].

Note: One gains a nice understanding of how (2.4) arise from (2.3) from the discretisation result of S.Cohen (§III.3 of [Coh]). Taking $0 = T^n_o < T^n_1 < \ldots < T^n_k \underset{k}{\to} \infty$ to be a sequence of stochastic partitions whose mesh tends to zero, it is shown that the solution of (2.4), up to its explosion time, is obtained by taking limits of the sequence defined by

$$\phi_0(t) = p$$
$$\phi_n(t) = \xi((X(t) - X(T^n_k-))\phi_n(T^n_k-) \quad \text{for } T^n_k \le t < T^n_{k+1}$$

In the case where $\dim \mathcal{L} < \infty$, (2.4) was solved in [ApKu] and it was shown that the solution defines a Lévy process on the transformation Lie group associated to $\mathcal{L}$ with $\sigma = \infty$ a.s. (see also [Est]). When $\mathcal{V}$ is compact equations of a similar type to (2.4) were studied in [Fuj] and the flat case $\mathcal{V} = \mathbb{R}^d$ can be found in [FuKu]. Further examples of classes of SDE's with jumps are investigated in [Rog] and [Coh].

Now consider (2.4) in the case $\mathcal{V} = \mathbb{R}^d$ and write $Y_j(x) = a^i_j(x)\partial_i$ for $1 \le j \le n$, $1 \le i \le d$ where each $a^i_j \in C^\infty(\mathbb{R}^d)$. The following result is established in §2.2 of [ApKu].

Theorem 2.1 Suppose that a^i_j, $\partial_k(a^i_j)$ and $\partial_k\partial_l(a^i_j)$ are bounded on $\mathbb{R}^d$ for all $1 \le j \le n$, $1 \le i,k,l \le d$, then (2.4) has a unique solution on $\mathbb{R}^d$.

The main result in this section is the following,

Theorem 2.2 There exists a unique maximal solution to (2.4) on $\mathcal{V}$.

Proof By Whitney's embedding theorem we can smoothly embed $\mathcal{V}$ into $\mathbb{R}^b$ where $b = 2d + 1$. Consider the corresponding extended version of (2.4) as an SDE on $\mathbb{R}^b$. Provided that the extension can be carried out in such a way that the hypothesis of theorem 2.1 is satisfied, we immediately have existence and uniqueness for the extended equation. It must now be shown that a solution of the extended SDE which has initial condition in V never leaves it. To do this we must construct the extension in a careful way. Our method is very closely based on that used by Elworthy in [Elw].

We begin by considering an equation closely related to (2.4)

$$f(\zeta(t)) = f(p) + \int_0^t (Z_0 f)(\zeta(s-))ds + \int_0^t (Z_j f)(\zeta(s-))\circ dB^j(s)$$

$$+ \int_0^{t+}\int_{|x|<1} [f(\xi(x)\zeta(s-)) - f(\zeta(s-))]\, \tilde{N}(ds,dx)$$

$$+ \int_0^{t+}\int_{|x|<1} [f(\xi(x)\zeta(s-)) - f(\zeta(s-)) - x^j Y_j f(\zeta(s-))]\, \nu(dx)\, ds$$

$$\ldots(2.5)$$

for $f \in C^\infty(V)$, $t \in \mathbb{R}^+$.

We will first show that (2.5) has a unique maximal solution on V. We recall some notation from [Elw]. Let $a: V \to \mathbb{R}^+$ be smooth and let $N(V)$ denote the normal bundle in $\mathbb{R}^b$ with base V and canonical projection π. If S is an open set in V, we define

$$M_a(S) = \bigcup_{q\in S} \{ y \in \mathbb{R}^b;\ ||y - q|| < a(q)\}$$

Note that $\{ y \in N(V),\ ||y|| < a(\pi(y))\}$ is diffeomorphic to the tubular neighborhood $M_a(V)$ in $\mathbb{R}^b$.

Now let G_0 be an open neighborhood of $p \in V$ with compact closure in $\mathbb{R}^b$ We denote as $B_R(0)$ the open ball of radius R about the origin in $\mathbb{R}^b$.

Choose $R > 0$ and (following [Elw]) define

$$a_R = \inf\{ a(q),\ q \in G_0 \cap B_{R+1}(0)\}.$$

Let $\lambda_R \in C^\infty(\mathbb{R}^b, \mathbb{R}^+)$ be such that $\mathrm{supp}(\lambda_R) \subseteq B_{R+1}(0)$ and $\lambda_R = 1$ in $B_R(0)$ and let $\mu_R \in C^\infty(\mathbb{R}^+)$ be such that $\mu_R(y) = 1$ if $|y| \leq \frac{1}{2} a_R^2$ and $\mu_R(y) = 0$ if $|y| > a_R^2$.

Let $\gamma : M_a(V) \to V$ be defined by $\gamma(y) = q$ where

$$||y - q|| = \inf\{ ||y - r||,\ r \in V\}.$$

Again as in [Elw], we extend the vector fields Y_j on G_0 to smoothvector fields $\overline{Y_j}$ on the whole of $\mathbb{R}^b$ $(1 \leq j \leq n)$ by

$$\overline{Y_j}(p) = 0, \quad p \notin M_a(G_0)$$

$$\overline{Y_j}(p) = \lambda_R(p)\ \mu(d(p,V)^2) Y_j(\gamma(p)), \quad p \in M_a(G_0)$$

where d is the usual (Euclidean) metric in $\mathbb{R}^b$.

Define the extended diffeomorphisms $\overline{\xi(x)} = \mathrm{Exp}(x^j\, \overline{Y_j})$. (For ease of notation, we have suppressed the dependence of $\overline{Y_j}$ and $\overline{\xi(x)}$ on R). We may now use theorem 2.1 to assert the existence and uniqueness of the extended SDE on $\mathbb{R}^b$.

Now choose S > R such that

$$S > \sup_{|x|<1} \sup_{q \in G_0} |\xi(x)\ G_0|$$

and let $g_S \in C^\infty(\mathbb{R}^b)$ be given by

$$g_S(y) = \lambda_S(y)\ \mu_S(d(y,V)^2)\ d(y,V)^2 \quad {}^1$$

Now consider equation (2.5) with $f = g_S$. As in [Elw], we have

$$\overline{Y_j}(q) g_S(q) = 0 \text{ for all } q \in B_R(0),\ 1 \leq j \leq n.$$

Define for $0 \leq t \leq 1$, $|x| < 1$ the automorphisms $j_t(x)$ of $C^\infty(\mathbb{R}^b)$ by

$$j_t(x)(f) = f \circ \overline{\xi(tx)}$$

[1] I am grateful to David Elworthy for this correction to [Elw].

then we have that

$$\frac{dj_t(x)f}{dt} = x^j j_t(x)(\overline{Y_j}(f))$$

However each $j_t(x)(\overline{Y_j}(g_s))(q) = \overline{Y_j}(\overline{\xi(tx)}q)\, g_s(\overline{\xi(tx)}q) = 0$ for $q \in B_R(0)$, since g_s is constant on the level sets of the map $q \to d(\overline{\xi(tx)}q, V)^2$ for each $|x| < 1$, $0 \le t \le 1$, thus

$$g_s(\overline{\xi(tx)}q) = g_s(q) \quad \text{for each } q \in B_R(0).$$

Hence from (2.5) we obtain $g_s(\zeta(t)) = 0$ whenever $p \in G_0$. As R is arbitrary, we find that $\zeta(t)$ remains in G_0 for all $0 \le t < \sigma_0$ where $\sigma_0 \le \sigma$.

Suppose that $\sigma_0 < \sigma$, then take G_1 to be an open neighborhood withcompact closure of $\zeta(\sigma_0)$ and repeat the above argument wherein $\zeta(\sigma_0)$ replaces p in (2.5). We thus obtain a new extension of the equation on $\mathbb{R}^b$ which yields a solution of (2.5) which lies in G_1 for $\sigma_0 \le t < \sigma_1 \le \sigma$. We continue in this fashion to obtain $\zeta(t) \in M$ for all $0 \le t < \sigma$.

This solution is clearly maximal. Uniqueness follows from that of the extended equation on each open neighborhood.

We conclude by constructing the unique maximal solution of (2.4).

Let $\rho = (\rho^1, \ldots, \rho^n)$ be the Poisson point process defined by

$$\rho^j(t) = \Delta\left(\int_{|x|\ge 1} x^j N(dt, dx) \right) \text{ for } 1 \le j \le n \text{ and let}$$

$\xi(\rho(t)) = \mathrm{Exp}(\rho^j(t)\, Y\ )$ for $t \in \mathbb{R}^+$. Define the random diffeomorphisms $\chi(t) = \xi(\rho(t)$ of V for $t \in \mathbb{R}^+$ and let $\{\tau_n;\ 0 \le n \le N\}$ be the jump times of ρ. We now proceed to construct ϕ as in [FuKu] p.84. Hence for $0 \le t < \tau_1$, define $\phi(t) = \zeta(t)$; for $t = \tau_1$, define $\phi(\tau_1) = \chi(\tau_1)(\zeta(\tau_1-))$ and for $\tau_1 < t < \tau_2$ we define $\phi(t) = \hat{\zeta}(t)$ where $\hat{\zeta}(t)$ is the solution of (2.5) with initial condition $\phi(\tau_1)$. We thus proceed inductively to define $\phi(t)$ for all $0 \le t \le \sigma$ □

Note: As indicated in the introduction, theorem 2.2 may also be proved by appealing to corollary 2 in §III.2 of [Coh] and taking the map Φ therein from $\mathbb{R}^n \times \mathcal{V} \times \mathbb{R}^n \to \mathcal{V}$ as

$$\Phi(a,p,b) = \xi(b - a)(p)$$

3 Horizontal Lévy Processes and Lévy Processes on Manifolds

We begin this section by collecting some geometrical facts which can all be found in [KoNo].

Let M be an n-dimensional connected, paracompact Riemannian manifold and denote by O(M) the bundle of orthonormal frames over M with canonical projection π: O(M) $\to$ M. Let $r = (r_1, \ldots, r_n) \in O(M)$ with $\pi(r) = p$ then we will also denote by π the induced linear map from $T_r(O(M))$ onto $T_p(M)$. Let $x = (x^1, \ldots, x^n) \in \mathbb{R}^n$, then r may be regarded as a linear map from $\mathbb{R}^n$ onto $T_p(M)$ with the action

$$r(x) = x^j r_j$$

We equip M with its unique Riemannian connection so that at each $r \in O(M)$ we have the decomposition

$$T_r(O(M)) = H_r(O(M)) \oplus V_r(O(M))$$

where H_r and V_r comprise the horizontal and vertical vectors at r (respectively). Note that each $\dim(H_r(O(M))) = n$. For each $x \in \mathbb{R}^n$ there exists a canonical horizontal vector field L(x) on O(M) which has the following properties

(i) L is smooth and each $L(x)(r) \in H_r(O(M))$,

(ii) $\pi(L(x)(r)) = r(x)$

We assume from now on that each L(x) is complete. M is then said to be geodesically complete. For each $p \in M$, let exp denote the exponential mapping from $T_p(M)$ into M, then if $\pi(r) = p$ we have

$$\pi(\text{Exp}(t\, L(x)(r))) = \exp(t\, r(x))(p) \qquad \ldots(3.1)$$

for all $t \in \mathbb{R}$. The right hand side of (3.1) is the unique geodesic through p in the direction $r(x) \in T_p(M)$.

We fix an orthonormal basis $e_1, \ldots, e_n$ in $\mathbb{R}^n$ and write $L_j = L(e_j)$ for $1 \le j \le n$ so that $L(x) = x^j L_j$ for each $x \in \mathbb{R}^n$. We now study the SDE (2.4) in the following context: take $\mathcal{V}$ to be O(M) and each $Y_j = L_j$ $(1 \le j \le n)$. From the above discussion we see immediately that each member of $\mathcal{T}$ is complete as is required.
By theorem 2.2 we can assert the existence of a unique maximal solution to the SDE

$$
\begin{aligned}
f(r(t)) &= f(r) + \int_0^t (Z_0 f)(r(s-))ds + \int_0^t (Z_j f)(r(s-))\circ dB^j(s) \\
&+ \int_0^{t+}\int_{|x|\ge 1} [f(\mathrm{Exp}(L(x))r(s-)) - f(r(s-))]\, N(ds,dx) \\
&+ \int_0^{t+}\int_{|x|<1} [f(\mathrm{Exp}(L(x))r(s-)) - f(r(s-))]\, \tilde{N}(ds,dx) \\
&+ \int_0^{t+}\int_{|x|<1} [f(\mathrm{Exp}(L(x))r(s-)) - f(r(s-)) - L(x)f(r(s-))]\, \nu(dx)\, ds
\end{aligned}
\qquad \ldots(3.2)
$$

for each $f \in C^\infty(O(M))$, $r \in O(M)$, $t \in \mathbb{R}^+$.

We call the solution $(r(t), 0 \le t \le \sigma)$ of (3.2) a <u>horizontal Lévy process</u> in O(M). It is interesting to compare the form of (3.2) with the formula obtained by A.Estrade and M.Pontier for the horizontal lift of a manifold-valued càdlàg semimartingale in proposition 4.3 of [EsPo].

We note that if $\nabla T = o$ where ∇ is the covariant derivative and T is the torsion tensor field then $\dim(\mathcal{L}) < \infty$ ([KoNo] p.137). In this case we have $\sigma = \infty$ a.s. by the results of [ApKu].
Now let $(T_t, t \in \mathbb{R}^+)$ be the Markov semigroup on $C_0(O(M))$

defined by

$$T_t(f)(r) = \mathbb{E}(f(r(t))/ r(0) = r) \qquad \ldots(3.3)$$

for $f \in C_0(O(M))$, $r \in O(M)$.

If $\mathcal{N}$ denotes its infinitesimal generator we have $C^\infty(O(M)) \subset \mathrm{Dom}(\mathcal{N})$ and a standard calculation yields

$$\mathcal{N}(f)(r) = m^j L_j(f)(r) + \frac{1}{2} a^{ij} L_i L_j(f)(r)$$

$$+ \int_{\mathbb{R}^n-\{0\}} [f(\mathrm{Exp}(L(x)(r)) - f(r) - \frac{x^j}{1 + |x|^2} L_j f(r)] \, \nu(dx) \qquad \ldots(3.4)$$

for $f \in C^\infty(O(M))$, $r \in O(M)$, where $a = (a^{ij})$ is the non-negative definite matrix $\sigma \sigma^T$ and for $1 \le j \le n$

$$m^j = c^j - \int_{|x|\ge 1} \frac{x^j}{1 + |x|^2} \nu(dx) + \int_{|x|<1} \frac{x^j |x|^2}{1 + |x|^2} \nu(dx)$$

We may regard (3.4) as a <u>horizontal Lévy-Khintchine-Hunt formulaon O(M)</u> (see [Hun], [ApKu]).
We now consider the càdlàg process $X = (X(t), 0 \le t < \sigma)$ on M defined by $X(t) = \pi(r(t))$. To investigate X, we define the linear operator $\mathcal{A}(r)$ on $C^\infty(M)$ by

$$\mathcal{A}(r) \, g(p) = \mathcal{N}(g \circ \pi)(r) \qquad \ldots(3.5)$$

where $g \in C^\infty(M)$, $r \in O(M)$ and $p = \pi(r)$.
Using (3.1), we then obtain the following for $g \in C^\infty(M)$, $p = \pi(r)$,

$$\mathcal{A}(r)(g)(p) = m^j R_j(g)(p) + \frac{1}{2} \Delta_a(g)(p)$$

$$+ \int_{\mathbb{R}^n-\{0\}} [g(\exp(x^j R_j)(p)) - g(p) - \frac{x^j}{1 + |x|^2} R_j(g)(p)] \, \nu(dx) \qquad \ldots(3.6)$$

where $R_j = r(e_j) \in T_p(M)$ and Δ_a is the linear operator on $C^\infty(M)$ defined by

$$\Delta_a \circ \pi = \pi \circ a^{ij} L_i L_j \qquad \ldots(3.7)$$

We call (3.6) the <u>Lévy-Khintchine-Hunt formula on the Riemannian manifold M</u>. $\mathcal{A}(r)$ is clearly independent of choice of orthonormal basis $(R_1, R_2, \ldots R_n)$ for $T_p(M)$ and hence is independent of the choice of lift r of p to O(M) in (3.5), however we still retain the frame r in the notation for reasons that will become clear below.
To get a clearer insight into the nature of Δ_a, we work in local co-ordinates in O(M). Let $r(p) = (p^i, e^i_j)$, and ∇_i be the covariant derivative in the direction ∂_i then a straightforward calculation yields (c.f [IkWa] p.260 - 274),

$$\Delta_a = g_a^{ij} \nabla_i \nabla_j$$

where $g_a^{ij} = a^{kl} e^i_k e^j_l$.

If the matrix a is positive-definite, $g_a = (g_a^{ij})$ is an (inverse) metric tensor on M. In this case we say that the generator $\mathcal{A}(r)$ is <u>non-degenerate</u>. We note that if σ is the matrix of an orthogonal transformation (so a is the identity matrix) g_a is the original (inverse) Riemannian metric on M and Δ_a is the Laplace-Beltrami operator.

The frame dependence of the operator $\mathcal{A}$ indicates that the semigroup $(T_t, t \in \mathbb{R}^+)$ on $C_0(O(M))$ does not, in general, project to a semigroup on $C_0(M)$ so that X is not, in general, a Markov process. Clearly one could define X to be a "Lévy process on a manifold" whenever it is indeed a Markov process. Alternatively, it might be argued that this is not a natural generalisation of the Euclidean case. Clearly further work on this question is required.

REFERENCES

[ApKu] D.Applebaum, H.Kunita, Lévy Flows on Manifolds and Lévy Processes on Lie Groups, J.Math. Kyoto Univ 33, 1103-23 (1993)

[Coh] S.Cohen, Géométrie Différentielle Stochastique Avec Sauts, ENPC CERMA Ph.D thesis (1992)

[Elw] K.D.Elworthy, Geometric Aspects of Diffusions on Manifolds in Springer LNM 1362 (1988)

[Eme] M.Emery, Stochastic Calculus in Manifolds, Springer-Verlag (1989)

[EsPo] A.Estrade, M.Pontier, Rélevement Horizontal d'une Semimartingale Càdlàg, Séminaire de Probabilités XXVI, 127-46 (Springer-Verlag), 1992

[Est] A.Estrade, Exponentielle Stochastique et Intégrale Multiplicative Discontinues, Ann.Inst.Henri Poincaré (Prob.Stat), 28, 107-29 (1992)

[FuKu] T.Fujiwara, H.Kunita, Stochastic Differential Equations of Jump Type and Lévy Processes in Diffeomorphisms Group, J.Math.Kyoto Univ 25, 71-106 (1985)

[Fuj] T.Fujiwara, Stochastic Differential Equations Of Jump Type on Manifolds and Lévy Flows, J.Math.Kyoto Univ. 31, 99-119 (1991)

[Hun] G.A.Hunt, Semigroups of Measures on Lie Groups, Trans. Amer. Math. Soc. 81, 264-93 (1956)

[IkWa] N.Ikeda, S.Watanabe, Stochastic Differential Equations and Diffusion Processes, North Holland-Kodansha (1981)

[Itô] K.Itô, On Stochastic Processes (I) (Infinitely Divisible Laws of Probability), Japanese J.Math. 18, 261-301 (1942)

[KoNo] S.Kobayashi , K.Nomizu, Foundations of Differential Geometry, Volume 1, Wiley (1963)

[Rog] S.J.Rogerson, Stochastic Dynamical Systems and Processes With Discontinuous Sample Paths, Warwick University Ph.D thesis (1981)

Some Markov properties of Stochastic Differential Equations with jumps

S. Cohen
CERMA-ENPC
La Courtine
URA CNRS 1502
93167 Noisy-le-Grand Cedex

May 9, 1995

1 Introduction

In previous articles [3] [4] we were interested in SDE's on manifolds driven by non continuous semimartingales; we got an extension of Meyer-Schwartz second order calculus [16], [19]. The main idea to deal with macroscopic jumps was to substitute 2-jets at x (only local behaviour is needed in the continuous case) by functions that are twice differentiable at x . In this setting a typical coefficient of SDE is a function φ from $V \times W \times V$ to W that describes how the solution Y jumps when the driving semimartingale X has a jump, we write :

$$Y_t = \varphi(X_{t^-}, Y_{t^-}, X_t) \quad \text{if } X_{t^-} \neq X_t.$$

In [3] the existence and uniqueness of a strong solution are obtained for such SDE's with locally Lipschitz coefficients, and the stochastic development of a càdlàg semimartingale in the tangent space of a Riemannian manifold was presented as an example of a SDE with jumps in [4]. Here we will be concerned with Markov properties of solutions, when the driving process is a Lévy process living in $\mathbf{R}^v$, more precisely we will compute their infinitesimal generator. This type of processes was already studied by Fujiwara, Applebaum [8] [1] but the techniques used were completely different; they construct Lévy flows on manifolds. Since we have not studied flows for SDE's with jumps we will not go further in this direction. In [18] Rogerson has defined the α-stable process with values in a Riemannian manifold M as a Brownian motion time-changed by a suitable subordinator; he has also constructed another process, called pseudo α-stable, as the stochastic development in M of a vector valued α-stable process in $T_{x_0}M$. These definitions agree when M is a Euclidean vector space but not in general. We will apply results on SDE's with jumps to compare pseudo and α-stable processes. In Section 2 we recall the existence and uniqueness theorem, and its application to stochastic development. Then we exhibit sufficient conditions for the solution to be Markovian, and compute its infinitesimal generator. In section 4 we give a probabilistic proof that pseudo α-stable and α-stable processes do not have the same law if the manifold is Riemannian with a pole and a rotationally invariant metric. We have to assume that $\Delta r < \frac{dim(M)-1}{r}$ (respectively $>$) where r represents the distance from the pole because we are studying the radial part of a Brownian motion. This evidently includes sphere in all dimensions. On the sphere in dimension 2, it is worth mentioning that $\Delta/2$ and the infinitesimal generator of the pseudo 1-stable process are linked via a formula involving a concave, piecewise affine function. As a consequence of this remark the pseudo 1-stable process on the 2-sphere is not a Brownian motion time-changed by a subordinator.

2 Summary on SDE's with jumps.

All filtered spaces $(\Omega, P, \mathcal{F}, \mathcal{F}_t)$ will verify the so called "usual conditions". If H is a real valued process, a new process starting at zero is defined by :

$$\mathcal{S}(H)_t = \sum_{0<s\leq t} H_s$$

if the sum is absolutely convergent, otherwise $\mathcal{S}(H)_t = \infty$.

All manifolds will possess a countable atlas and we will use Einstein's convention for summation.

To introduce SDE's with jumps we define first constrained coefficients of SDE's with jumps.

Definition 1 *Suppose that C is a closed submanifold of $V \times W$, such that the projection p_1 from C to V be onto and submersive. A measurable application φ from $C \times V$ to W will be called a constrained coefficient of SDE's with jumps if*

- *for each $z = (x, y)$ in C, $\varphi(z, x) = y$*
- *φ is C^3 in a neighborhood of $\{(z, p_1(z)) \,/\, z \in C\}$*
- *$\forall x \in V, \forall z \in C;\ (x, \varphi(z, x)) \in C$.*

Remark 1 *The particular case when $C = V \times W$ can be interpreted as the unconstrained case .*

The following existence and uniqueness result was shown in [3].

Theorem 1 *Take a constrained coefficient φ from C on to V, a semimartingale X living on V. We will note a SDE with jumps*

$$\begin{cases} \overset{\Delta}{d}Y &= \Phi(Y, \overset{\Delta}{d}X) \\ Y_0 &= y_0, \end{cases} \tag{1}$$

Equation (1) admits a unique solution (Y_{sol}, η), where η is the previsible stopping time of explosion of Y_{sol}, with the meaning that if $(x^i)_{i=1\ to\ v}$, $(y^\alpha)_{\alpha=1\ to\ w}$ are two C^2

imbeddings from V *to* $\mathbf{R}^v$*, respectively from* W *to* $\mathbf{R}^w$

$$\forall \alpha = 1 \text{ to } w \quad Y^\alpha = y_0^\alpha + \int \frac{\partial \varphi^\alpha}{\partial x^i}((X_-,Y_-),X_-)dX^i + 1/2 \int \frac{\partial^2 \varphi^\alpha}{\partial x^i \partial x^j}((X_-,Y_-),X_-)d\langle (X^i)^c,(X^j)^c\rangle + \mathrm{S}(\varphi^\alpha((X_-,Y_-),X) - Y_-^\alpha - \frac{\partial \varphi^\alpha}{\partial x^i}((X_-,Y_-),X_-)\Delta X^i). \tag{2}$$

Remark 2 *Because of equation (2) it is clear that the solution jumps only when* X *has a jump, and* $Y_t = \varphi(X_{t^-}, Y_{t^-}, X_t)$ *if* $X_{t^-} \neq X_t$*, which gives an intuitive interpretation of the coefficient* φ*.*

As an example of such SDE's stochastic development of semimartingales with jumps has been presented in [4]. For continuous processes this method had been used to obtain a Brownian motion on a Riemannian manifold from a flat Brownian motion. Stochastic development has been extended to discontinuous driving process in [18] [6]. We suppose that M is a complete connected Riemannian manifold with a C^3 atlas and dimension m, $O(M)$ will be the orthonormal frame bundle, and x_0 a reference point of M. With these notations $R_0 \in O_{x_0}(M)$ will be an isomorphism from $T_{x_0}M$ to $\mathbf{R}^m$. Two steps are necessary to develop a deterministic or random curve : solve a SDE between $V = \mathbf{R}^m$ and $W = O(M)$, then project the solution onto M. Let us describe the coefficient we need for the first step when the driving process may have jumps. The process $R_0(X_t)$ is the driving semimartingale in $\mathbf{R}^m$, and the coefficient of this SDE is constructed with the Riemannian exponential. If x, x' are two points in $\mathbf{R}^m$ and if R belongs to $O_\xi(M)$ we define $\varphi(x,(\xi,R),x')$ as the result of parallel transporting R along the geodesic $exp_\xi(tR^{-1}(x-x'))$ until time $t=1$. In this example no constraint is needed. The equation

$$\begin{cases} \overset{\triangle}{d}Y &= \Phi(Y, \overset{\triangle}{d}X) \\ Y_0 &= (x_0, R_0) \end{cases}$$

has a solution Y until the explosion time η. We call this process Y the horizontal lift of X and $\pi(Y) = y$ is the developed curve associated to X where $\pi : O(M) \longmapsto M$ is the trivial bundle projection. If X is smooth enough y is nothing else than the usual developed curve of textbooks in Mechanics. This setting includes continuous stochastic development of Brownian motions that leads to horizontal Brownian motion as in [5]. In Section 4 we will study with some care what happens when X is the α-stable process, extending results of [18].

3 Markov solutions to SDE's with Jumps

In a vector space, the pair of the driving process and the solution is Markovian, when the driving process is Markovian. But if you want the solution alone to be Markovian we "practically" have to suppose that the driving process is a Lévy process. A precise formulation of this is given by Theorem 32 in [17], and by [14] for the converse. Therefore in this geometric setting we will suppose that the driving Lévy process lives in a vector space. Actually only the group structure is needed. Furthermore if X jumps from X_{t^-} to X_t, solution will not explicitly depend on X_{t^-} but on jumps $X_t - X_{t^-}$.

Let us introduce classical notations for Markov processes. Consider an homogeneous Markov V-valued process X_t and its filtration $\mathcal{F}_t = \sigma(X_s, s \leq t)$. Suppose the existence of a family of transition probabilities $P(t, x, A)$ defined for all x, $0 \leq s \leq t$, and Borel sets A such that

$$E(X_t \in A|\mathcal{F}_s) = E(X_t \in A|X_s) = P(t-s, X_s, A) \;\; P \;\; ps$$

for all s, t $0 \leq s \leq t$ and Borel sets A. Usual hypothesis is assumed for transition probabilities. Because of the possible explosion of solutions of SDE's we have to consider submarkov transition probabilities. Adding to V a cemetery point is the usual trick to deal with such problem. Last we state the classical representation for Lévy processes see [15].

Theorem 2 *Let X be a càdlàg Lévy process living in $\mathbf{R}^v$, then X has a decomposition*

$$X_t = at + \sigma B_t + \int_{\{|x|<1\}\times[0,t]} \{\nu(dx, ds) - \pi(dx)ds\} + \mathrm{S}_t(\Delta X 1_{\{|\Delta X|\geq 1\}}) \qquad (3)$$

where a is a vector in $\mathbf{R}^v$, σ belongs to $\mathcal{L}(\mathbf{R}^{\sqsubseteq}, \mathbf{R}^{\sqsubseteq})$, B is a $\mathbf{R}^v$ Brownian motion; $\nu(dx, ds)$ a vector Poisson measure and π a measure on $\mathbf{R}^v$ such that

$$E(\nu(A \times [0,t])) = t\pi(A) \;\; avec \;\; A \in \mathcal{B}(\mathbf{R}^{\sqsubseteq}).$$

In the next proposition we exhibit sufficient conditions for the solution of a SDE to be a Markov process. The emphasis is put on the non-linear treatment of jumps in the infinitesimal generator.

Proposition 1 *Consider a unconstrained coefficient φ. Suppose that the driving process lives in $\mathbf{R}^v$, y_0 is a random variable on W, φ is a function of the increment $\varphi(x, y, z) = \psi(y, z - x)$, and the driving semimartingale is a Lévy process with a decomposition as in Theorem 2. The solution of*

$$\begin{cases} \overset{\Delta}{d}Y &= \Phi(Y, \overset{\Delta}{d}X) \\ Y_0 &= y_0 \end{cases}$$

is an homogeneous Markov process with transition probability

$$P(t, x, C) = P(Y_t^y \in C)$$

where $y \in W$ and Y^y verifies

$$\begin{cases} \overset{\Delta}{d}Y^y &= \Phi(Y^y, \overset{\Delta}{d}X) \\ Y_0^y &= y. \end{cases}$$

Furthermore if $f \in C^2(W)$, $f(Y) - f(Y_0) - \int_0 Af(Y_s)ds$ is a local martingale and A is defined by

$$\begin{aligned} Af(y) = &\langle df(y), \frac{\partial\psi}{\partial x}(y,0)a\rangle + 1/2\; Tr(Hess(f).(\frac{\partial\psi}{\partial x}(y,0)\sigma)(\frac{\partial\psi}{\partial x}(y,0)\sigma)^*) \\ &+ \int_{\mathbf{R}^v} \{f(\psi(y,x)) - f(y) - df(y)\frac{\partial\psi}{\partial x}(y,0)1\{|x|<1\}\}\pi(dx) \end{aligned} \qquad (4)$$

where $\frac{\partial\psi}{\partial x}(y,0)$ stands for $\frac{\partial\psi}{\partial x}(y,x)|_{x=0}$ and maps $\mathbf{R}^v$ to V.

Proof : Choose an imbedding $(y^\alpha)_{\alpha=1 \text{ to } w}$, and recall the system (2). For all $\alpha =$ 1 to w on $[[0,\eta[[$ we get

$$\begin{aligned} Y^\alpha &= y_0^\alpha + \int_0 \frac{\partial \psi^\alpha}{\partial x^i}(Y_{s^-},0)dX_s^i \\ &+1/2 \int_0 \frac{\partial^2 \psi^\alpha}{\partial x^i \partial x^j}(Y_{s^-},0)d\langle (X^i)^c,(X^j)^c\rangle_s \\ &+\mathrm{S}(\psi^\alpha(Y_-,0) - Y_-^\alpha - \frac{\partial \psi^\alpha}{\partial x^i}(Y_-,0)\Delta X^i). \end{aligned} \tag{5}$$

Replace X by its decomposition to obtain

$$\begin{aligned} &\forall \alpha = 1 \text{ to } w \quad \text{on} \quad [[0,\eta[[\\ &Y^\alpha = y_0^\alpha + \int_0 \frac{\partial \psi^\alpha}{\partial x^i}(Y_{s^-},0)a^i ds + \int_0 \frac{\partial \psi^\alpha}{\partial x^i}(Y_{s^-},0)(\sigma dB)_s^i \\ &+1/2 \int_0 \frac{\partial^2 \psi^\alpha}{\partial x^i \partial x^j}(Y_{s^-},0)(\sigma\sigma^*)_{i,j} ds \\ &+\int_{[0,.]\times\{|x|<1\}} \frac{\partial \psi^\alpha}{\partial x^i}(Y_{s^-},0)x^i\{\nu(dx,ds) - \pi(dx)ds\} \\ &+\mathrm{S}(\frac{\partial \psi^\alpha}{\partial x^i}(Y_-,0)\Delta X^i 1_{|\Delta X|\geq 1}) \\ &+\mathrm{S}(\psi^\alpha(Y_-,0) - Y_-^\alpha - \frac{\partial \psi^\alpha}{\partial x^i}(Y_-,0)\Delta X^i). \end{aligned} \tag{6}$$

We can write this expression with Poisson measures

$$\begin{aligned} Y^\alpha &= y_0^\alpha + \int_0 \frac{\partial \psi^\alpha}{\partial x^i}(Y_{s^-},0)a^i ds + \int_0 \frac{\partial \psi^\alpha}{\partial x^i}(Y_{s^-},0)(\sigma dB)_s^i \\ &+1/2 \int_0 \frac{\partial^2 \psi^\alpha}{\partial x^i \partial x^j}(Y_{s^-},0)(\sigma\sigma^*)_{i,j} ds \\ &+\int_{[0,.]\times\{|x|<1\}} \frac{\partial \psi^\alpha}{\partial x^i}(Y_{s^-},0)x^i\{\nu(dx,ds) - \pi(dx)ds\} \\ &+\int_{[0,.]\times\{|x|<1\}} \{\psi^\alpha(Y_-,0) - Y_-^\alpha - \frac{\partial \psi^\alpha}{\partial x^i}(Y_-,0)\Delta X^i)\}\nu(dx,ds). \end{aligned} \tag{7}$$

SDE's with jumps are translated in Rogerson's language, they become SDE's with Poisson jumps, and Rogerson [18] proves in theorem 3 p 4.4 that the solution is a homogeneous Markov process. He proceeds as in the classical Gihman Skorokhod's demonstration [10], the main steps are similar to those for diffusions : first verify the measurability of Y_t^y with respect to (y,t), and conclude using the flow property of the solution. However there is an additional technical problem with the explosion. The computation of the infinitesimal generator is an immediate consequence of Itô's formula. ◇

4 Pseudo α-stable and α-stable processes

Stochastic development of jump processes was presented in Section 2 . Rogerson in [18] tried to develop symmetric α-stable processes in order to find a Markov process with infinitesimal generator $-1/2\ (-\Delta)^{\alpha/2}$ on the Riemannian manifold. But what works for continuous processes like Brownian motion fails when a jump occurs. Rogerson has depicted the situation with a non-commutative diagram : the vertical arrows correspond to the stochastic development, the horizontal one to subordination. A classical construction of a Markov process with the infinitesimal generator $-1/2\,(-\Delta)^{\alpha/2}$ $\alpha\in]0,2[$ is achieved in [7] with a pair of independent processes (B, T^α), where B is a Brownian motion, and T^α a one sided $\alpha/2$- stable $\mathbf{R}^+$-valued process. The subordination is simply the time-change consisting in taking B at random time T_t^α, and it works on any state space whatsoever as soon as a Laplacian is defined on it.

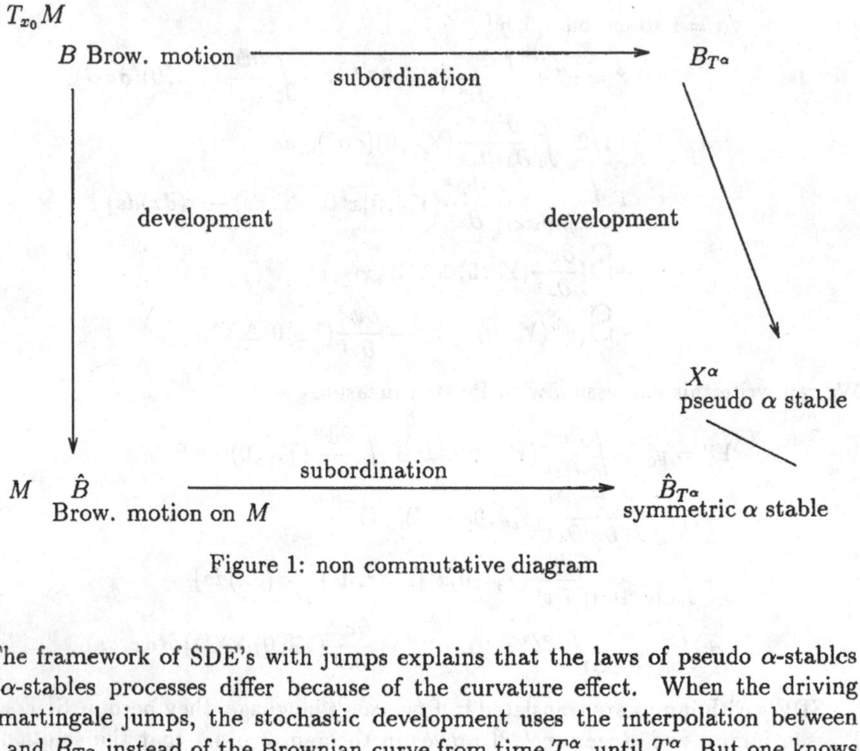

Figure 1: non commutative diagram

The framework of SDE's with jumps explains that the laws of pseudo α-stablcs and α-stables processes differ because of the curvature effect. When the driving semimartingale jumps, the stochastic development uses the interpolation between $B_{T^\alpha_{t-}}$ and $B_{T^\alpha_t}$ instead of the Brownian curve from time T^α_{t-} until T^α_t. But one knows that two curves that possess the same ending points are not necessarily mapped to curves with the same ending points. We can state this claim precisely as soon as we can compare how fast the Brownian motion goes to infinity in a vector space and on the particular Riemannian manifold. The radial part of the Brownian motion will be the basic tool of the next result.

Theorem 3 *Take a Riemannian connected manifold V with dimension $v > 1$. Suppose that V has a pole x_0 : in polar geodesic coordinates centered at x_0, the metric has the form $ds^2 = dr^2 + g^2(r)d\theta$, suppose also that $\Delta_y d(x,y) < \frac{v-1}{d(x,y)}$ for all x, y in a normal neighborhood of x_0 (respectively $>$). The laws of pseudo α-stable and α-stable processes are distinct.*

Proof : Assume that exp_{x_0} is a diffeomorphism from the ball $B_\rho(T_{x_0}\mathbf{R}^v)$ with center x_0 and diameter $\rho < \infty$ onto $\mathcal{V}$, we will choose ρ small enough such that there exists no (x, y) belonging to $\mathcal{V}$ and y conjugate to x. A Brownian motion B^V living on V with a starting point x_0 is constructed with a stochastic development, and we call τ the first exit time of B^V out of $\mathcal{V}$. Taking a one-sided α-stable process T^α independent of B^V, we define a sequence of stopping times

$$T_n = \inf\{t \geq 0 / d^V(B^V_{T^\alpha_{t-}}, B^V_{T^\alpha_t}) > 1/n\},$$

where d^V is the geodesic distance on V. The limit of T_n is almost surely 0, the sequence of events $A_n = \{\tau > T^\alpha_{T_n}\}$ is non decreasing, and $P(\bigcup_n A_n) = 1$. Applying the Itô formula to the function $d^V(B^V_{(T^\alpha_{T_n})^-}, x)$ when $s > (T^\alpha_{T_n})^-$, leads to

$$dr^V_s = dW_s + 1/2\frac{(v-1)g'(r^V_s)}{g(r^V_s)}ds \tag{8}$$

if $(T^\alpha_{T_n})^- < s < \tau$ and $r^V_s = d^V(B^V_{(T^\alpha_{T_n})^-}, B^V_s)$. In equation (8) W is a real valued Brownian motion, and one can construct a $T_{x_0}V$ Brownian motion, with $r_s = d(B_{(T^\alpha_{T_n})^-}, B_s)$ satisfying

$$dr_s = dW_s + 1/2\frac{(v-1)}{r_s}ds. \tag{9}$$

Therefore thanks to the comparison Theorem of solutions of SDE's, we know that

$$d^V(B^V_{(T^\alpha_{T_n})^-}, B^V_.) < d(B_{(T^\alpha_{T_n})^-}, B_.)$$

on $]](T^\alpha_{T_n})^-, \tau[[$ and

$$d^V(B^V_{(T^\alpha_{T_n})^-}, B^V_{T^\alpha_{T_n}}) < d(B_{(T^\alpha_{T_n})^-}, B_{T^\alpha_{T_n}}) \text{ on } A_n.$$

But when $B_{(T^\alpha)}$ jumps, the pseudo α-stable process X^α jumps too and remark 2 implies $d^V(X^\alpha_{T_n^-}, X^\alpha_{T_n}) = d(B_{(T^\alpha_{T_n})^-}, B_{T^\alpha_{T_n}})$. Hence almost surely for n big enough

$$d^V(B^V_{(T^\alpha_{T_n})^-}, B^V_{T^\alpha_{T_n}}) < d^V(X^\alpha_{T_n^-}, X^\alpha_{T_n}).$$

Consequently if $\tilde{T}_n = \inf\{t \geq 0 / d^V(X^\alpha_{t-}, X^\alpha_t) > 1/n\}$, we know that $\tilde{T}_n \leq T_n$ and when $\tilde{T}_n = T_n$ then $d^V(B^V_{(T^\alpha_{T_n})^-}, B^V_{T^\alpha_{T_n}}) < d^V(X^\alpha_{\tilde{T}_n^-}, X^\alpha_{\tilde{T}_n})$. If $P(\exists n \in \mathbf{N}, \tilde{T}_n < T_n) > 0$ we take the conditional probability $P(\ |\exists n \in \mathbf{N}, \tilde{T}_n < T_n)$, $B^V_{T^\alpha}$ and X^α do not have the same conditional law, since $\tilde{T}_n$ is obtained with the same construction from X^α as T_n from $B^V_{T^\alpha}$. But if we consider $P_2 = P(\ |\forall n \in \mathbf{N}, \tilde{T}_n = T_n)$ assuming $P(\forall n \in \mathbf{N}, \tilde{T}_n = T_n) > 0$, then $B^V_{T^\alpha}$ and X^α do not have the same law on $(\Omega, P_2, \mathcal{F})$ because $d^V(B^V_{(T^\alpha_{T_n})^-}, B^V_{T^\alpha_{T_n}}) < d^V(X^\alpha_{\tilde{T}_n^-}, X^\alpha_{\tilde{T}_n})$. So the laws of pseudo α-stable and α-stable processes are distinct. The same proof works when inequalities are reversed.
$\diamond$

5 Spectral comparison between pseudo α-stable and α-stable processes

Rogerson proved the same theorem in the special case of sphere in dimension 2, but he used different techniques : he compared the spectra of the infinitesimal generator of both processes. In this particular case the spherical harmonic functions are the eigenfunctions for the pseudo α-stable infinitesimal generator A_α. This phenomenon is quite general and it is a consequence of a geometrical remark. It is clear that both processes have their law invariant under a rotation preserving the starting point. We should consider this remark as a hint to study those processes on symmetric spaces. As a conclusion we will exhibit a fairly strange geometrical property of the spectrum of the pseudo α-stable process that prevents it to be a subordinated process of a Brownian motion.

The next proposition computes the infinitesimal generator of the pseudo α-stable process, it is a mere consequence of Proposition 1 applied to the first step in stochastic development. For the second step we use the rotational invariance of the Lévy measure of an α-stable process to get the Markov property of the projected process from $O(V)$ onto V.

Proposition 2 *If* $f \in \mathcal{D}(\mathcal{A}_\alpha) \bigcap \mathcal{C}_l^\epsilon(\mathcal{V})$

$$A_\alpha f(x) = \int_{T_x V} \{f(exp_x(u)) - f(x) - df(x).u1_{\|u\|<1}\}\pi(du) \tag{10}$$

where $\pi(du) = \frac{C(\alpha,v)}{\|u\|^{\alpha+v}}du$ *is the Lévy measure associated to α-stable symmetric process in* $\mathbf{R}^v$.

Geometers working on symmetric spaces usually introduce the spherical functions as the eigenfunctions of all differential operators that are invariant by the action of the isometry group. We can show that the spherical functions are also eigenfunctions of A_α, although A_α is not a differential operator. We first have to recall basic facts and notation for globally symmetric spaces.

Definition 2 *Let V be an analytic Riemannian manifold; V is called a symmetric space if each $x \in V$ is an isolated fixed point of an involutive isometry s_x.*

Symmetric spaces are analytically diffeomorph to the quotient of the connected component $G = I_0(V)$ of the isometry group which contains the identity by the subgroup K of G for which x_0 is a fixed point of V

$$\begin{aligned} G/K &\longmapsto V \\ gK &\longmapsto g(x_0). \end{aligned}$$

On the other hand G acts on G/K by $(g, hK) \longmapsto \tau(g)(hK)$ where $\tau(g) : hK \longmapsto ghK$, and $D(G/K)$ represents the space of differential operators on G/K invariant under the action of G. We get an easy example with the Laplace Beltrami operator of V when it is identified to G/K with its Riemannian structure.

If *proj* is the projection of G onto G/K recall the definition of spherical functions in [13].

Definition 3 *Let φ be a complex-valued function on G/K of class C^∞ which satisfies $\varphi(proj(e)) = 1$, φ is called a spherical function if*

(i) $\varphi^{\tau(k)} = \varphi \quad \forall k \in K$.

(ii) $D\varphi = \lambda_D \varphi \quad \forall D \in D(G/K)$.

Although A_α is not a differential operator, its law is invariant under the action of K, and the spherical functions are still eigenfunctions.

Proposition 3 *Let φ be a spherical function which belongs to the domain of A_α. We get*

$$A_\alpha(\varphi) = \lambda_{A_\alpha}\varphi.$$

Proof : Take $x_0 = proj(e)$

$$A_\alpha(\varphi)(x_0) = \int_{T_{x_0}V} \{\varphi(exp_{x_0}(u)) - 1 - d\varphi(x_0).u1_{\|u\|<1}\}\pi(du)$$

thanks to symmetry of π we can write

$$A_\alpha(\varphi)(x_0) = 1/2 \int_{T_{x_0}V} \{\varphi(exp_{x_0}(u)) + \varphi(exp_{x_0}(-u)) - 2\}\pi(du).$$

We have then to solve the problem in G. Theorem 3.3 in [12] explains that Riemannian exponential of a symmetric space may be expressed with Exponential of the Lie group G. Take the notations of [12], $\mathfrak{G}$ is the Lie algebra associated to G, $\mathfrak{T}$ with K, and $\mathfrak{P}$ satisfies $\mathfrak{G} = \mathfrak{T} \oplus \mathfrak{P}$, moreover $d(proj)_e$is an isomorphism from $\mathfrak{P}$ to $T_{x_0}V$. If we note $\tilde{\varphi} = \varphi \circ proj$, we express $A_\alpha(\varphi)(x_0)$ as an integral on $\mathfrak{P}$ with

$$A_\alpha(\varphi)(x_0) = 1/2 \int_{\mathfrak{P}} \{\tilde{\varphi}(Exp(u)) + \tilde{\varphi}(Exp(-u)) - 2\tilde{\varphi}(e)\}\pi(du).$$

With a technical transformation it appears as

$$A_\alpha(\varphi)(x_0) = 1/2 \int_K \int_{\mathfrak{P}} \{\tilde{\varphi}(kExp(u)) + \tilde{\varphi}(kExp(-u)) - 2\tilde{\varphi}(e)\}\pi(du)dk$$

where dk stands for the Haar measure on K. If we want to compute $A_\alpha(\varphi)(x)$ where $x = proj(h)$, we apply an isometry h which maps x_0 onto x and we get

$$A_\alpha(\varphi)(x) = 1/2 \int_K \int_{\mathfrak{P}} \{\tilde{\varphi}(hkExp(u)) + \tilde{\varphi}(hkExp(-u)) - 2\tilde{\varphi}(h)\}\pi(du)dk.$$

One knows that the spherical functions are solutions of the following functional equation (Proposition 2.2 [13])

$$\forall g, h \in G \quad \int_K \tilde{\varphi}(gkh)dk = \tilde{\varphi}(g)\tilde{\varphi}(h). \tag{11}$$

So we get

$$A_\alpha(\varphi)(x) = (A_\alpha(\varphi)(x_0))\tilde{\varphi}(h)$$

as we claimed. ◇

A probabilistic approach of this problem is related to the study of semi groups that are invariant under action of K as presented in [9][11]. Let (X_t^x) be a Markov process starting from x the law of X^x is assumed to be invariant under the action of the subgroup of isometries fixing x. The commutation property of the semi groups (P_t) and (Q_s) corresponding to another process Y^y of same type can be expressed as follows :

$$P_t(Q_s(f))(x) = E_\omega(Q_s(f)(X_t^x(\omega))) = E_\omega(E_{\omega'}(f(Y_s^{X_t^x(\omega)}(\omega')))).$$

But if G/K is put for V, the law of $X_t^{Y_s^x}$ is nothing but the law of $\tau(Y_s^x).(X_t^x)$ on G/K, this law is called in [9] the convolution of Y_s^x by X_t^x. The convolution commutes on the symmetric spaces (we can find the proof in [9][11]), and it implies the commutation of P_t , Q_s. It is the same phenomenon than in Proposition 3 expressed on the law of random variables and not on infinitesimal generators.

Nevertheless more can be deduced from the spectral study of those processes. We will show that there is no subordinator T such that the law of the pseudo 1-stable process on the sphere in dimension 2 is the same as that of B_T where B is a S^2-Brownian motion and (B,T) are independent. We will introduce the Bernstein functions by considering the Laplace transform of subordinators.

Definition 4 *If T_t is a real non decreasing càdlàg Lévy process*

$$E(e^{-pT_t}) = e^{-\psi(p)t}$$

where ψ is a Bernstein function.

We get in [2] two other characterizations of the Bernstein functions.

Definition 5 *A function $\psi \in C^\infty(]0,\infty[,\mathbf{R})$ is called a Bernstein function if*

$$\psi \geq 0,\ (-1)^p \psi^{(p)} \leq 0 \quad \forall p \geq 1\,;$$

but there is also an integral representation of these functions built on the same pattern as the Lévy Khintchine formula.

Definition 6 *A function ψ is called a Bernstein function if and only if there exist two positive constants a, b, and a positive measure μ on $]0,\infty[$ with $\int_0^\infty \frac{s}{1+s} d\mu(s) < \infty$ such that*

$$\psi(x) = a + bx + \int_0^\infty (1 - e^{xs}) d\mu(s) \quad \forall x > 0\,;$$

the triple (a, b, μ) is uniquely determined by ψ.

The Bernstein functions give the correspondence between the eigenvalues of $(-\Delta/2)$ and those of the infinitesimal generator of subordinated processes. Take a process B_T subordinated to a Brownian motion with a subordinator T, which corresponds to the Bernstein function ψ. If φ is a bounded spherical function for the eigenvalue λ with respect to $(-\Delta/2)$, it will be associated to the eigenvalue $-\psi(\lambda)$ as eigenfunction of the infinitesimal generator of B_T. Actually if P_t is the semi group of B_T

$$\begin{aligned} P_t(\varphi)(x) &= E(\varphi(B_{T_t}^x)) \\ &= E(E(\varphi(B_{T_t}^x)|T_t = a)) \\ &= \int_0^\infty e^{-a\lambda} \varphi_{\lambda(x)} dP_{T_t}(a) \\ &= exp(-t\psi(\lambda))\varphi_\lambda(x). \end{aligned}$$

Let us go back to the special case of the sphere in dimension 2. We can fix as reference point the north pole with Cartesian coordinates $PN = (0,0,1)$, we write (r,θ) for the polar geodesic coordinates from the north pole. The spherical functions can be expressed with Legendre polynomials of first kind

$$f_n(x) = P_n(cos(r(x))) \text{ and } (-\Delta/2))f_n(PN) = \frac{n(n+1)}{2} = \lambda_n.$$

On the other hand if A is the infinitesimal generator associated to the pseudo 1-stable process, formula (10) leads to

$$Af_n(PN) = 2\pi \int_0^\infty \frac{P_n(cos(r)) - 1}{r^2} C(1,2)dr = \mu_n.$$

But on the graph of $(\lambda_n, -\mu_n)$ for $n = 1$ to 10 we remark that those points are aligned three by three. Since Bernstein functions are either affine or strictly concave thanks to definition 6. we know that there is no Bernstein function such that $\psi(\lambda_n) = -\mu_n$. Hence

Proposition 4 *The pseudo* 1*-stable on the* 2*-sphere is not a Brownian motion time-changed by a subordinator : more precisely it is not equal in law to B_T where B is a Brownian motion and T a subordinator independent of B.*

6 Exact computation of eigenvalues

In this part we would like to present the computations of eigenvalues which allow us to prove the last result. They have been obtained using MAPLE symbolic software. It is the reason why we can claim that points $(\lambda_n, -\mu_n)$ are aligned and not nearly aligned as the answer given by classical numerical program. The same result could have been obtained without a computer, but we are not sure that we would have tried it because it is a lot of work, and you do not know at the beginning if it will be useful. We first recall the expression of μ_n as

$$\mu_n = 2\pi \int_0^\infty \frac{P_n(cos(r)) - 1}{r^2} C(1,2)dr.$$

In the next table we present on the first row the eigenvalue corresponding to $-(\Delta/2)$ $\lambda_n = \frac{n(n+1)}{2}$, on second line $-\mu_n$, and we can read that points $(\lambda_n, -\mu_n)$ are aligned three by three since $\delta_n = \frac{\mu_n - \mu_{n+1}}{\lambda_{n+1} - \lambda_n}$ is printed on the third line.

n	1	2	3	4	5
λ_n	1	3	6	10	15
μ_n	$(-1/4)\pi$	$(-3/8)\pi$	$(-9/16)\pi$	$(-45/64)\pi$	$(-225/256)\pi$
δ_n	$(-1/16)\pi$	$(-1/16)\pi$	$(-9/256)\pi$	$(-9/256)\pi$	$(-25/1024)\pi$

n	6	7	8	9
λ_n	21	28	36	45
μ_n	$(-525/512)\pi$	$(-1225/1024)\pi$	$(-11025/8192)\pi$	$(-99225/65536)\pi$
δ_n	$(-25/1024)\pi$	$(-1225/65536)\pi$	$(-1225/65536)\pi$	$(-3969/262144)\pi$

This phenomenon was first suggested by the graph where you put λ_n on the X axis and $-\mu_n$ on the Y axis. It can also be seen on the graph that the points are close to the parabola

$$y = 1/2(x^{1/2}),$$

which is a graphic representation of the idea that the pseudo α-stable processes are perturbated α-stable processes in a spectral sense.

References

[1] David Applebaum. Stochastic flows of diffeomorphism on manifolds driven by Lévy processes. Preprint Department of Mathematics, Statistics and Operation Research, Nottingham Polytechnic, Burton Streeet, Nottingham NG1 4BU England.

[2] C. Berg and G. Forst. *Potential theory on Locally Compact Abelian Groups.* Ergebnisse der Mathematik und ihrer Grenzgebiete. Springer Verlag, 1975.

[3] S. Cohen. Géométrie différentielle stochastique avec sauts 1. accepté pour publication dans Stochastics Reports, 1993.

[4] S. Cohen. Géométrie différentielle stochastique avec sauts 2 : discrétisation et applications des eds avec sauts. préprint, 1993.

[5] K.D. Elworthy. *Stochastic Differential Equations on Manifolds.* Cambridge University Press, 1982.

[6] A. Estrade and M. Pontier. Relèvement horizontal d'une semi-martingale cadlag. *Séminaire de Probabilités XXVI*, Lecture Notes in Mathematics(1526):127–145, 1991.

[7] W Feller. *An Introduction to Probability Theory and its Applications*, volume 2. Wiley, 1950.

[8] T. Fujiwara. Stochastic differential equation of jump type on manifolds and Lévy flows. *J. Math. Kyoto Univ.*, 31:99–119, 1991.

[9] Gangolli. Isotropic infinitely divisible measures on symmetric spaces. *Acta. Math.*, (111):213–246, 1964.

[10] I.I. Gihman and A.V. Skorokhod. *Stochastic Differential Equations*, volume 72 of *Ergebnisse der Mathematik und ihrer Grenzgebiete.* Springer-Verlag, 1972.

[11] R. Godement. Introductions aux travaux de A.Selberg. *Séminaire Bourbaki*, (144), 1957.

[12] S. Helgason. *Differential Geometry Lie Groups and Symmetric Spaces*, volume 80. Academic Press, second edition, 1978.

[13] S. Helgason. *Group and Geometric Analysis*, volume 113. Academic Press, 1984.

[14] J. Jacod and P. Protter. Une remarque sur les équations différentielles stochastiques à solutions markoviennes. *Séminaire de Probabilités XXV*, 1992.

[15] J. Jacod and A. N. Shiryaev. *Limit theorems for stochastic processes.* Springer, 1987.

[16] P.A. Meyer. Géométrie stochastique sans larmes. *Séminaire de Probabilités XV*, Lecture Notes in Mathematics(850), 1981.

[17] P. Protter. *Stochastic Integration and Differential Equations.* Springer-Verlag, 1990.

[18] S. Rogerson. Stochastic differential equation with discontinuous sample path and differential geometry. Control Theory Center report 101 Warwick University.

[19] L. Schwartz. Géométrie différentielle du deuxième ordre, semi-martingales et équations différentielles sur une variété différentielle. *Séminaire de Probabilités XVI*, Lecture Notes in Mathematics(921), 1982.

CHAOS MULTIPLICATIF : UN TRAITEMENT SIMPLE ET COMPLET DE LA FONCTION DE PARTITION

J. FRANCHI

Laboratoire de Probabilités - Université Paris VI
(4, Place Jussieu , Tour 56, 3ème étage - 75252 Paris Cedex 05)
et Université Paris XII

I. Introduction

Le modèle dit du chaos multiplicatif a été introduit comme modèle de turbulence par B. Mandelbrot [M] : soient d un entier > 1 , W une variable aléatoire positive non constante d'espérance 1 , et n un entier positif; à chaque intervalle $I_{k,n} = [kd^{-n},(k+1)d^{-n}[$, où $0 \le k < d^n$, on associe une variable aléatoire $W_{k,n}$ de même loi que W, toutes ces variables étant choisies indépendantes; soit f_n la fonction aléatoire sur $[0,1[$ qui vaut $W_{k,n}$ sur $I_{k,n}$ pour $k \in \{0,..,d^n-1\}$; soit μ_n la mesure aléatoire sur $[0,1[$ admettant pour densité le produit $f_1 f_2 .. f_n$.
Le théorème de convergence des martingales positives montre que p.s. μ_n converge vaguement vers μ_∞ lorsque $n \to \infty$.

Kahane ([K],[KP]) a calculé $\mathbb{E}\left(\mu_\infty([0,1[)\right)$, et a caractérisé l'appartenance de $\mathbb{E}\left((\mu_\infty([0,1[))^h\right)$ à $\mathbb{R}_+^*$;

Peyrière ([KP]) a calculé la dimension de Hausdorff du support de μ_∞ .

Le succès de ce modèle est attesté par son intervention dans plusieurs branches de la physique théorique : verres de spin, croissance de polymères ([DS],[CD]) , thermodynamique ([CK]) .
La quantité à laquelle les physiciens théoriciens s'intéressent le plus est la fonction de partition : $\sigma_n(\beta) = d^n \mu_n^\beta([0,1[)$, où β est un réel >0 (représentant l'inverse de la température) et où μ_n^β est formée à partir de W^β comme μ_n l'était à partir de W ; cette fonction de partition est la masse d'une certaine mesure de Gibbs aléatoire; le comportement asymptotique de $Z_n(\beta) = n^{-1}\mathrm{Log}(\sigma_n(\beta))$ est étudié dans [DS],[CD], puis [CK].

Dans [DS] et [CD], le modèle de Mandelbrot est présenté sous la forme isomorphe d'arbre, chaque branche codant une suite décroissante d'intervalles $I_{k_n,n}$, et la variable W y est notée e^X ; alors

$\sigma_n(\beta) = \Sigma\, e^{\beta S_b}$, où la somme porte sur les branches b de longueur n et où S_b est la somme des variables $X_{b'}$ rencontrées le long de la branche b .

On trouve dans [CD](page 513) une expression générale de $Z_\infty(\beta)$ lorsque d=2, tandis que [CK] calculent $Z_\infty(\beta)$ pour d quelconque mais pour X bornée et β_c finie.

Le but de cet article (qui est repris de la prépublication [F]) est de présenter simplement, élémentairement et rapidemment le détail du calcul de la pression $Z_\infty(\beta)$ dans le cas général, en mettant en évidence que l'argument crucial est l'un de ceux de Kahane ($\delta\Rightarrow\beta$ du théorème 1 de [KP], mais on peut aussi bien utiliser l'argument principal de [K]), qui assure en fait l'intégrabilité uniforme de $\mu_n^\beta([0,1[)$ sous le seuil critique; (la réciproque $\beta\Rightarrow\delta$ du théorème 1 de [KP] est en revanche une conséquence du calcul de $Z_\infty(\beta)$).

De plus on relie simplement le seuil critique à la fonctionnelle des grandes déviations de X, et on caractérise simplement la finitude du seuil critique.

L'origine de ce travail est l'intérêt porté (dans le cas de X gaussienne) par A. Benassi à la limite de $n^{-1}\mathrm{Max}\{\, S_n(t) \mid |t|=n \,\}$, qui est déduite ici de $Z_\infty(\beta)$.

La présentation du modèle retenue ici est celle de l'arbre comme dans [DS] et [CD], tandis que la démarche est pour une part adaptée de [CK].

II. Notations

X est une variable aléatoire réelle (non p.s. constante) ayant tous ses moments exponentiels finis ;

$\bar{x} = \text{ess-sup } X = \sup\{x\in\mathbb{R} \mid \mathbb{P}(X<x)<1\}$; $\underline{x} = \text{ess-inf } X$; $m=\mathbb{E}(X)$;

$\gamma(\beta) = \mathrm{Log}\left(\mathbb{E}(e^{\beta X})\right)$ et $g(\beta) = \beta\gamma'(\beta)-\gamma(\beta)$, pour β réel ;

$I(\theta) = \mathrm{Sup}\{\, \theta\beta-\gamma(\beta) \mid \beta\in\mathbb{R} \,\}$, pour θ réel ;

$I^{-1}(z) = \mathrm{Inf}\{\, \theta>m \mid I(\theta)>z \,\}$, pour z dans $\mathbb{R}_+$;

d est un entier ≥ 2 ;

$\{\, X_b \mid b\in\mathcal{A} \,\}$ est une suite de variables aléatoires indépendantes distribuées comme X, indexées par les branches (finies) d'un arbre d-adique $\mathcal{A}$ (de chaque branche sont issus d segments); notons $\angle$ l'ordre naturel sur $\mathcal{A}$, et $|b|$ la longueur, i.e. le nombre de minorants, de la branche b de l'arbre ; (de sorte que $\mathcal{A}$ compte d^n branches de longueur n);

pour $n\in\mathbb{N}^*$ et $b\in\mathcal{A}$, $S_b = \sum_{b'\angle b} X_{b'}$ et $S_n^* = \mathrm{Max}\{\, S_b \mid |b| = n \,\}$.

$\sigma_n(\beta) = \sum_{|b|=n} e^{\beta S_b}$, $Y_n(\beta) = \sigma_n(\beta).\mathbb{E}^{-1}(\sigma_n(\beta))$, et

$Z_n(\beta) = n^{-1}.\mathrm{Log}(\sigma_n(\beta))$, pour n dans $\mathbb{N}^*$ et β dans $\mathbb{R}_+$.

Le théorème qui suit, dont l'essentiel est le point (ii), énonce ce qui est établi dans cet article. La démonstration est ensuite découpée en 20 courtes étapes.

III. Théorème

i) Il existe une unique valeur critique $\beta_c \in]0,+\infty]$ telle que $g(\beta)<\mathrm{Log}\,d$ si $0\le\beta<\beta_c$ et $g(\beta)>\mathrm{Log}\,d$ si $\beta>\beta_c$; β_c est finie si et seulement si $d.\mathbb{P}(X=\bar{x}) < 1$; (auquel cas $\beta_c = g^{-1}(\mathrm{Log}\,d)$);

ii) $Z_n(\beta)$ converge lorsque n tend vers $+\infty$, p.s. et dans L^p pour $1\le p<\infty$, vers $Z_\infty(\beta)$ qui vaut : $\mathrm{Log}\,d+\gamma(\beta)$ si $0\le\beta\le\beta_c$ et $\beta\gamma'(\beta_c)$ si $\beta\ge\beta_c$;

iii) $n^{-1}S^*_n$ converge p.s. et dans L^p pour $1\le p<\infty$ vers $\gamma'(\beta_c) = I^{-1}(\mathrm{Log}\,d)$.

Remarques 1) Biggins ([B], corollaire (3.4)) obtient déjà (de façon moins élémentaire) une expression de la limite presque sûre du point (iii).

2) Z_∞ est convexe, de classe C^1 sur $\mathbb{R}_+$, C^∞ hors de β_c , mais pas C^2 en β_c : c'est ce que les physiciens théoriciens nomment une transition de phase.

Exemples 1) Si X est gaussienne centrée de variance V, alors $g(\beta)=\gamma(\beta)=V\beta^2/2$, $I(\theta)=\theta^2/2V$, $\beta_c= (2V^{-1}\mathrm{Log}\,d)^{1/2}$ et $I^{-1}(\mathrm{Log}\,d)= (2V\mathrm{Log}\,d)^{1/2}$;

2) Si X vaut +1 et -1 avec probabilité 1/2 , alors $I(\pm 1)=\mathrm{Log}\,2$, $\beta_c= \infty$, et $I^{-1}(\mathrm{Log}\,d)=1$.

IV. Démonstration du Théorème

Commençons par quelques propriétés de γ,g,I et I^{-1} ;

1) γ est C^∞ strictement convexe, et $\gamma(0)=0$, $\gamma'(0)=m$;

en effet $\gamma''(\beta) = e^{-2\gamma(\beta)}.\left(\mathbb{E}(X^2e^{\beta X})\mathbb{E}(e^{\beta X})-\mathbb{E}^2(Xe^{\beta X})\right)$ est ≥ 0 par l'inégalité de Schwarz, et >0 car X est p.s. non constante .

2) g est strictement croissante sur $\mathbb{R}_+$;

car $g'(\beta) = \beta.\gamma''(\beta) > 0$ sur $\mathbb{R}^*_+$.

3) $\gamma'(+\infty) = \bar{x} = \lim_{\beta\to\infty} \gamma(\beta)/\beta$; si $\bar{x}<\infty$, $\lim_{\beta\to\infty}(\gamma(\beta)-\beta\bar{x}) = \mathrm{Log}(\mathbb{P}(X=\bar{x}))$;

preuve: clairement $\gamma'(\beta)\le\bar{x}$ et $\gamma(\beta)\le\beta\bar{x}$ pour tout $\beta\ge 0$; réciproquement fixons $m<y<z<\bar{x}$; on a $\mathbb{E}((X-m)e^{\beta X}) \ge (y-m)\mathbb{E}(e^{\beta X}(1-1_{\{X<y\}}))$ d'où

$\gamma'(\beta) \ge y -(y-m)\mathbb{E}(e^{\beta X} 1_{\{X<y\}})\mathbb{E}^{-1}(e^{\beta X}) \ge y - (y-m)e^{\beta(y-z)}\mathbb{P}^{-1}(X\ge z)$

qui tend vers y lorsque β tend vers $+\infty$, ce qui montre que $y<\bar{x} \Rightarrow \gamma'(\infty)\ge y$; de même, $\gamma(\beta)/\beta \ge y + \beta^{-1}\mathrm{Log}(\mathbb{P}(X\ge y)) \to y$; et si $\bar{x}<\infty$,

$\gamma(\beta)-\beta\bar{x} = \mathrm{Log}\left(\mathbb{E}\left(e^{\beta(X-\bar{x})}\right)\right)$ tend vers $\mathrm{Log}\left(\mathbb{E}\left(1_{\{X=\bar{x}\}}\right)\right)$ lorsque β tend vers ∞.

4) Si $\underline{x}<\theta<\bar{x}$, alors $I(\theta) = \int_m^\theta (\gamma')^{-1} = g\circ(\gamma')^{-1}(\theta)$;

en effet 1) et 3) montrent que $\theta\beta-\gamma(\beta)$ passe par un maximum strict pour $\beta=\gamma'^{-1}(\theta)$, lorsque $\theta\in]\underline{x},\bar{x}[$, et donc $I(\theta)=\theta\gamma'^{-1}(\theta)-\gamma(\gamma'^{-1}(\theta))=g(\gamma'^{-1}(\theta))$; enfin $g\circ\gamma'^{-1}(m)=0$ et $(g\circ\gamma'^{-1})'=\gamma'^{-1}$.

5) I est finie sur $]\underline{x},\bar{x}[$, et vaut ∞ sur $]\bar{x},\infty[$; $I(\bar{x})=-\mathrm{Log}\left(\mathbb{P}(X=\bar{x})\right)$;

preuve: pour $\varepsilon>0$ on a $\mathbb{E}(e^{\beta X}) \le e^{\beta(\bar{x}-\varepsilon)} + e^{\beta\bar{x}}\mathbb{P}(X>\bar{x}-\varepsilon)$ et donc
$I(\bar{x}) \ge \beta\bar{x} - \mathrm{Log}(\mathbb{E}(e^{\beta X})) \ge -\mathrm{Log}\left(\mathbb{P}(X>\bar{x}-\varepsilon) + e^{-\beta\varepsilon}\right) \xrightarrow[\beta\to\infty]{} -\mathrm{Log}\left(\mathbb{P}(X>\bar{x}-\varepsilon)\right)$;
d'autre part $I(\bar{x}) \le \mathrm{Sup}_\beta\{\beta\bar{x} - \mathrm{Log}(e^{\beta\bar{x}}\mathbb{P}(X=\bar{x}))\} = -\mathrm{Log}(\mathbb{P}(X=\bar{x}))$;
enfin si $\theta>\bar{x}$: $I(\theta) \ge \theta\beta - \beta\bar{x} \xrightarrow[\beta\to\infty]{} \infty$.

6) I est convexe sur $\mathbb{R}$, et bijective croissante C^∞ de $[m,\bar{x}[$ sur $[0,I(\bar{x})[$;

en effet d'après 4) et 5) $I'= \gamma'^{-1}$ croît strictement sur $]\underline{x},\bar{x}[$; de plus I est s.c.i. et croît sur $\mathbb{R}_+$, d'où la convexité et $I(\bar{x}-)=I(\bar{x})$; d'où :

7) I^{-1} est concave croissante sur $\mathbb{R}_+$, égale à $\bar{x}$ sur $[I(\bar{x}),\infty[$;

8) $I(\bar{x}-)=I(\bar{x})=g(\infty)$ et $\gamma'(\beta_c) = I^{-1}(\mathrm{Logd}) = \mathrm{Inf}\{ (\gamma(\beta)+\mathrm{Logd})/\beta \mid \beta>0 \}$;

preuve: $I(\bar{x}-)=g(\infty)$ découle de 4) et 3) ; si $\beta_c=\infty$, alors $\mathrm{Logd} \ge g(\infty) = I(\bar{x})$ et donc $I^{-1}(\mathrm{Logd}) = \bar{x} = \gamma'(\beta_c)$; de plus dans ce cas
$\frac{\partial}{\partial\beta}\left((\gamma(\beta)+\mathrm{Logd})/\beta\right) = (g(\beta)-\mathrm{Logd})/\beta^2 < 0$ montre que
$\mathrm{Inf}\{ (\gamma(\beta)+\mathrm{Logd})/\beta \mid \beta>0 \} = \lim_{\beta\to\infty} \gamma(\beta)/\beta = \bar{x}$ aussi ;
enfin si $\beta_c<\infty$, alors $(\gamma(\beta)+\mathrm{Logd})/\beta$ admet un minimum strict en β_c , égal à $\gamma'(\beta_c)$, et d'après 4) $I^{-1}(\mathrm{Logd}) = \gamma'\circ g^{-1}(\mathrm{Logd}) = \gamma'(\beta_c)$.

9) Le point i) ainsi que l'égalité du point iii) du théorème découlent de 2), 5) et 8) , car $\beta < \infty \Leftrightarrow g(\infty) > \mathrm{Logd}$.

L'estimation classique de Chernoff permet de dominer simplement S^*_n :

10) $\mathbb{P}(S^*_n>n\theta) \le d^n e^{-nI(\theta)}$ pour tout $\theta>0$;

en effet $\mathbb{P}(S_b>n\theta) \le \mathbb{E}\left(e^{\beta(X_{b_1}+..+X_{b_n})-\beta n\theta}\right) = e^{-\beta n\theta}\ \mathbb{E}^n(e^{\beta X}) = e^{-n(\theta\beta-\gamma(\beta))}$
$= e^{-nI(\theta)}$ pour $\beta>0$ bien choisi
et pour $\theta>0$, d'où $\mathbb{P}(S^*_n>n\theta) \le \sum_{|b|=n} \mathbb{P}(S_b>n\theta) \le d^n e^{-nI(\theta)}$.

11) $\limsup_{n\to\infty} n^{-1}S^*_n \le I^{-1}(\mathrm{Logd})$ p.s. ;

en effet $\sum_{n>0} \mathbb{P}(S^*_n > n\theta) \le \sum_{n>0} e^{n(\mathrm{Logd} - I(\theta))} < \infty$ dès que $I(\theta) > \mathrm{Logd}$; donc $\theta > I^{-1}(\mathrm{Logd}) \Rightarrow \limsup_{n\to\infty} n^{-1}S^*_n \le \theta$ p.s. , d'où le résultat.

12) $\mathbb{E}(\sigma_n(\beta)) = d^n e^{n\gamma(\beta)}$.

13) $Y_n(\beta) \xrightarrow[n\to\infty]{p.s.} Y_\infty(\beta) \in \mathbb{R}_+$, et $\mathbb{E}(Y_\infty(\beta)) \le 1$;

preuve: si $|b| = n$ et $|b'| = p$, notant bb' la branche de longueur (n+p) commençant par b et finissant par b', on a : $S_{bb'} = S_b + S^b_{b'}$, où $S^b_{b'}$ est indépendante de S_b et a la loi de $S_{b'}$;
Soit $\mathfrak{F}_n$ la tribu engendrée par les variables $\{X_b \,|\, |b| \le n\}$;
utilisant 12), on a :

$$\mathbb{E}\left(Y_{n+p}(\beta)\middle|\mathfrak{F}_n\right) = \sum_{|b|=n}\left(e^{\beta S_b}\ \mathbb{E}\left(\sum_{|b'|=p} e^{\beta S^b_{b'}}\right)\right).d^{-p}e^{-p\gamma(\beta)}.d^{-n}e^{-n\gamma(\beta)} = Y_n(\beta) \ ;$$

donc $Y_n(\beta)$ est une martingale; comme elle est >0 et d'espérance 1, elle converge p.s. selon le résultat classique de Doob, et le lemme de Fatou assure que $\mathbb{E}(Y_\infty(\beta)) \le 1$.

14) $\mathbb{P}(Y_\infty(\beta)>0) = 0$ ou 1 ;

car la loi du 0-1 s'applique à $\mathfrak{F}_n$ et à $\{Y_\infty(\beta)>0\}$; on a en effet

$$Y_\infty(\beta) = \sum_{|b|=n}\left(e^{\beta S_b}\ \mathbb{E}^{-1}(\sigma_n(\beta))\right)Y^b_\infty(\beta) \quad \text{p.s.}$$

, où $Y^b_\infty(\beta)$ est relative au sous-arbre débutant à l'extrémité de la branche b , ce qui montre que $\{Y_\infty(\beta)>0\} = \bigcup_{|b|=n} \{Y^b_\infty(\beta)>0\}$ est indépendant de $\mathfrak{F}_n$.

Passons maintenant au cœur de la démonstration, qui consiste à obtenir l'intégrabilité uniforme des variables $Y_n(\beta)$ sous la valeur critique β_c . 15) ci-dessous est repris de Kahane ([KP], théorème 1 , $\delta\Rightarrow\beta$).

15) $0 \le \beta < \beta_c \Rightarrow Y_\infty(\beta) > 0$ p.s. ;

preuve: on remarque que $\left(\sum_{i=1}^d x_i\right)^h \ge \sum_{i=1}^d (x_i)^h - 2(1-h).\sum_{i<j}(x_i x_j)^{h/2}$

pour $h\in]h_0,1[$ et $x_i>0$, ce qui appliqué à $Y_n(\beta) = d^{-1}\sum_{i=1}^d e^{\beta X_i - \gamma(\beta)}\, Y^i_{n-1}(\beta)$
entraîne :

$$d^h.\mathbb{E}\left((Y_n(\beta))^h\right) \ge d.\mathbb{E}\left(e^{h(\beta X-\gamma(\beta))}\right).\mathbb{E}\left((Y_{n-1}(\beta))^h\right) - d(d-1)(1-h).\mathbb{E}^2\left(e^{h(\beta X-\gamma(\beta))/2}\right).\mathbb{E}^2\left((Y_{n-1}(\beta))^{h/2}\right)$$

puis via l'inégalité de Jensen :

$$\mathbb{E}\left((Y_n(\beta))^h\right).\left(\exp[(1-h)\mathrm{Logd}-h\gamma(\beta)+\gamma(h\beta)] - 1\right) \le (1-h)d^{1-h}(d-1).\mathbb{E}^2\left((Y_{n-1}(\beta))^{h/2}\right)$$

d'où en faisant croître h vers 1 après division par (1-h) :
$0 < \mathrm{Logd} - g(\beta) \le (d-1)\,\mathbb{E}^2\left((Y_{n-1}(\beta))^{1/2}\right) \xrightarrow[n\to\infty]{} (d-1)\,\mathbb{E}^2\left((Y_\infty(\beta))^{1/2}\right)$,
puisque $(Y_n(\beta))^{1/2}$, bornée dans L^2, est équiintégrable; le résultat découle maintenant de 14).

D'après 12) on a $Z_n(\beta) = \mathrm{Logd} + \gamma(\beta) + n^{-1}\mathrm{Log}(Y_n(\beta))$,
et donc 15) donne l'expression de Z_∞ sous la valeur critique :

$$16)\quad 0 \le \beta < \beta_c \Rightarrow \left(Z_n(\beta) \xrightarrow[n\to\infty]{p.s.} \mathrm{Logd} + \gamma(\beta) \right.$$

On va maintenant déduire dans 17) et 18) le comportement lorsque $\beta>\beta_c$ du comportement sous β_c , à l'aide pour la minoration d'un argument de convexité et pour la majoration de la domination de S^*_n .

$$17)\quad \beta \ge \beta_c \Rightarrow \liminf_{n\to\infty} Z_n(\beta) \ge \beta.\gamma'(\beta_c) \quad \text{p.s.} ;$$

preuve $\frac{\partial^2}{\partial\beta^2} \mathrm{Log}\left(\sum_{i=1}^n (x_i)^\beta\right) =$

$$\left(\sum_{i=1}^n (x_i)^\beta\right)^{-2} . \left(\left(\sum_{i=1}^n (x_i)^\beta \mathrm{Log}^2 x_i\right)\left(\sum_{i=1}^n (x_i)^\beta\right) - \left(\sum_{i=1}^n (x_i)^\beta \mathrm{Log} x_i\right) \right) \ge 0$$

par inégalité de Schwarz , pour tous $x_i>0$, ce qui montre que Z_n est p.s. convexe ; on a donc pour $0 < \beta_1 < \beta_2 < \beta_c \le \beta$:
$Z_n(\beta) \ge Z_n(\beta_2) + (\beta-\beta_2)(\beta_2-\beta_1)^{-1}(Z_n(\beta_2)-Z_n(\beta_1))$ p.s. , d'où par 16) :
$\liminf_{n\to\infty} Z_n(\beta) \ge \gamma(\beta_2) + \mathrm{Logd} + (\beta-\beta_2)(\beta_2-\beta_1)^{-1}(\gamma(\beta_2)-\gamma(\beta_1))$ p.s. ;
faisant ensuite croître β_1 vers β_2 , puis β_2 vers β_c , on en déduit :
$\liminf_{n\to\infty} Z_n(\beta) \ge \mathrm{Logd} + \gamma(\beta_c) + (\beta-\beta_c)\gamma'(\beta_c)$ p.s. ;
enfin on utilise que $g(\beta_c)=\mathrm{Logd}$, i.e. que $\beta_c\gamma'(\beta_c) = \mathrm{Logd} + \gamma(\beta_c)$.

$$18)\quad \beta \ge \beta_c \Rightarrow \limsup_{n\to\infty} Z_n(\beta) \le \beta.\gamma'(\beta_c) \quad \text{p.s.} ;$$

preuve: On a pour $0 < \beta' < \beta_c \le \beta$:
$Z_n(\beta) = Z_n(\beta') + \left[Z_n(\beta'+(\beta-\beta')) - Z_n(\beta') \right]$

$$= Z_n(\beta') + n^{-1}\mathrm{Log}\left(\left[\sum_{|b|=n} e^{\beta' S_b} e^{(\beta-\beta')S_b}\right].\left[\sum_{|b|=n} e^{\beta' S_b}\right]^{-1}\right)$$

$\le Z_n(\beta') + (\beta-\beta')n^{-1}S^*_n$, d'où par 11) et 16) :
$\limsup_{n\to\infty} Z_n(\beta) \le \mathrm{Logd} + \gamma(\beta') + (\beta-\beta').I^{-1}(\mathrm{Logd})$ p.s. , d'où le résultat en faisant croître β' vers β_c et en utilisant 8) .

19) La convergence presque sûre du point ii) du théorème est établie par 16),17) et 18). Remarquons ensuite l'encadrement évident :

$$n^{-1}S^*_n \le \beta^{-1}Z_n(\beta) \le n^{-1}S^*_n + \beta^{-1}\mathrm{Logd} \;;$$

si l'on y fait tendre n vers $+\infty$, pour $\beta>\beta_c$ si $\beta_c<\infty$, puis β vers $+\infty$, on obtient la convergence presque sûre du point iii) du théorème, avec l'aide de 3) si $\beta_c = \infty$.

Enfin les convergences dans L^p découlent aussitôt de :

20) $\{ \exp|n^{-1}S_n^*| \mid n\in\mathbb{N}^* \}$ est bornée dans L^p pour $1\le p<\infty$, et donc $\{ |n^{-1}S_n^*|^p \mid n\in\mathbb{N}^* \}$ et $\{ |Z_n(\beta)|^p \mid n\in\mathbb{N}^*, 0\le\beta\le B \}$ sont uniformément intégrables ;

preuve: par convexité de I il existe a et u >0 tels que $\theta\ge u \Rightarrow I(\theta)>a\theta$; fixons p>0 , et $\alpha = \mathrm{Max}\{e^{up},d^{p/a}\}$; on a alors pour tout $n > (1+p)/a$:

$$\mathbb{E}\left(e^{pn^{-1}S_n^*}\right) = \int_0^\infty \mathbb{P}\left(e^{pn^{-1}S_n^*}>r\right) dr = \int_0^\infty \mathbb{P}\left(S_n^* > np^{-1}\mathrm{Log}r\right) dr$$

$$\le \alpha + d^n.\int_\alpha^\infty e^{-nI(p^{-1}\mathrm{Log}r)} dr = \alpha + pd^n.\int_{p^{-1}\mathrm{Log}\alpha}^\infty e^{-nI(\theta)}e^{p\theta} d\theta$$

$$\le \alpha + pd^n.\int_{p^{-1}\mathrm{Log}\alpha}^\infty e^{(p-na)\theta} d\theta = \alpha + p\alpha(na-p)^{-1}e^{n(\mathrm{Log}d-ap^{-1}\mathrm{Log}\alpha)}$$

$$\le \alpha(1+p) < \infty ;$$

enfin tout ceci étant vrai avec (-X) comme avec X , on peut remplacer S_n^* par $|S_n^*|$.■

BIBLIOGRAPHIE

[B] Biggins J.D. *Chernoff's Theorem in the Branching Random Walk.* J. Appl. Prob. 14, 630-636, 1977.

[CD] Cook J.- Derrida B. *Finite-Size Effects in Random Energy Models and in the Problem of Polymers in a Random Medium.* J. of Stat. Physics, vol 63, 505-539, 1991.

[CK] Collet P.- Koukiou F. *Thermodynamics of the multiplicative chaos.* Preprint du centre de physique théorique de l'école polytechnique, 1992.

[DS] Derrida B.- Spohn H. *Polymers on Disordered Trees, Spin Glasses, and Traveling Waves.* J. of Stat. Physics, vol 51, 817-840, 1988.

[F] Franchi J. *Chaos Multiplicatif: un Traitement Simple et Complet de la Fonction de Partition.* Prépublication n° 148 du Laboratoire de Probabilités de Paris 6, Janvier 1993.

[K] Kahane J.P. *Sur le Modèle de Turbulence de B. Mandelbrot.* C.R.A.S. 278, série A, 621-623, 1974.

[KP] Kahane J.P.- Peyrière J. *Sur certaines Martingales de B. Mandelbrot.* Adv. in Math. 22, 131-145, 1976.

[M] Mandelbrot B. *Multiplications Aléatoires Itérées et Distributions Invariantes par Moyenne Pondérée Aléatoire.* C.R.A.S. 278, série A, 289-292, 1974.

On the Hypercontractivity of Ornstein–Uhlenbeck Semigroups with Drift*

Zhongmin Qian and Sheng-Wu He

In the framework of white noise analysis we study an Ornstein–Uhlenbeck semigroup with drift, which is a self-adjoint operator. Let $(S) \subset (L^2) \subset (S)^*$ be the Gel'fand's triple over white noise space $(S'(R), \mathcal{B}(S'(R)), \mu)$. Let H be a strictly positive self-adjoint operator in $L^2(R)$. Then

$$P_t^H \varphi(x) = \int_{S'(R)} \varphi(e^{-tH}x + \sqrt{1 - e^{-2tH}}y)\mu(dy), \varphi \in (S), t \geq 0,$$

determines a diffusion semigroup in $(L^p), p \geq 1$, called the Ornstein–Uhlenbeck semigroup with drift operator H. We shall show that the Bakry-Emery's curvature of $(P_t^H)_{t\geq 0}$ is bounded below by

$$\alpha = \inf_{0 \neq \xi \in S(R)} \frac{(H\xi, H\xi)}{(H\xi, \xi)}.$$

In particular if $\alpha > 0$, then (P_t^H) is hypercontractive : for any $p \geq 1$, $q(t) = 1 + (p-1)e^{2\alpha t}$ and nonnegative $f \in (L^p)$,

$$||P_t^H f||_{q(t)} \leq ||f||_p.$$

The importance of hypercontractivity for classical Ornstein–Uhlenbeck semigroup in the constructive quantum field theory has already been shown by E. Nelson (cf. [13], [14], [20] and [21]). Since then it became an active research field (cf. [6] and [20]). Moreover, it is clear recently that there are connections between hypercontractivity and spectral theory , and other aspects of operator theory (cf. [2], [6] and [19]). In his famous paper [9], L. Gross established the equivalence between logarithmic Sobolev inequality and hypercontractivity of diffusion semigroups. In recent, D. Bakry and M. Emery ([3]) gave a local criterion (i.e., only involved with the generator of a diffusion semigroup) for hypercontractivity (cf. [2] and references there). Thus one way to establish a hypercontractivity criterion for the semigroup $(P_t^H)_{t\geq 0}$ is to identify the Dirichlet

*The project supported by National Natural Science Foundation of China and in part by a Royal Society Fellowship for Zhongmin Qian.

space associated with the semigroup $(P_t^H)_{t\geq 0}$. In this paper, however, we computer the Bakry-Emery's curvature of the semigroup $(P_t^H)_{t\geq 0}$.

A brief introduction to white noise analysis is given in section 1. More materials on white noise analysis may be obtained from [11] or [22]. Ornstein–Uhlenbeck semigroup with drift is defined in section 2. A detailed discussion on Ornstein–Uhlenbeck semigroup may be found in [10]. A lower bound of the Bakry-Emery's curvature of the semigroup $(P_t^H)_{t\geq 0}$, then a hypercontractivity criterion are established in section 3.

1. White noise space. Let $S(R)$ be the Schwartz space of rapidly decreasing functions on R .Denote by A the self-adjoint extention of the harmonic oscillator operator in $L^2(R)$:

$$Af(u) = -f''(u) + (1+u^2)f(u), \quad f \in S(R).$$

Put

$$e_n(u) = (-1)^n(\pi^{1/2}2^n n!)^{-1/2} e^{u^2/2} \frac{d^n}{du^n} e^{-u^2}, \quad n \geq 0.$$

Then $e_n \in S(R)$ is the eigenfunction of A, corresponding to eigenvalue $2n+2$, and $\{e_n, n \geq 0\}$ is an orthogonal normed basis of $L^2(R)$. Define

$$|f|_{2,p}^2 = |A^p f|_2^2 = \sum_{n=0}^{\infty} (2n+2)^{2p} |\langle f, e_n\rangle|^2, \quad f \in L^2(R),$$

$$S_p(R) = \mathcal{D}(A^p) = \{f \in L^2(R) : |f|_{2,p}^2 < \infty\}, \qquad p \geq 0,$$

where $|\cdot|_2$ denotes the norm of $L^2(R)$. With $\{|\cdot|_{2,p},\ p \geq 0\}$ $S(R)$ is a nuclear space. Let $S'(R)$ be its dual space. Set

$$S_p(R) = \{f \in S'(R) : |f|_{2,p}^2 = \sum_{n=0}^{\infty} (2n+2)^{2p} |\langle f, e_n\rangle|^2 < \infty\}, \qquad p \in R,$$

where $\langle \cdot, \cdot \rangle$ denotes the pairing between $S(R)$ and $S'(R)$. Then

$$S(R) = \bigcap_{p \in R} S_p(R), \qquad S'(R) = \bigcup_{p \in R} S_p(R).$$

The famous Minlos theorem states that there exists a unique probability measure μ on $\mathcal{B}(S'(R))$, the σ-field generated by cylinder sets, such that

$$\int_{S'(R)} e^{i\langle x, \xi\rangle} \mu(dx) = \exp\Big\{-\frac{1}{2}|\xi|_2^2\Big\}, \quad \xi \in S(R).$$

The measure μ is called the white noise measure, and the probability space $(S'(R), \mathcal{B}(S'(R)), \mu)$ is called the white noise space. Set

$$X_\xi(x) = \langle x, \xi \rangle, \qquad x \in S'(R),\ \xi \in S(R).$$

$\{X_\xi,\ \xi \in S(R)\}$ is called the canonical process on the white noise space. Under μ the canonical process is a Gaussian process with zero mean and covariance $C(\xi,\eta) = \langle \xi,\eta\rangle,\ \xi,\eta \in S(R)$. On white noise space one can define a Brownian motion $B = \{B_t, -\infty < t < \infty\}$ such that $X_\xi = \int_{-\infty}^{\infty} \xi(t)dB_t$ and $\mathcal{B}(S'(R)) = \sigma\{B_t, -\infty < t < \infty\}$. Each $\varphi \in (L^2) = L^2(S'(R), \mathcal{B}(S'(R)), \mu)$ has chaotic representation:

$$\varphi = \sum_{n=0}^{\infty} \int \cdots \int \varphi^{(n)}(t_1, ..., t_n) dB_{t_1} ... dB_{t_n}, \tag{1.1}$$

$$\|\varphi\|_2^2 = \sum_{n=0}^{\infty} n! |\varphi^{(n)}|_2^2,$$

where $\varphi^{(n)} \in \widehat{L}^2(R^n)$ (the symmetric subspace of $L^2(R^n)$), $\|\cdot\|_2$ denotes the norm of (L^2). We denote (1.1) also by $\varphi \sim (\varphi^{(n)})$ simply. If for all n, $\varphi^{(n)} \in \mathcal{D}(A^{\otimes n})$, and $\sum_{n=0}^{\infty} n! |A^{\otimes n}\varphi^{(n)}|_2^2 < \infty$, define

$$\Gamma(A)\varphi \in (L^2), \qquad \Gamma(A)\varphi \sim (A^{\otimes n}\varphi^{(n)}). \tag{1.2}$$

$\Gamma(A)$ is a self-adjoint linear operator in (L^2), and is called the second quantization of A. For $p \geq 0$, set

$$(S)_p = \mathcal{D}(\Gamma(A)^p),$$

$$\|\varphi\|_{2,p}^2 = \|\Gamma(A)^p\varphi\|_2^2 = \sum_{n=0}^{\infty} n! |\varphi^{(n)}|_{2,p}^2, \quad \varphi \sim (\varphi^{(n)}) \in (S)_p.$$

$$(S) = \bigcap_{p \geq 0} (S)_p.$$

With $\{\|\cdot\|_{2,p},\ p \geq 0\}$ (S) is also a nuclear space, each element of (S) is called a test functional. Denote by $(S)_{-p}$ the dual of $(S)_p, p \geq 0$, by $(S)^*$ the dual of (S), then

$$(S)^* = \bigcup_{p \geq 0} (S)_{-p}.$$

Each element of $(S)^*$ is called a generalized Wiener functional or Hida distribution. (S) is an algebra, and each $\varphi \in (S)$ has a continuous version (in the strong topology of $S'(R)$), thus each member of (S) is assumed continuous in the sequel (cf. [23]).

For $\xi \in L^2(R)$, exponential functional $\mathcal{E}(\xi)$ is defined as

$$\mathcal{E}(\xi) = \exp\Big\{\langle \cdot, \xi\rangle - \frac{1}{2}|\xi|_2^2\Big\} \sim \Big(\frac{1}{n!}\xi^{\otimes n}\Big).$$

If $\xi \in S(R)$, then $\mathcal{E}(\xi)$ is a test functional. Let $F \in (S)^*$. The S-transform of F is defined as

$$(SF)(\xi) = \langle\!\langle F, \mathcal{E}(\xi)\rangle\!\rangle, \qquad \xi \in S(R),$$

where $\langle\!\langle \cdot, \cdot \rangle\!\rangle$ denotes the pairing between $(S)^*$ and (S).

A functional U on $S(R)$ is called a U-functional, if

1) for each $\xi \in S(R)$ the mapping $\lambda \to U(\lambda\xi)$ has analytic continuation, denoted by $u(z,\xi)$, on the whole plane;
2) for each $n \geq 1$

$$U_n(\xi_1 \otimes \cdots \otimes \xi_n) = \frac{1}{n!}\sum_{k=1}^{n}(-1)^{n-k} \sum_{l_1<\cdots<l_k} \frac{d^n}{dz^n}u(0, \xi_{l_1} + \cdots + \xi_{l_k})$$

is multilinear in $(\xi_1, \cdots, \xi_n) \in (S)^n$;
3) there exist constants $C_1 > 0$, $C_2 > 0$, $p \in R$ such that for all z and ξ

$$|u(z,\xi)| \leq C_1 \exp\{C_2 |z|^2 |\xi|_{2,-p}^2\}.$$

Potthoff and Streit (cf. [15]) have proved that a functional on $S(R)$ is the S-transform of a Hida distribution if and only if it is a U-functional. Each Hida distribution is uniquely determined by its S-transform. For any $F, G \in (S)^*$ there exists a unique Hida distribution, denoted by $F : G$ and called the Wick product of F and G, such that $S(F : G) = S(F)S(G)$.

Let ν be a probability measure on $(S'(R), \mathcal{B}(S'(R)))$. If under ν the canonical process $X = \{X_\xi, \xi \in S(R)\}$ is a Gaussian process, we call ν a Gaussian measure (cf. [10]). In this case, the mean functional

$$\langle m_\nu, \xi\rangle = \int X_\xi d\nu, \qquad \xi \in S(R),$$

is a generalized function, i.e., $m_\nu \in S'(R)$, and the covariance functional

$$C_\nu(\xi,\eta) = \int X_\xi X_\eta d\nu - \langle m_\nu, \xi\rangle\langle m_\nu, \eta\rangle, \qquad \xi, \eta \in S(R),$$

is a nonnegative-definite continuous bilinear functional on $S(R) \times S(R)$. The characteristic functional of Gaussian measure ν is

$$\int e^{i\langle x,\xi\rangle}\nu(dx) = \exp\Big\{i\langle m_\nu, \xi\rangle - \frac{1}{2}C_\nu(\xi,\xi)\Big\}, \quad \xi \in S(R),$$

and it is not difficult to see

$$\int \mathcal{E}(\xi) d\nu = \exp\Big\{-\frac{1}{2}|\xi|_2^2 + \langle m_\nu, \xi\rangle + \frac{1}{2}C_\nu(\xi,\xi)\Big\} \tag{1.3}$$

is a U-functional. For any affine transform T on $S'(R)$, νT^{-1} remains a Gaussian measure (see Theorem 2 in [10]).

Let $y \in S'(R)$ and $\varphi \in (S)$. The derivative $D_y\varphi$ of φ in direction y is defined by

$$D_y\varphi = \lim_{t\to 0} \frac{\varphi(\cdot + ty) - \varphi}{t},$$

where the limit is taken in (S). For any $F \in (S)^*$

$$\langle\langle F, D_y\varphi\rangle\rangle = \langle\langle F{:}I_1(y), \varphi\rangle\rangle, \tag{1.4}$$

where $I_1(y) \sim (0, y, 0, \cdots) \in (S)^*$. For any φ, $\psi \in (S)$

$$D_y(\varphi\psi) = (D_y\varphi)\psi + \varphi(D_y\psi), \quad D_y(\varphi{:}\psi) = (D_y\varphi){:}\psi + \varphi{:}(D_y\psi). \tag{1.5}$$

2. Ornstein-Uhlenbeck semigroup. Let H be a strictly positive self-adjoint operator in $L^2(R)$. Set

$$M_t = e^{-tH}, \qquad T_t = \sqrt{1 - e^{-2tH}} = \sqrt{1 - M_{2t}}, \quad t \geq 0. \tag{2.1}$$

We make the following assumptions:

(H_1) $S(R) \subset \mathcal{D}(H)$ and H is a continuous mapping from $S(R)$ into itself.

(H_2) $\forall t > 0$ M_t and T_t are continuous operators from $S(R)$ into itself.

Then M_t and T_t, $t > 0$, can be extended onto $S'(R)$: $\forall x \in S'(R)$, $\xi \in S(R)$,

$$\langle M_t x, \xi\rangle = \langle x, M_t\xi\rangle, \quad \langle T_t x, \xi\rangle = \langle x, T_t\xi\rangle. \tag{2.2}$$

Now for all $t \geq 0, x \in S'(R)$ and $\varphi \in (S)$ define

$$P_t^H\varphi(x) = \int \varphi(M_t x + T_t y)\mu(dy) = \int \varphi(y)\mu^H_{x,t}(dy), \tag{2.3}$$

where $\mu^H_{x,t}$ is a Gaussian measure with mean functional $\langle M_t x, \xi\rangle$ and covariance fuctional $\langle (1 - e^{-2tH})\xi, \eta\rangle$. Hence the definition (2.3) makes sense.

Let $\Gamma(e^{-tH}) = \Gamma(M_t)$ be the second quantization of M_t. Then we have

$$P_t^H = \Gamma(e^{-tH}) = e^{-td\Gamma(H)}, \qquad t \geq 0, \tag{2.4}$$

where $d\Gamma(H)$ is a self-adjoint operator in (L^2):

$$d\Gamma(H) = \sum_{n=1}^{\infty} \oplus\{\underbrace{H \otimes I \otimes \cdots \otimes I}_{n \text{ factors}} + \underbrace{I \otimes H \otimes I \otimes \cdots \otimes I}_{n \text{ factors}} + \cdots + \underbrace{I \otimes \cdots \otimes H}_{n \text{ factors}}\}$$

i.e., $\{P_t^H, t \geq 0\}$ is a Markov semigroup with infinitesimal generator $L_H = -d\Gamma(H)$. $\{P_t^H, t > 0\}$ is called the Ornstein-Uhlenbeck semigroup with drift operator H . When $H = I$, the identity operator, it reduces to ordinary infinite dimensional Ornstein-Uhlenbeck semigroup (Refer to Theorem 8 in [10]). To help the understanding the definition of semigroup $(P_t^H)_{t\geq 0}$, the reader may think of its finite dimensional analogue. In this case, the Hilbert space $L^2(R)$ is replaced by R^n, μ is the standard normal measure on R^n and H is a positive symmetric matrix, e.g., $Hx = \sum_{i=1}^n \lambda_i \langle x, e_i \rangle e_i$, where (e_i) is the standard base of R^n, so that

$$P_t^H f(x) = \int_{R^n} f\left(e^{-Ht}x + \sqrt{1 - e^{-2Ht}}y\right) \mu(dy)$$

and

$$L_H = \frac{1}{2}\Delta - \sum_{i=1}^n \lambda_i x_i \frac{\partial}{\partial x_i}.$$

The following properties of Ornstein–Uhlenbeck semigroup are immediate.

Proposition 2.1. *For any φ, $\psi \in (S)$ and $t \geq 0$*
1) $\|P_t^H \varphi\|_2 \leq \|\varphi\|_2$,
2) $\int \varphi(P_t^H \psi) d\mu = \int (P_t^H \varphi)\psi d\mu$,
3) $\lim_{t\to 0} \|P_t^H \varphi - \varphi\|_2 = 0$,
4) $\lim_{t\to\infty} \|P_t^H \varphi - \int \varphi d\mu\|_2 = 0$.
In particular, for any $p \geq 1$, $(P_t^H)_{t\geq 0}$ can be uniquely extended to be a μ-symmetric, contractive, strongly continuous semigroup on (L^p), and the above properties remain true.

We need also the properties of operator $d\Gamma(H)$. For any $n \geq 1$, let $H^{(n)}$ be the self-adjoint operator in $\widehat{L}^2(R^n)$ such that

$$H^{(n)}\xi^{\otimes n} = \underbrace{H\xi\widehat{\otimes}\xi\cdots\widehat{\otimes}\xi}_{n \text{ factors}}, \tag{2.5}$$

and $H^{(0)} = I$. Then for any $\varphi \sim (\varphi^{(n)}) \in \mathcal{D}(d\Gamma(H))$ by definition we have

$$d\Gamma(H)\varphi \sim (nH^{(n)}\varphi^{(n)}). \tag{2.6}$$

In particular, for any $\xi \in S(R)$

$$d\Gamma(H)\mathcal{E}(\xi) \sim (\frac{1}{(n-1)!}(H\xi)\widehat{\otimes}\xi^{\otimes n-1}) = I_1(H\xi){:}\,\mathcal{E}(\xi). \tag{2.7}$$

Proposition 2.2 $(S) \subset \mathcal{D}(d\Gamma(H))$.

Proof. Let $p \geq 0$. Since H is a continuous mapping from $S(R)$ into itself, there are $q \geq p$ and $C_p > 0$ such that for all $\xi \in S(R)$

$$|H\xi|_{2,p} \leq C_p|\xi|_{2,q}.$$

Let $\varphi \sim (\varphi^{(n)}) \in (S)$. Then

$$\begin{aligned}|H^{(n)}\varphi^{(n)}|_{2,p} &\leq \sum_{i_1,\cdots,i_n} |(\varphi^{(n)}, e_{i_1} \otimes \cdots \otimes e_{i_n})||H^{(n)}e_{i_1} \otimes \cdots \otimes e_{i_n}|_{2,p} \\ &\leq \sum_{i_1,\cdots,i_n} |(\varphi^{(n)}, e_{i_1} \otimes \cdots \otimes e_{i_n})|nC_p \prod_{k=0}^{n} (2i_k+2)^q \\ &\leq nC_p\Big[\sum_{k=0}^{\infty}(2k+2)^{-2}\Big]^{n/2}|\varphi^{(n)}|_{2,q+1},\end{aligned}$$

$$\|d\Gamma(H)\varphi\|_{2,p}^2 \leq C_p^2 \sum_{n=0}^{\infty} n!n^2\Big[\sum_{k=0}^{\infty}(2k+2)^{-2}\Big]^n|\varphi^{(n)}|_{2,q+1}^2 \leq \|\varphi\|_{2,q+1+\alpha}^2,$$

where $\alpha > 0$ is taken such that for all n

$$nC_p 2^{-n\alpha}\Big[\sum_{k=0}^{\infty}(2k+2)^{-2}\Big]^n < 1.$$

Thus $d\Gamma(H)\varphi \in (S)$.

From the definition of $d\Gamma(H)$ it is easy to verify directly the following

Proposition 2.3. *For any $\varphi, \psi \in \mathcal{D}(d\Gamma(H))$, we have $\varphi\colon\psi \in \mathcal{D}(d\Gamma(H))$ and*

$$d\Gamma(H)(\varphi\colon\psi) = (d\Gamma(H)\varphi)\colon\psi + \varphi\colon(d\Gamma(H)\psi).$$

Lemma 2.4. *Let $\varphi \in (S)$. Then*

$$[S(d\Gamma(H)\varphi)](\xi) = \langle\langle D_{H\xi}(\varphi), \mathcal{E}(\xi)\rangle\rangle. \tag{2.8}$$

Proof. By the symmetry of $d\Gamma(H)$, (2.7) and (1.4)

$$\begin{aligned}[S(d\Gamma(H)\varphi)](\xi) &= \langle\langle d\Gamma(H)\varphi, \mathcal{E}(\xi)\rangle\rangle = \langle\langle \varphi, d\Gamma(H)\mathcal{E}(\xi)\rangle\rangle \\ &= \langle\langle \varphi, I_1(H\xi)\colon\mathcal{E}(\xi)\rangle\rangle = \langle\langle D_{H\xi}(\varphi), \mathcal{E}(\xi)\rangle\rangle.\end{aligned}$$

Corollary 2.5. *For any $\xi, \eta, \zeta \in S(R)$*

$$[S(\mathcal{E}(\eta)d\Gamma(H)\mathcal{E}(\zeta))](\xi) = (H\zeta, \eta+\xi)e^{(\xi,\eta)+(\eta,\zeta)+(\zeta,\xi)}, \tag{2.9}$$

that is

$$\mathcal{E}(\eta)d\Gamma(H)\mathcal{E}(\zeta)=\{(H\zeta,\eta)+I_1(H\zeta)\}:\{\mathcal{E}(\eta)\mathcal{E}(\zeta)\}. \tag{2.10}$$

Proof. Note that for any ξ, $\eta \in S(R)$

$$\mathcal{E}(\xi)\mathcal{E}(\eta)=\mathcal{E}(\xi+\eta)e^{(\xi,\eta)}, \tag{2.11}$$

and by (2.7)

$$[S(d\Gamma(H)\mathcal{E}(\eta))](\xi)=(H\eta,\xi)e^{(\xi,\eta)}. \tag{2.12}$$

Now from (2.11) and (2.12) we have

$$\begin{aligned}[S(\mathcal{E}(\eta)d\Gamma(H)\mathcal{E}(\zeta))](\xi)&=\langle\!\langle \mathcal{E}(\eta)d\Gamma(H)\mathcal{E}(\zeta),\mathcal{E}(\xi)\rangle\!\rangle\\&=\langle\!\langle d\Gamma(H)\mathcal{E}(\zeta),\mathcal{E}(\xi+\eta)\rangle\!\rangle e^{(\xi,\eta)}\\&=(H\zeta,\eta+\xi)e^{(\xi,\eta)+(\eta,\zeta)+(\zeta,\xi)}.\end{aligned}$$

Then (2.10) follows from (2.9) and (2.11).

Denote $\mathcal{A}=\mathrm{sp}\{\mathcal{E}(\xi),\,\xi\in S(R)\}$.

Lemma 2.6. *For any $\varphi\in\mathcal{A}$ we have*

$$d\Gamma(H)\varphi^3=3\varphi d\Gamma(H)\varphi^2-3\varphi^2 d\Gamma(H)\varphi. \tag{2.13}$$

Proof. At first, note that for any positive integer k and $\xi\in S(R)$

$$\mathcal{E}(\xi)^k=\mathcal{E}(k\xi)e^{\frac{1}{2}(k^2-k)|\xi|_2^2}. \tag{2.14}$$

It can be shown by induction and (2.11). By means of (2.9) and (2.14) it is easy to calculate

$$[S(d\Gamma(H)\mathcal{E}(\eta)^3)](\xi)=3(H\eta,\xi)\exp\{3|\eta|_2^2+3(\eta,\xi)\} \tag{2.15}$$
$$[S(\mathcal{E}(\eta)d\Gamma(H)\mathcal{E}(\eta)^2)](\xi)=2(H\eta,\eta+\xi)\exp\{3|\eta|_2^2+3(\eta,\xi)\} \tag{2.16}$$
$$[S(\mathcal{E}(\eta)^2d\Gamma(H)\mathcal{E}(\eta))](\xi)=(H\eta,2\eta+\xi)\exp\{3|\eta|_2^2+3(\eta,\xi)\} \tag{2.17}$$

Let $\varphi=\sum_{i=1}^n c_i\mathcal{E}(\eta_i)$, $\eta_i\in S(R)$, $c_i\in R$, $1\le i\le n$. Then (2.13) follows from (2.15)–(2.17) by straightforward computation.

Since the strong topology of $S'(R)$ is generated by $\mathcal{A}$, by making use of a Bakry and Emery's result in [3] from Lemma 2.6 we get the following

Theorem 2.7. *The semigroup* (P_t^H) *is a diffusion semigroup, i.e., for any* $\varphi_1, \cdots, \varphi_n \in \mathcal{D}(L_H)^n$ *and* $\Phi \in C_b^2(R^n)$ *with* $\Phi(\varphi_1, \cdots, \varphi_n) \in \mathcal{D}(L_H)$ *we have*

$$L_H\Phi(\varphi_1, \cdots, \varphi_n) = \sum_{i=1}^{n} \frac{\partial \Phi}{\partial x_i}(\varphi_1, \cdots, \varphi_n) L_H\varphi_i$$
$$+ \frac{1}{2}\sum_{i,j=1}^{n} \frac{\partial^2}{\partial x_i \partial x_j}(\varphi_1, \cdots, \varphi_n)[L_H(\varphi_i\varphi_j) - \varphi_i(L_H\varphi_j) - (L_H\varphi_i)\varphi_j].$$

If we denote by $(\mathcal{E}, \mathcal{F})$ the Dirichlet form associated with the μ-symmetric semigroup (P_t^H), then $(\mathcal{E}, \mathcal{F})$ is local (cf. [5], [8] and [12]). It is not difficult to show that there is a diffusion process $X = (X_t, P^x)$ with transition semigroup (P_t^H). Then for any bounded $\varphi \in \mathcal{D}(L_H)$

$$M_t^\varphi = \varphi(X_t) - \varphi(X_0) - \int L_H\varphi(X_s)ds$$

is a P^x-martingale for any x, and $d\langle M^\varphi, M^\varphi\rangle_t \ll dt$ (cf. [8] and [12]). In fact, we have

$$\langle M^\varphi, M^\varphi\rangle_t = \int_0^t [L_H(\varphi^2) - 2\varphi(L_H\varphi)](X_s)ds.$$

3. The hypercontractivity of (P_t^H). Define

$$\Gamma(\varphi, \psi) = \frac{1}{2}\{L_H(\varphi\psi) - \varphi(L_H\psi) - (L_H\varphi)\psi\}, \quad (\varphi, \psi) \in \mathcal{D}(\Gamma),$$

where

$$\mathcal{D}(\Gamma) = \{(\varphi, \psi) : \varphi, \psi, \varphi\psi \in \mathcal{D}(L_H)\},$$

Obviously, by Proposition 2.2 we have $(S) \times (S) \subset \mathcal{D}(\Gamma)$, since (S) is an algebra. Γ is called the square field operator of the semigroup (P_t^H). Define

$$\Gamma_2(\varphi, \psi) = \frac{1}{2}\{L_H\Gamma(\varphi, \psi) - \Gamma(L_H\varphi, \psi) - \Gamma(\varphi, L_H\psi)\}, \quad \varphi, \psi \in \mathcal{D}(\Gamma_2),$$

where

$$\mathcal{D}(\Gamma_2) = \{\varphi : \varphi, \varphi^2, L_H\varphi, \varphi L_H\varphi, L_H\varphi^2 \in \mathcal{D}(L_H)\}.$$

By the same reason we still have $(S) \subset \mathcal{D}(\Gamma_2)$. Γ_2 is called the iterated square field operator of the semigroup (P_t^H) or the Bakry-Emery's curvature of the diffusion operator L_H, and was introduced by D. Bakry ([1]).

Lemma 3.1. *For any* $\eta, \zeta \in S(R)$

$$\Gamma(\mathcal{E}(\eta), \mathcal{E}(\zeta)) = (H\eta, \zeta)\mathcal{E}(\eta)\mathcal{E}(\zeta). \tag{3.1}$$

Proof. Let $\xi \in S(R)$. For convenience, denote

$$a(\xi, \eta, \zeta) = e^{(\xi, \eta)+(\eta, \zeta)+(\zeta, \xi)} = [S(\mathcal{E}(\eta)\mathcal{E}(\zeta))](\xi). \tag{3.2}$$

We are to check the S-transforms of the two sides of (3.1) are equal. By (2.12) we have

$$\begin{aligned}[S(L_H(\mathcal{E}(\eta)\mathcal{E}(\zeta))](\xi) &= [S(L_H\mathcal{E}(\eta+\zeta))](\xi)e^{(\eta,\zeta)} \\ &= -(H(\eta+\zeta), \xi)e^{(\xi,\eta+\zeta)}e^{(\eta,\zeta)} \\ &= \{-(H\eta, \xi) - (H\zeta, \xi)\}a(\xi, \eta, \zeta).\end{aligned} \tag{3.3}$$

Similarly, by (2.9) we have

$$[S(\mathcal{E}(\eta)L_H\mathcal{E}E(\zeta))](\xi) = \{-(H\zeta, \eta) - (H\zeta, \xi)\}a(\xi, \eta, \zeta), \tag{3.4}$$

$$[S(\mathcal{E}(\zeta)L_H\mathcal{E}(\eta))](\xi) = \{-(H\eta, \zeta) - (H\eta, \xi)\}a(\xi, \eta, \zeta). \tag{3.5}$$

Noting that H is symmetric, from (3.3), (3.4) and (3.5) we get

$$\begin{aligned}[S\Gamma(\mathcal{E}(\eta), \mathcal{E}(\zeta))](\xi) &= (H\eta, \zeta)a(\xi, \eta, \zeta) \\ &= [S((H\eta, \zeta)\mathcal{E}(\eta)\mathcal{E}(\zeta))](\xi).\end{aligned}$$

Thus (3.1) is verified.

Lemma 3.2. *For any* $\eta, \zeta \in S(R)$

$$\Gamma_2(\mathcal{E}(\eta), \mathcal{E}(\zeta)) = \{(H\eta, H\zeta) + (H\eta, \zeta)^2\}\mathcal{E}(\eta)\mathcal{E}(\zeta). \tag{3.6}$$

Proof. Samely, we need only to verify for $\xi \in S(R)$

$$[S\Gamma_2(\mathcal{E}(\eta), \mathcal{E}(\zeta))](\xi) = \{(H\eta, H\zeta) + (H\eta, \zeta)^2\}a(\xi, \eta, \zeta). \tag{3.7}$$

At first, by (2.12) and Lemma 3.1 we have

$$\begin{aligned}&[SL_H\Gamma(\mathcal{E}(\eta), \mathcal{E}(\zeta))](\xi) = (\eta, \zeta)e^{(\eta,\zeta)}[SL_H\mathcal{E}E(\eta+\zeta)](\xi) \\ &= -(H\eta, \zeta)\{(H\xi, \eta) + (H\xi, \zeta)\}a(\xi, \eta, \zeta).\end{aligned} \tag{3.8}$$

Secondly, we are to calculate the S-transform of $\Gamma(L_H\mathcal{E}(\eta), \mathcal{E}(\zeta))$. By (2.8) and Proposition 2.3

$$[SL_H(\mathcal{E}(\zeta)L_H\mathcal{E}(\eta))](\xi) = \langle\langle D_{-H\xi}[\mathcal{E}(\zeta)L_H\mathcal{E}(\eta)], \mathcal{E}(\xi)\rangle\rangle$$
$$= \langle\langle -(H\xi, \zeta)\mathcal{E}(\zeta)L_H\mathcal{E}(\eta), \mathcal{E}(\xi)\rangle\rangle + \langle\langle \mathcal{E}(\zeta)D_{-H\xi}L_H\mathcal{E}(\eta), \mathcal{E}(\xi)\rangle\rangle$$
$$= -(H\xi, \zeta)\langle\langle L_H\mathcal{E}(\eta), \mathcal{E}(\xi+\zeta)\rangle\rangle e^{(\xi,\zeta)} + \langle\langle D_{-H\xi}L_H\mathcal{E}(\eta), \mathcal{E}(\xi+\zeta)\rangle\rangle e^{(\xi,\zeta)}$$
$$= (H\xi, \zeta)(H\eta, \xi+\zeta)a(\xi, \eta, \zeta) + \langle\langle -I_1(H\xi){:}\,\mathcal{E}(\xi+\zeta), L_H\mathcal{E}(\eta)\rangle\rangle e^{(\xi,\zeta)}$$
$$= (H\xi,\zeta)(H\eta, \xi+\zeta)a(\xi, \eta, \zeta) - \{\langle\langle (L_HI_1(H\xi)){:}\,\mathcal{E}(\xi+\zeta), \mathcal{E}(\eta)\rangle\rangle$$
$$+ \langle\langle I_1(H\xi){:}\,(L_H\mathcal{E}(\xi+\zeta)), \mathcal{E}(\eta)\rangle\rangle\}e^{(\xi,\zeta)}$$
$$= (H\xi, \zeta)(H\eta, \xi+\zeta)a(\xi, \eta, \zeta) + \langle\langle I_1(H^2\xi), \mathcal{E}(\eta)\rangle\rangle a(\xi, \eta, \zeta)$$
$$- (H\xi, \eta)\langle\langle L_H\mathcal{E}(\xi+\zeta), \mathcal{E}(\eta)\rangle\rangle e^{(\xi,\zeta)}$$
$$= \{(H\xi, \zeta)(H\eta, \xi+\zeta) + (H\xi, H\eta) + (H\xi,\eta)(H\eta, \xi+\zeta)\}a(\xi, \eta, \zeta)$$
$$= \{(H\xi, H\eta) + (H\xi, \eta+\zeta)(H\eta, \xi+\zeta)\}a(\xi, \eta, \zeta). \tag{3.9}$$

Using the symmetry of L_H and $L_H\mathcal{E}(\eta) \in (S)$ we get

$$[S(L_H\mathcal{E}(\eta)L_H\mathcal{E}(\zeta))](\xi) = \langle\langle L_H\mathcal{E}(\eta)L_H\mathcal{E}(\zeta), \mathcal{E}(\xi)\rangle\rangle$$
$$= \langle\langle L_H\mathcal{E}(\zeta), \mathcal{E}(\xi)L_H\mathcal{E}(\eta)\rangle\rangle = \langle\langle L_H(\mathcal{E}(\xi)L_H\mathcal{E}(\eta)), \mathcal{E}(\zeta)\rangle\rangle$$
$$= \{(H\zeta, H\eta) + (H\zeta, \xi+\eta)(H\eta, \xi+\zeta)\}a(\xi, \eta, \zeta), \tag{3.10}$$

where the last equality comes from (3.9). By using (2.7) repeatedly we have

$$L_H^2\mathcal{E}(\eta) = I_1(H^2\xi){:}\,\mathcal{E}(\eta) + I_1(H\eta){:}\,I_1(H\eta){:}\,\mathcal{E}(\eta).$$

Thus

$$[S(\mathcal{E}(\zeta)L_H^2\mathcal{E}(\eta))](\xi) = \langle\langle L_H^2\mathcal{E}(\eta), \mathcal{E}(\xi+\zeta)\rangle\rangle e^{(\xi,\zeta)}$$
$$= \{(H\eta, H(\xi+\zeta)) + (H\eta, \xi+\zeta)^2\}a(\xi, \eta, \zeta). \tag{3.11}$$

Combining (3.9), (3.10) and (3.11), we get

$$[S\Gamma(L_H\mathcal{E}(\eta), \mathcal{E}(\zeta))](\xi)$$
$$= \frac{1}{2}\{(H\xi, H\eta) + (H\xi, \eta+\zeta)(H\eta, \xi+\zeta) - (H\eta, \zeta)$$
$$- (H\zeta, \xi+\eta)(H\eta, \xi+\zeta) - (H\eta, H\xi+H\zeta) - (H\eta, \xi+\zeta)^2\}a(\xi, \eta, \zeta)$$
$$= \{-(H\eta, H\zeta) - (H\eta, \zeta)(H\eta, \xi) - (H\eta, \zeta)^2\}a(\xi, \eta, \zeta). \tag{3.12}$$

(3.12) may also apply to $[S\Gamma(\mathcal{E}(\eta), L_H\mathcal{E}(\zeta))](\xi)$. Now (3.7) follows from (3.8) and (3.12) :

$$[S\Gamma_2(\mathcal{E}(\eta), \mathcal{E}(\zeta))](\xi)$$
$$= \frac{1}{2}\{-(H\eta, \zeta)(H\xi, \eta+\zeta) + (H\eta, H\zeta) + (H\eta, \xi)(H\eta, \zeta) + (H\eta, \zeta)^2$$
$$+ (H\eta, H\zeta) + (H\zeta, \xi)(H\zeta, \eta) + (H\zeta, \eta)^2\}a(\xi, \eta, \zeta)$$
$$= \{(H\eta, H\zeta) + (H\eta, \zeta)^2\}a(\xi, \eta, \zeta).$$

Set

$$\alpha = \inf_{0\neq\xi\in S(R)} \frac{(H\xi, H\xi)}{(H\xi, \xi)}. \tag{3.13}$$

Theorem 3.3. *For all $\varphi \in \mathcal{A}$, we have following Bakry-Emery's curvature inequality,*

$$\Gamma_2(\varphi, \varphi) \geq \alpha\Gamma(\varphi, \varphi). \tag{3.14}$$

Proof. Let $\varphi = \sum_{i=1}^n c_i\mathcal{E}(\eta_i)$, $\eta \in S(R)$, $c_i \in R$, $1 \leq i \leq n$. By Lemma 3.1, 3.2 and the condition (3.13) we know

$$\begin{aligned}&\Gamma_2(\varphi, \varphi) - \alpha\Gamma(\varphi, \varphi)\\ &= \sum_{i,j=1}^n c_ic_j\mathcal{E}(\eta_i)\mathcal{E}(\eta_j)\{(H\eta_i, H\eta_j) - \alpha(H\eta_i, \eta_j) + (H\eta_i, \eta_j)^2\} \geq 0,\end{aligned}$$

noting that $\big((H\eta_i, \eta_j)\big)$ and $\big((H\eta_i, \eta_j)^2\big)$ are nonnegative-definite.

L. Gross ([9]) and D. Bakry – M. Emery ([3]) proved that (P_t^H) is hypercontractive if there is a dense algebra $\mathcal{B}$ such that it is stable under C^∞-maps, $\mathcal{B} \times \mathcal{B} \subset \mathcal{D}(\Gamma_2) \cap \mathcal{D}(\Gamma)$, and (3.14) holds for every $\varphi \in \mathcal{B}$. Unfortunately, the algebra $\mathcal{A}$ is not stable under C^∞-maps. However, the following Theorem 3.5 permits us to establish a hypercontractivity criterion for the semigroup (P_t^H) along the lines of L. Gross and D. Bakry – M. Emery.

Lemma 3.4. *Let $(a_{i,j})_{1\leq i,j\leq n}$ be a symmetric nonnegative-definite matrix and c_i, $1 \leq i \leq n$, be arbitrary reals. Then*

$$\sum_{i,j,k,l=1}^n c_ic_jc_kc_l(a_{i,j}a_{k,l} + a_{k,l}^2 - a_{i,j}a_{i,k} - a_{i,j}a_{j,k}) \geq 0.$$

Proof. Let $(X_1, \cdots, X_n)$ and $(Y_1, \cdots, Y_n)$ be two independent random vectors with the same normal law $N(0, (a_{i,j}))$. Denote $X = \sum_{i=1}^n c_iX_i$, $Y = \sum_{i=1}^n c_iY_i$, $Z = \sum_{i=1}^n c_iX_iY_i$, $c = \sum_{i=1}^n c_i$. Then

$$
\begin{aligned}
&\sum_{i,j,k,l=1}^{n} c_i c_j c_k c_l \{a_{i,j} a_{k,l} + a_{k,l}^2 - a_{i,j} a_{i,k} - a_{i,j} a_{j,k}\} \\
= &\sum_{i,j,k,l=1}^{n} c_i c_j c_k c_l \{\mathsf{E}(X_i X_j)\mathsf{E}(Y_k Y_l) + \mathsf{E}(X_k X_l)\mathsf{E}(Y_k Y_l) \\
&\qquad - \mathsf{E}(X_i X_j)\mathsf{E}(Y_i Y_k) - \mathsf{E}(X_i X_j)\mathsf{E}(Y_j Y_k)\} \\
= &\sum_{i,j,k,l=1}^{n} c_i c_j c_k c_l \mathsf{E}(X_i X_j Y_k Y_l + X_k X_l Y_k Y_l - X_i X_j Y_i Y_k - X_i X_j Y_j Y_k) \\
= &\ \mathsf{E}(X^2 Y^2 + c^2 Z^2 - cZXY - cXZY) \\
= &\ \mathsf{E}(XY - cZ)^2 \geq 0.
\end{aligned}
$$

Theorem 3.5. *For any $\varphi \in \mathcal{A}$ with $\varphi > 0$ we have*

$$
\Gamma_2(\ln\varphi, \ln\varphi) \geq \alpha\Gamma(\ln\varphi, \ln\varphi). \tag{3.15}
$$

Proof. Since (P_t^H) is a diffusion semigroup, we have

$$
\Gamma(\ln\varphi, \ln\varphi) = \frac{1}{\varphi^2}\Gamma(\varphi, \varphi),
$$

$$
\Gamma_2(\ln\varphi, \ln\varphi) = \frac{1}{\varphi^2}\Gamma_2(\varphi, \varphi) + \frac{1}{\varphi^4}\Gamma(\varphi, \varphi)^2 - \frac{1}{\varphi^3}\Gamma(\varphi, \Gamma(\varphi, \varphi)).
$$

Let $\varphi = \sum_{i=1}^{n} c_i \mathcal{E}(\eta_i)$, $\eta_i \in S(R)$, $c_i \in R$, $1 \leq i \leq n$. Denote

$$
\begin{aligned}
\psi &= \varphi^4\Gamma_2(\ln\varphi, \ln\varphi) - \alpha\varphi^4\Gamma(\ln\varphi, \ln\varphi) \\
&= \varphi^2\Gamma_2(\varphi, \varphi) + \Gamma(\varphi, \varphi)^2 - \varphi\Gamma(\varphi, \Gamma(\varphi, \varphi)) - \alpha\varphi^2\Gamma(\varphi, \varphi).
\end{aligned}
$$

Observing that

$$
\Gamma\big(\mathcal{E}(\eta_k), \Gamma(\mathcal{E}(\eta_i, \mathcal{E}(\eta_j))\big) = (H\eta_i, \eta_j)(H\eta_k, \eta_i + \eta_j)\mathcal{E}(\eta_i)\mathcal{E}(\eta_j)\mathcal{E}(\eta_k),
$$

by Lemma 3.1 and 3.2 we have

$$
\begin{aligned}
\psi = &\sum_{i,j,k,l,=1}^{n} c_i c_j c_k c_l \{\mathcal{E}(\eta_k)\mathcal{E}(\eta_l)\Gamma_2(\mathcal{E}(\eta_i), \mathcal{E}(\eta_j)) \\
&+ \Gamma(\mathcal{E}(\eta_i), \mathcal{E}(\eta_j))\Gamma(\mathcal{E}(\eta_k), \mathcal{E}(\eta_l)) - \mathcal{E}(\eta_l)\Gamma(\mathcal{E}(\eta_k), \Gamma(\mathcal{E}(\eta_i), \mathcal{E}(\eta_j))) \\
&- \alpha\mathcal{E}(\eta_k)\mathcal{E}(\eta_l)\Gamma(\mathcal{E}(\eta_i), \mathcal{E}(\eta_j))\} \\
= &\sum_{i,j,k,l,=1}^{n} c_i c_j c_k c_l \mathcal{E}(\eta_i)\mathcal{E}(\eta_j)\mathcal{E}(\eta_k)\mathcal{E}(\eta_l)\{(H\eta_i, \eta_j)(H\eta_k, \eta_l) \\
&\qquad + (H\eta_k, \eta_l)^2 - (H\eta_i, \eta_j)(H\eta_i, \eta_k) - (H\eta_i, \eta_j)(H\eta_j, \eta_k) \\
&\qquad + (H\eta_k, H\eta_l) - \alpha(H\eta_k, \eta_l)\}.
\end{aligned}
$$

By Lemma 3.4

$$\sum_{i,j,k,l,=1}^{n} c_i c_j c_k c_l \mathcal{E}(\eta_i)\mathcal{E}(\eta_j)\mathcal{E}(\eta_k)\mathcal{E}(\eta_l)\{(H\eta_i,\eta_j)(H\eta_k,\eta_l)+(H\eta_k,\eta_l)^2$$
$$-(H\eta_i,\eta_j)(H\eta_i,\eta_k)-(H\eta_i,\eta_j)(H\eta_j,\eta_k)\}\geq 0,$$

and by the condition (3.13)

$$\sum_{i,j,k,l,=1}^{n} c_i c_j c_k c_l \mathcal{E}(\eta_i)\mathcal{E}(\eta_j)\mathcal{E}(\eta_k)\mathcal{E}(\eta_l)\{(H\eta_k,H\eta_l)-\alpha(H\eta_k,\eta_l)\}\geq 0.$$

Hence $\psi \geq 0$. So (3.15) is verified.

Now starting from Theorem 3.5 and by making use of the similar arguments in D. Bakry and M. Emery ([4]), we may get the following results consecutively. We omit the details of the proofs.

Lemma 3.6. *Let $\varphi \in \mathcal{A}, \varphi > 0$ and $\alpha > 0$. Then for any $t \geq 0$*

$$P_t^H(\varphi \ln \varphi) - (P_t^H\varphi)\ln(P_t^H\varphi) \leq \frac{1}{2\alpha}(1-e^{-2\alpha t})P_t^H\Big(\frac{1}{\varphi}\Gamma(\varphi,\varphi)\Big). \quad (3.16)$$

Proposition 3.7. *If $\varphi \in \mathcal{A}$ and $\alpha > 0$, then*

$$\int \varphi^2 \ln \varphi^2 d\mu - \Big(\int \varphi^2 d\mu\Big)\ln\Big(\int \varphi^2 d\mu\Big) \leq \frac{2}{\alpha}\int \Gamma(\varphi,\varphi)d\mu. \quad (3.17)$$

Theorem 3.8. *Assume*

$$\alpha = \inf_{0\neq\xi\in S(R)} \frac{(H\xi,H\xi)}{(H\xi,\xi)} > 0.$$

Then for any $p \geq 1$, $q(t) = 1+(p-1)e^{2\alpha t}$, $t \geq 0$ and $f \in (L^p)$ with $f \geq 0$ we have

$$\|P_t^H f\|_{q(t)} \leq \|f\|_p. \quad (3.18)$$

This hypercontractivity criterion for (P_t^H) is our main result. The equivalence of (3.17) and (3.18) was estabished by L. Gross ([9]).

Acknowledgement. The first draft of this paper is completed by Zhongmin Qian when he was at Mathematics Institute, University of Warwick. He is grateful to Professor K. D. Elworthy for helpful comments and encouragement.

REFENRENCES

[1] Bakry, D., Etude probabiliste des transformees de Riesz et de l'espace H^1 sur les spheres, Sem. Prob. XVIII, Lecture Notes in Math. **1059**, 197-218, Springer, 1984.

[2] Bakry, D., L'hypercontractivite et son utilisation en theorie des semi-groupes, Preprint, 1993.

[3] Bakry, D. and Emery, M., Diffusions hypercontractives, Sem. Prob. XIX, Lecture Notes in Math. **1123**, 177-206, Springer, 1985.

[4] Bakry, D. and Emery, M., Propaganda for Γ_2, in *From Local Times to Global Geometry, Control and Physics*, 39-46, K. D. Elworthy (ed.), Longman Sci. Tech., 1986.

[5] Bouleau, N. and Hirsh, F., *Dirichlet Forms and Analysis on Wiener Space*, Walter de Gruyter, 1991.

[6] Davis, E. B., *Heat Kernels and Spectral Theory*, Cambridge Univ. Press, 1989.

[7] Dellacherie, C. and Meyer, P. A., *Probabilites et Potentiel* IV, Hermann, 1991.

[8] Fukushima, M., *Dirichlet Forms and Markov Processes*, North Holland, 1980.

[9] Gross, L., Logarithmic Sobolev inequalities, Amer. J. Math. **97** (1975), 1061-1083.

[10] He, S.W. and Wang, J.G., Gaussian measures on white noise space, Preprint, 1993.

[11] Hida, T., Kuo, H. H., Potthoff, J. and Streit, L., *White Noise – An Infinite Dimensional Calculus*, Kluwer Academic Publ., 1993.

[12] Ma, Z. and Röckner, M., *An Introduction to Non-symmetric Dirichlet Forms*, Springer, 1992.

[13] Nelson, E., A quadratic interaction in two dimension, In *Mathematical Theory of Elementary Particles*, R. Goodman and I. Segal (eds.), M.I.T. Press, 1966.

[14] Nelson, E., The free Markov field, J. Funct. Anal. **12**(1973), 211-227.

[15] Potthoff, J. and Streit, L., A characterization on Hida distribution, J. Funct. Anal. **101**(1991), 212-229.

[16] Potthoff, J. and Yan, J. A., Some results about test and generalized functionals of white noise, In *Proc. Singapore Prob. Conf.*, L.H.Y. Chen et al. (eds.), Walter de Gruyter, 1992.

[17] Qian, Z. M., On the Martin boundary of the Ornstein-Uhlenbeck operator on the white noise space, Preprint, 1993.

[18] Röckner, M., On the parabolic Martin boundary of the Ornstein–Uhlenbeck operator on Wiener space, Ann. Prob. (1992), 1063-1085.

[19] Rothaus, O., Analytic inequalities, isoperimetric inequalities and logarithmic Sobolev inequalities, J. Funct. Anal. **64**(1985), 296-313.

[20] Reed, M. and Simon, B., *Methods of Modern Mathematical Physics*, Academic Press, 1985.

[21] Simon, B., *The $P(\varphi)_2$ Euclidean (Quantum) Field Theory*, Princeton Univ. Press, 1974.
[22] Yan J. A., Some recent developments in white noise analysis. In *Probability and Statistics*, A. Badrikian et al. (eds.), World Scientific, 1993.
[23] Yokoi, Y., Positive generalized functionals, Hiroshima Math. J. **20**(1990), 137-157.

Zhongmin Qian,
Institute of Applied Mathematics,
Shanghai Institute of Railway Technology,
Shanghai 200333, China.

Sheng-Wu He,
Department of Mathematical Statistics,
East China Normal University,
Shanghai 200062, China.

On the differentiability of functions of an operator

by YaoZhong HU

Introduction. Let f be a continuous function on $\mathbb{R}$. Then it is well known how to define $f(A)$ when A is a bounded self-adjoint operator on a Hilbert space $\mathcal{H}$, using the spectral decomposition of A. But if A, H are two non-commuting self-adjoint operators, no explicit computation of $f(A+tH)$ is known. Our problem here is to study the regularity of $f(A+tH)$ under some regularity assumptions on f. We will assume that $\mathcal{H}$ is finite-dimensional. This note is a complement to our article "Some operator inequalities" in volume XXVIII[1], and answers a question of P.A. Meyer.

Notation. Given real numbers $\lambda_i \neq \lambda_j$, we put (divided differences : see the first pages of Donoghue [1] for more detail)

$$\{\lambda_2, \lambda_1\} f = \frac{f(\lambda_2) - f(\lambda_1)}{\lambda_2 - \lambda_1}$$

$$\{\lambda_{k+1}, \ldots, \lambda_1\} f = \{\lambda_{k+1}, \lambda_k\}\{\cdot, \lambda_{k-1} \ldots, \lambda_1\} f$$

LEMMA. *Let f have continuous derivatives up to of order $n+1$. Let $R_{n+1}(a,b)$ be the corresponding Taylor remainder*

$$R_{n+1}(a,b)\, f = f(b) - f(a) - (b-a) f'(a) - \tfrac{(b-a)^{n+1}}{(n+1)!} f^{(n+1)}(a)\ .$$

Then $\{\cdot, \lambda\} f$ has derivatives up to order n, and we have

$$\frac{d^n}{dt^n}\{t, \lambda\} f = \frac{n!}{(\lambda - x)^{n+1}}\, R_{n+1}(\lambda, x)\, f\ .$$

One can deduce that, if f is of class C^k, the function $\{\lambda_{k+1}, \ldots, \lambda_1\} f$ can be extended by continuity to $\mathbb{R}^{k+1}$ including the diagonals, and that

$$|\{\lambda_{k+1}, \ldots, \lambda_1\} f| \leq \gamma(k)\, \| f \|_{k;T}$$

where the last norm is the C^k norm of f on any interval T containing $\lambda_1, \ldots \lambda_{k+1}$. Then many results proved for polynomials $f(t) = t^d$ can be extended by density to C^k functions. In particular, the divided differences are symmetric in all their arguments from the following lemma, which is the crucial point of the calculation.

LEMMA. *When $f(t) = t^d$, we have*

$$\{\lambda_k, \ldots, \lambda_1\} f = \sum_{m_k + \ldots + m_1 = d-k+1} \lambda_k^{m_k} \ldots \lambda_1^{m_1}\ .$$

PROOF. By induction on k.

[1] La rédaction du Séminaire regrette ce retard de publication, dû à une erreur de transmission du manuscrit.

Computation of derivatives. When f is a polynomial, the operator function $f(A)$ is infinitely differentiable at every A and we can write its partial derivatives

$$\frac{\partial^k}{\partial t_1 \dots \partial t_k} f(A + t_1 H^1 + \dots + t_k H^k)\big|_{t_1 = \dots = t_k = 0} = \Phi(A\,;\, H^1, \dots, H^k)$$

where $\Phi(A\,;\,\cdot)$ is a symmetric k-linear functional. The problem is to give a uniform estimate of these derivatives knowing the C^k norm of f, which will allow us to extend the result from polynomials to C^k functions. Since Φ arises from polarization of the function $\Phi(H, \dots, H)$, it will be sufficient to estimate this function.

When $f(t) = t^d$ we have

$$\frac{d^k}{dt^k}(A + tH)^d\big|_{t=0} = \sum_{m_1 + \dots + m_{k+1} = d-k} A^{m_1} H \dots H A^{m_{k+1}}$$

Choose a basis in which A is diagonal with eigenvalues λ_i. Then the matrix of this operator D is

$$D_{ij} = \sum_{u_1, \dots, u_{k+1}} \delta_{iu_1} h_{u_1 u_2} \dots h_{u_k u_{k+1}} \delta_{u_{k+1} j} \sum_{m_1 + \dots + m_{k+1} = d-k} \lambda_{u_1}^{m_1} \dots \lambda_{u_{k+1}}^{m_{k+1}}$$

and this last coefficient is $\{\lambda_{u_{k+1}}, \dots, \lambda_{u_1}\} f$, in which the explicit form of f no longer appears. It follows that, if f is a polynomial, whenever the spectrum of A is contained in some interval T, we have a domination in Hilbert-Schmidt norm

$$\| \frac{d^k}{dt^k} f(A + tH)\big|_{t=0} \|_{HS} \leq C \| f \|_{k,T} \| H^k \|_{HS}$$

This is no longer basis dependent, and shows that, approximating locally a C^k function by polynomials in C^k norm, the function $f(A+tH)$ is k-times continuously differentiable in t.

This reasoning suggests that in infinite dimensions $f(A)$ is differentiable at A bounded, but along Hilbert-Schmidt directions.

REFERENCES

[1] DONOGHUE (W.F.). *Monotone Matrix Functions and Analytic Continuation*, Springer (Grundlehren 207), 1974.

[2] NÖRLUND (N.). Vorlesungen über Differenzenrechnung, Berlin 1924.

Institute of Math. Science, Academia Sinica,
Wuhan 430071, China
and
University of Oslo, POB 1053 Blindern N-0316 Oslo

The gap between the past supremum and the future infimum of a transient Bessel process

By

Davar Khoshnevisan

Department of Mathematics

University of Utah

Salt Lake City, UT. 84112

davar@math.utah.edu

Summary. This paper consists of some remarks on an earlier paper of the author with T.M. Lewis and W.V. Li regarding some almost sure path properties of the future infima of transient Bessel processes. In particular, we study the a.s. asymptotic discrepancy between $\sup_{s\le t} X_s$ and $\inf_{s\ge t} X_s$.

1. Introduction. Let $\{X_t; t \ge 0\}$ denote a transient Bessel process starting at zero. This means that X is a positive diffusion with a (strong) infinitesimal generator given by the following:

$$\mathcal{L}f(x) = \frac{1}{2}f''(x) + \frac{(d-1)}{2x}f'(x).$$

By transience, we must have $d > 2$ and this condition is assumed throughout this article without further mention. Moreover, the domain of the above generator is exactly all real-valued functions, f, which are twice continuously differentiable on $(0, \infty)$ and have the following boundary behavior: $|f(\varepsilon) - f(0)| = o(\varepsilon^{2-d})$, as $\varepsilon \to 0$.

Let $I_t \triangleq \inf_{s\ge t} X_s$ and $M_t \triangleq \sup_{s\le t} X_s$. Future infima processes such as I occur quite naturally in a variety of situations: see Aldous [A] for an application to random walks on trees. Furthermore, when $d = 3$, I appears quite naturally in Pitman's theorem: $2I_t - X_t$ is a Brownian motion. For the latter, see Revuz and Yor [RY, Thm. VI.(3.5), p. 234]; a surprising extension to all $d > 2$ appears in Chapter 12 of Yor [Y]. General extensions of Pitman's theorem to transient diffusions appear in the work of Saisho and Tanemura [ST].

In an earlier paper ([KLL]), together with T.M. Lewis and W.V. Li, we proved results about the asymptotic behavior of the process I with respect to the process X.

This, in turn, gives some information on the nature of transience of the underlying Bessel process X. Motivated by a question of K. Burdzy, this note is concerned with the size of the "gap" between the processes I and M. More precisely, we offer the following results:

Theorem 1.1. *Suppose* $\varphi : \mathbb{R}^1_+ \mapsto (0,\infty)$ *is increasing with* $\lim_{t\to\infty} \varphi(t) = \infty$. *Suppose further that* $t \mapsto \varphi(t)$ *is slowly varying and that*

$$\int_1^\infty \frac{dt}{t\varphi(t)} = \infty.$$

Then with probability one,

$$\liminf_{t\to\infty} \varphi(t) \cdot \left(1 - \frac{I_t}{M_t}\right) = 0.$$

Theorem 1.1 has a "converse" which is the following:

Proposition 1.2. *Suppose* $\varphi : \mathbb{R}^1_+ \mapsto (0,\infty)$ *is increasing with* $\lim_{t\to\infty} \varphi(t) = \infty$. *If*

$$\int_1^\infty \frac{dt}{t\varphi(t)} = \infty,$$

then with probability one,

$$\limsup_{t\to\infty} \varphi(t) \cdot \left(1 - \frac{I_t}{M_t}\right) = \infty.$$

We next mention a related result which has to do with the difference between M_t and I_t instead of their ratio.

Theorem 1.3. *Suppose* $\varphi : \mathbb{R}^1_+ \mapsto (0,\infty)$ *is decreasing. Suppose further that* $t \mapsto \varphi(t)$ *is slowly varying and*

$$\int_1^\infty \varphi(t)\frac{dt}{t} = \infty.$$

Then with probability one,

$$\liminf_{t\to\infty} \frac{M_t - I_t}{\varphi(t)} = 0.$$

An obvious consequence of Theorem 1.1 is that almost surely,

$$\liminf_{t\to\infty} (\ln t \ln\ln t) \cdot (1 - I_t \cdot M_t^{-1}) = 0,$$

and correspondingly by Theorem 1.3,

$$\liminf_{t\to\infty} (\ln t \ln\ln t) \cdot (M_t - I_t) = 0.$$

It may at first seem surprising that the above rates are independent of the dimension, $d > 2$; but this is not so, since we believe the results above are far from being sharp. To illustrate the problem, we point out that our techniques cannot establish that almost surely,

$$\liminf_{t\to\infty} f(t) \cdot (1 - I_t \cdot M_t^{-1}) = \infty,$$

even for a function such as $f(x) = \exp(e^x)$, which we believe ought to do the job. Thus it is important to find a more robust method of handling this gap.

The above results are partial attempts at estimating the size of the gap when it is small; the gap is small when $I_t \simeq M_t$. Below we provide the following theorem which gives a complete characterization of the size of the gap when it is large, i.e., when I_t is much smaller then M_t.

Theorem 1.4. *Suppose $\psi : (0,\infty) \mapsto (0,\infty)$ is decreasing to zero and is slowly varying. Then with probability one,*

$$\liminf_{t\to\infty} \frac{I_t}{M_t \psi(t)} = \begin{cases} 0, & \text{if } J(\psi) = \infty \\ \infty & \text{if } J(\psi) < \infty, \end{cases}$$

where

$$J(\psi) \triangleq \int_1^\infty (\psi(t))^{d-2} \frac{dt}{t}.$$

Therefore, almost surely,

$$\liminf_{t\to\infty} (\ln t)^a \frac{I_t}{M_t} = \begin{cases} 0, & \text{if } a \le \frac{1}{d-2} \\ \infty, & \text{if } a > \frac{1}{d-2} \end{cases}.$$

From now on, $\{\mathcal{F}_t; t \ge 0\}$ denotes the natural filtration of the process X and for any measurable $A \subseteq C([0,\infty))$, $\mathbb{P}^x(A)$ is a nice version of the probability of A conditional on $\{\omega : X_0 = x\}$. Unimportant finite positive constants are denoted as K_0 and K and their value may vary from line to line.

I wish to thank Chris Burdzy for introducing me to this problem as well as for interesting conversations on this topic. Also many thanks are due to Yuval Peres, Russ Lyons, Marc Yor and an anonymous referee for several useful suggestions.

2. The Proofs. Define the first hitting times, $\sigma(t)$, by the following:

$$\sigma(t) \triangleq \inf\{s > 0 : X_s = t\}.$$

Supposing that $\psi : \mathbb{R}^1_+ \mapsto (0,1)$ increases to one as $t \to \infty$, let us define measurable events, $E(t) = E_\psi(t)$ by

$$E(t) \triangleq \{\omega : \ I_{\sigma(t)} \geq t\psi(t)\}. \tag{2.1}$$

Lemma 2.1. *In the above notation,*

$$\mathbb{P}\big(E(t)\big) = 1 - \psi^{d-2}(t).$$

Proof. First condition on $\mathcal{F}_{\sigma(t)}$ and then use the gambler's ruin problem for X, using the fact that X^{2-d} is a continuous martingale. □

Lemma 2.2. *Let ψ be as in the statement of Lemma 2.1 and let $E(t)$ be defined by (2.1). Suppose that $t > s > 0$ are such that $t\psi(t) \geq s$. Then*

$$\mathbb{P}\big(E(t) \cap E(s)\big) = \mathbb{P}\big(E(t)\big)\mathbb{P}\big(E(s)\big) \times \frac{1}{1 - \big(s\psi(s)/t\big)^{d-2}}.$$

Proof. By a gambler's ruin calculation,

$$\begin{aligned}
\mathbb{P}^s\big(\sigma(t) < \sigma(s\psi(s))\big) &= \frac{\big(s\psi(s)\big)^{2-d} - s^{2-d}}{\big(s\psi(s)\big)^{2-d} - t^{2-d}} \\
&= \frac{1 - \psi^{d-2}(s)}{1 - \big(s\psi(s)/t\big)^{d-2}} \\
&= \mathbb{P}\big(E(s)\big) \times \frac{1}{1 - \big(s\psi(s)/t\big)^{d-2}},
\end{aligned} \tag{2.2}$$

by Lemma 2.1. On the other hand, by another gambler's ruin calculation,

$$\begin{aligned}
\mathbb{P}^t\big(\sigma(t\psi(t)) = \infty\big) &= 1 - \psi^{d-2}(t) \\
&= \mathbb{P}\big(E(t)\big),
\end{aligned} \tag{2.3}$$

by Lemma 2.1. By the strong Markov property,

$$\mathbb{P}\big(E(s) \cap E(t)\big) = \mathbb{P}^s\big(\sigma(t) < \sigma(s\psi(s))\big) \cdot \mathbb{P}^t\big(\sigma(t\psi(t)) = \infty\big).$$

The lemma follows from (2.2) and (2.3). □

Proof of Theorem 1.1. Without loss of generality, we shall assume that $\varphi(t) \geq 1$ for all $t > 0$. Otherwise, we let $t' \triangleq \inf\{s : \varphi(s) \geq 1\}$ and shift everything by t'. With this in mind, fix $\varepsilon \in (0,1)$ and define $\psi(t) \triangleq \left(1 - \varepsilon\varphi^{-1}(t)\right)$. (Note that $\psi(t) \geq 1 - \varepsilon$ since $\varphi \geq 1$.) Also define

$$t_n \triangleq e^n \qquad \text{and} \qquad \psi_n \triangleq \psi(t_n).$$

Recalling the definition of $E(t)$ from (2.1), it follows from Lemma 2.1 that as $n \to \infty$, $P\big(E(t_n)\big) \sim \varepsilon(d-2) \cdot \varphi^{-1}(t_n)$. In particular, there exists some constant, K, such that for all $n \geq 1$, $\mathbb{P}\big(E(t_n)\big) \geq K/\varphi(t_n)$. Hence,

$$\begin{aligned}
\sum_{n=0}^{\infty} \mathbb{P}\big(E(t_n)\big) &\geq K \sum_{n=0}^{\infty} \int_{t_n}^{t_{n+1}} \frac{dt}{(t_{n+1} - t_n)\varphi(t_n)} \\
&= \frac{K}{e-1} \sum_{n=0}^{\infty} \int_{t_n}^{t_{n+1}} \frac{dt}{t_n \varphi(t_n)} \\
&\geq \frac{K}{e-1} \int_1^{\infty} \frac{dt}{t\varphi(t)} = \infty, \qquad (2.4)
\end{aligned}$$

by assumption. On the other hand, since $\varphi(t_k) \to \infty$, for all integers k and all integers n large enough,

$$\frac{t_{n+k}\psi_{n+k}}{t_n} = e^k \big(1 - \varepsilon/\varphi(t_{n+k})\big) \geq 1,$$

since $\varepsilon \in (0,1)$. This means that Lemma 2.2 applies. More precisely by Lemma 2.2, for all $k \geq 1$ and all n large,

$$\begin{aligned}
\mathbb{P}\big(E(t_n) \cap E(t_{n+k})\big) &= \mathbb{P}\big(E(t_n)\big)\mathbb{P}\big(E(t_{n+k})\big) \times \frac{1}{1 - \left(t_n\psi_k/t_{n+k}\right)^{d-2}} \\
&\leq \left(1 - e^{2-d}\right)^{-1} \mathbb{P}\big(E(t_n)\big)\mathbb{P}\big(E(t_{n+k})\big),
\end{aligned}$$

since $d > 2$ and $\psi_k \leq 1$ for all $k \geq 1$. By the Kochen–Stone lemma ([KS]), (2.4) and the above together imply that

$$\mathbb{P}\big(E(t_n), \text{ i.o.}\big) \geq \left(1 - e^{2-d}\right).$$

By Kolmogorov's 0–1 law, $\mathbb{P}\big(E(t_n), \text{ i.o.}\big) = 1$. In other words, with probability one,

$$I_{\sigma(t_n)} \geq t_n \psi(t_n), \qquad \text{i.o.} \qquad .$$

Hence for every $\varepsilon \in (0,1)$, almost surely,

$$\varphi(t_n)\cdot\left(1-\frac{I_{\sigma(t_n)}}{t_n}\right)\le \varepsilon, \qquad \text{i.o.}$$

Since $t \mapsto M_t$ is a.s. continuous, $M_{\sigma(t)} = t$. Substituting M_t for t,

$$\liminf_{n\to\infty}\varphi(M_{t_n})\cdot\left(1-\frac{I_{t_n}}{M_{t_n}}\right)=0, \qquad \text{a.s.} \tag{2.5}$$

It, therefore, remains to prove that for some $K>0$,

$$\liminf_{n\to\infty}\frac{\varphi(M_{t_n})}{\varphi(t_{n+1})}\ge\frac{1}{K}, \qquad \text{a.s.} \tag{2.6}$$

For if we proved (2.6), (2.5) implies the following:

$$\begin{aligned}
\liminf_{t\to\infty}\varphi(t)\cdot\left(1-\frac{I_t}{M_t}\right) &\le \liminf_{n\to\infty}\varphi(t_n)\cdot\left(1-\frac{I_{t_n}}{M_{t_n}}\right)\\
&= \liminf_{n\to\infty}\varphi(e^{-1}t_{n+1})\cdot\left(1-\frac{I_{t_n}}{M_{t_n}}\right)\\
&\le K_0\liminf_{n\to\infty}\varphi(t_{n+1})\cdot\left(1-\frac{I_{t_n}}{M_{t_n}}\right)\\
&\le K_0K\cdot\liminf_{n\to\infty}\varphi(M_{t_n})\cdot\left(1-\frac{I_{t_n}}{M_{t_n}}\right)=0,
\end{aligned}$$

for some $K_0>1$, since φ is slowly varying. As this proves Theorem 1.1, it suffices to prove (2.6). It is pointed out to us by Marc Yor that (2.6) can be obtained as a consequence of Chung's law of the iterated logarithm. We shall provide a direct proof for the sake of completeness.

Recall that for any $c>0$ there exists some $K=K(c)>0$ and $T(c)>0$, so that for all $t\ge T(c)$, $\varphi(t^c)\ge K^{-1}\varphi(t)$. Since φ is increasing, by standard calculations for any $c\in(1/2,1)$, and all $t_n\ge T(c)$,

$$\begin{aligned}
\mathbb{P}\big(\varphi(M_{t_n})\le K^{-1}\varphi(t_{n+1})\big) &\le \mathbb{P}\big(M_{t_n}\le t_{n+1}^c\big)\\
&\le \left(\frac{t_n^{1/2}}{t_{n+1}^c}\right)\\
&\le e^c\cdot\exp\big(-n(c-0.5)\big),
\end{aligned}$$

which sums. An application of the easy half of the Borel–Cantelli lemma proves (2.6) and hence finishes the proof of the theorem. □

Proof of Proposition 1.2. By considering $\varphi_0(t)\triangleq\varphi(t)\wedge(\ln t)$, we might as well assume that $\varphi(t)\ge\ln t$ for all $t>1$. By Theorem 4.1 (2) of Khoshnevisan et al. [KLL],

$$\limsup_{t\to\infty}\frac{X_t-I_t}{\sqrt{2t\ln\ln t}}=1, \qquad \text{a.s.}$$

Hence for any $\varepsilon > 0$,

$$\frac{M_t - I_t}{\sqrt{2t \ln \ln t}} \geq (1-\varepsilon), \qquad \text{i.o., a.s.} \tag{2.7}$$

By the usual law of the iterated logarithm for Bessel processes (see Revuz and Yor [RY, Ex. XI.(1.20), p. 419], for instance),

$$\limsup_{t\to\infty} \frac{M_t}{\sqrt{2t \ln \ln t}} = 1,$$

almost surely. Hence almost surely, $M_t \leq \sqrt{2t \ln t}$, eventually. By (2.7), the above immediately implies Proposition 1.2. □

Proof of Theorem 1.3. The proof of Theorem 1.3 is very similar to that of Theorem 1.1; the main difference is the choice of the subsequence along which one can use the Borel–Cantelli lemma. To this end, define

$$\psi(t) \triangleq 1 - \frac{\varphi(t)}{t}.$$

Recalling (2.1), we see from Lemma 2.1 that

$$\begin{aligned} \mathbb{P}\big(E(n)\big) &= 1 - \left(1 - \frac{\varphi(n)}{n}\right)^{d-2} \\ &\sim (d-2)\frac{\varphi(n)}{n}. \end{aligned} \tag{2.8}$$

Therefore,

$$\begin{aligned} \sum_n \mathbb{P}\big(E(n)\big) &\geq K \sum_n \frac{\varphi(n)}{n} \\ &\geq K \int_1^\infty \varphi(t)\frac{dt}{t} = \infty, \end{aligned} \tag{2.9}$$

by assumption. It now follows from Lemma 2.2 and (2.8) that

$$\begin{aligned} \mathbb{P}\big(E(n)\cap E(n+k)\big) &= \mathbb{P}\big(E(n)\big)\mathbb{P}\big(E(n+k)\big)\left[1 - \left(\frac{n-\varphi(n)}{n+k}\right)^{d-2}\right]^{-1} \\ &\sim (d-2)^3 \frac{\varphi(n)}{n}\cdot\frac{\varphi(n+k)}{n+k}\cdot\frac{n+k}{k+\varphi(n)} \\ &\leq (d-2)^3 \frac{\varphi(n)}{n}\cdot\frac{\varphi(n+k)}{k} \\ &\leq (d-2)^3 \frac{\varphi(n)}{n}\cdot\frac{\varphi(k)}{k} \\ &\sim (d-2)\mathbb{P}\big(E(n)\big)\mathbb{P}\big(E(k)\big), \end{aligned}$$

where the penultimate line follows from the fact that $t \mapsto \varphi(t)$ is assumed to be decreasing. This development implies the existence of some $\varepsilon_N \to 0$, such that

$$\sum_{n=1}^{N}\sum_{k=1}^{N-n} \mathbb{P}(E(n)\cap E(n+k)) \le (d-2)(1+\varepsilon_N)\sum_{n=1}^{N}\sum_{k=1}^{N-n}\mathbb{P}(E(n))\mathbb{P}(E(k))$$

$$\le (d-2)(1+\varepsilon_N)\Big(\sum_{n=1}^{N}\mathbb{P}(E(n))\Big)^2. \tag{2.10}$$

By the lemma of Kochen and Stone [KS], (2.9) and (2.10) together imply

$$\mathbb{P}\big(E(n),\ \text{i.o.}\ \big) \ge (d-2)^{-1}.$$

Hence, by Kolmogorov's 0–1 law, $\mathbb{P}\big(E(n),\ \text{i.o.}\ \big) = 1$. Since $M_{\sigma(n)} = n$, we have shown that with probability one,

$$M_t - I_t \le \varphi(M_t),\ \text{i.o..}$$

By Chung's law of the iterated logarithm, for each $\varepsilon > 0$, almost surely we have: $M_t \ge t^{1/2-\varepsilon}$, eventually. By the assumed properties of φ, for each $K > 1$,

$$M_t - I_t \le K\cdot\varphi(t),\ \text{i.o.}$$

Replacing $\varphi(\cdot)$ by $\varepsilon\varphi(\cdot)$ and letting $\varepsilon \to 0$, the result follows. □

Proof of Theorem 1.4. Let $t_n \triangleq e^n$, and $\psi_n \triangleq \psi(t_n)$. Since $\psi(t) \to 0$ as $t\to\infty$, we might as well assume that $\psi(t) < 1$. We begin with the elementary observation that $J(\psi) < \infty$ if and only if $\sum_n \psi_n^{d-2} < \infty$. Arguing as in Lemma 2.1, there exists some $K > 1$ such that for all $n \ge 1$,

$$\mathbb{P}\big(I_{\sigma(t_n)} < t_{n+1}\psi_{n+1}\big) \le K\psi_{n+1}^{d-2}.$$

Hence, if $J(\psi) < \infty$, then by the Borel–Cantelli lemma,

$$I_{\sigma(t_n)} > t_{n+1}\psi_{n+1}, \qquad \text{eventually,}$$

almost surely. Now suppose t is large. Then there exists some large n such that $t_n \le t \le t_{n+1}$. Since $t \mapsto I_{\sigma(t)}$ and $t \mapsto t\psi(t)$ are both increasing, it follows that with probability one, $I_{\sigma(t)} > t\psi(t)$, eventually. Now if $J(\psi)$ converges, $J(K\psi)$ also converges no matter the value of $K > 0$. Hence applying the above to $K\cdot\psi(t)$ instead, it follows that with probability one: $I_{\sigma(t)} > K\cdot\psi(t)$, eventually. In other words, we have argued that if $J(\psi) < \infty$, then almost surely,

$$\lim_{t\to\infty}\frac{I_t}{M_t\psi(M_t)} = \infty. \tag{2.11}$$

Now supppose $J(\psi) = \infty$ and define,

$$E_n \triangleq \{\omega : \ I_{\sigma(t_n)} < t_n \psi_n\}.$$

As in Lemma 2.1, it follows that there exists some $K > 1$ such that for all $n \geq 1$,

$$K^{-1} \psi_n^{d-2} \leq \mathbb{P}(E_n) \leq K \psi_n^{d-2}.$$

Since $J(\psi) < \infty$ if and only if $\sum_n \psi_n^{d-2} < \infty$, we have shown that $\sum_n \mathbb{P}(E_n) = \infty$. Suppose we could prove the following:

$$\liminf_{N \to \infty} \frac{\sum_{n=1}^{N} \sum_{k=1}^{N-n} \mathbb{P}(E_n \cap E_{n+k})}{\left(\sum_{n=1}^{N} \mathbb{P}(E_n)\right)^2} < \infty. \tag{2.12}$$

By the Kochen–Stone lemma ([KS]) and Kolmogorov's 0–1 law, it would then follow that almost surely, $I_{\sigma(t)} \leq t\psi(t)$, infinitely often. However, $J(\psi)$ diverges if and only if $J(\varepsilon\psi)$ diverges, for any choice of $\varepsilon > 0$. Hence, applying the above discussion to $t \mapsto \varepsilon\psi(t)$, we see that $J(\psi) = \infty$ implies the following:

$$\liminf_{t \to \infty} \frac{I_t}{M_t \psi(M_t)} = 0, \qquad \text{a.s.}$$

Together with (2.11), this shows that if we could prove (2.12), then we have shown the following:

$$\liminf_{t \to \infty} \frac{I_t}{M_t \psi(M_t)} = \begin{cases} 0, & \text{if } J(\psi) = \infty \\ \infty, & \text{if } J(\psi) < \infty \end{cases}. \tag{2.13}$$

Supposing (2.12) for the time being, let us see how (2.13) implies Theorem 1.4. For any $\theta > 0$, let $\psi_\theta(t) \triangleq \psi(t^\theta)$. Note that ψ_θ satisfies the conditions of the theorem if and only if ψ does. Moreover, $J(\psi_\theta) = \theta^{-1} J(\psi)$, and hence $J(\psi_\theta) = \infty$ if and only if $J(\psi) = \infty$.

Suppose $J(\psi) = \infty$. By (2.13), for any $\theta > 2$,

$$\liminf_{t \to \infty} \frac{I_t}{M_t \psi(M_t^\theta)} = 0, \qquad \text{a.s.}$$

Refining the proof of (2.6) (alternatively, using Chung's LIL), $M_t^\theta \geq t$, eventually, a.s.. Since $t \mapsto \psi(t)$ is decreasing, it follows that

$$\text{if} \quad J(\psi) = \infty \quad \text{then} \quad \liminf_{t \to \infty} \frac{I_t}{M_t \psi(t)} = 0, \qquad \text{a.s.} \tag{2.14}$$

On the other hand, if $J(\psi) < \infty$, then $J(\psi_\theta) < \infty$ for $\theta \in (0,2)$. Applying (2.13) to J_θ for such a θ, it follows that

$$\liminf_{t\to\infty} \frac{I_t}{M_t\psi(M_t^\theta)} = \infty, \qquad \text{a.s.}$$

Another argument (e.g, the LIL or a refinement of the argument leading to (2.6)) shows that almost surely: $M_t^\theta \le t$, eventually. Therefore we have shown that (a.s.),

$$\text{if} \qquad J(\psi) < \infty \qquad \text{then} \qquad \liminf_{t\to\infty} \frac{I_t}{M_t\psi(t)} = \infty.$$

This and (2.14) together prove the theorem. It therefore remains to prove (2.12).

In the course of the proof of (2.12), there are potentially two seperate cases to consider: (1) when k is so large that $t_{n+k}\psi_{n+k} \in [t_n, t_{n+k}]$, and (2) when $t_{n+k}\psi_{n+k} \in [t_n\psi_n, t_n]$. Both estimates follow the guidelines of the proof of Lemma 2.2 and one gets the same estimates (modulo some constant multiples) in both cases. Therefore, we shall be content to handle case (1) only. In this case, by the gambler's ruin problem (cf. Lemma 2.1) and the strong Markov property, for all $n, k \ge 1$,

$$\begin{aligned}
\mathbb{P}(E_n \cap E_{n+k}) &= \mathbb{P}\big(I_{\sigma(t_n)} < t_n\psi_n, I_{\sigma(t_{n+k})} \le t_n\psi_n\big) \\
&\quad + \mathbb{P}\big(I_{\sigma(t_n)} < t_n\psi_n \ , \ t_n\psi_n < I_{\sigma(t_{n+k})} < t_{n+k}\psi_{n+k}\big) \\
&= \left(\frac{t_n\psi_n}{t_{n+k}}\right)^{d-2} + \mathbb{P}^{t_n}\big(\sigma(t_n\psi_n) < \sigma(t_{n+k})\big) \cdot \mathbb{P}^{t_{n+k}}\big(\sigma(t_{n+k}\psi_{n+k}) < \infty\big) \\
&\qquad \cdot \mathbb{P}^{t_{n+k}\psi_{n+k}}\big(\sigma(t_n\psi_n) = \infty\big) \\
&\triangleq T_1 + T_2.
\end{aligned}$$

Evidently, there exists some $K > 0$ such that for all $n \ge 1$, $T_1 \le Ke^{-k(d-2)}\mathbb{P}(E_n)$. Likewise, T_2 is estimated as follows,

$$\begin{aligned}
T_2 &= \frac{t_{n+k}^{d-2} - t_n^{d-2}}{t_{n+k}^{d-2} - (t_n\psi_n)^{d-2}} \psi_n^{d-2}\psi_{n+k}^{d-2}\big(1 - (t_n\psi_n)^{d-2}\psi_{n+k}^{2-d}\big) \\
&\le 2\psi_n^{d-2}\psi_{n+k}^{d-2} \\
&\le K\mathbb{P}(E_n)\mathbb{P}(E_{n+k}).
\end{aligned}$$

Hence there exists some $K > 1$ such that for all k and n satisfying case (1), we have $\mathbb{P}(E_n \cap E_{n+k}) \le K \cdot \mathbb{P}(E_n)\big(1 + \mathbb{P}(E_{n+k})\big)$. Since a similar estimate holds for k and n satisfying case (2), (2.12) follows from the fact that $\sum_n \mathbb{P}(E_n) = \infty$. This proves Theorem 1.4. $\square$

References.

[A] D. Aldous (1992). Greedy search on the binary tree with random edge-weight. *Comb., Prob. & Computing* **1**, pp. 281–293

[KLL] D. Khoshnevisan, T.M. Lewis and W.V. Li (1993). On the future infima of some transient processes. To appear in *Prob. Theory and Rel. Fields*

[KS] S.B. Kochen and C.J. Stone (1964). A note on the Borel–Cantelli problem. *Ill. J. Math.* **8**, pp. 248–251

[RY] D. Revuz and M. Yor (1991). *Continuous Martingales and Brownian Motion.* Springer Verlag. Grundlehren der mathematischen Wissenschaften #293. Berlin–Heidelberg

[ST] Y. Saisho and H. Tanemura (1990). Pitman type theorem for one–dimensional diffusion processes. *Tokyo J. Math.* Vol.13, No.2, pp. 429–440

[Y] M. Yor (1993). *Some Aspects of Brownian Motion. Part II: Some Recent martingale Problems.* Forthcoming Lecture Notes

THE LEVEL SETS OF ITERATED BROWNIAN MOTION

BY

KRZYSZTOF BURDZY[1] AND DAVAR KHOSHNEVISAN

University of Washington & University of Utah

ABSTRACT. We show that the Hausdorff dimension of every level set of iterated Brownian motion is equal to 3/4.

§1. Introduction and Main Result. Suppose $(\Omega, \mathcal{F}, \mathbb{P})$ is a probability space, rich enough to carry three independent Brownian motions, X^+, X^- and Y, all starting from the origin. Iterated Brownian motion (IBM) is the process defined by $Z(t) = X(Y(t))$, where $X(t) = X^+(t)1_{[0,\infty)}(t) + X^-(-t)1_{(-\infty,0)}(t)$. The probabilistic and analytical properties of IBM and related processes have been the subject of recent vigorous investigations; see BERTOIN [B], BURDZY [B1,B2], CSÁKI ET AL. [CsCsFR1,CsCsFR2], DEHEUVELS AND MASON [DM], FUNAKI [F], HU ET AL. [HPS], HU AND SHI [HS], KHOSHNEVISAN AND LEWIS [KL1,KL2] and SHI [S] together with their combined references. Define the set–valued x–level set process, $\mathcal{L}_x(t)$, by

$$\mathcal{L}_x(t) = \{0 \le s \le t : Z(s) = x\}, \qquad \text{for all } x \in \mathbb{R}^1. \tag{1.1}$$

The main result of this paper is the following analogue of Paul Lévy's well-known result for Brownian motion (see ITÔ AND MCKEAN [IM] and ADLER [A]):

(1.2) Theorem. *Let* $\dim_H$ *denote Hausdorff dimension. Then, outside a single null set,*

$$\dim_H\big(\mathcal{L}_x(t)\big) = \tfrac{3}{4},$$

simultaneously for all $t \ge 0$ *and all* x *in the interior of* $Z([0,t])$.

Here and throughout, if $f : \mathbb{R}^1 \mapsto \mathbb{R}^1$ is Borel measurable and $A \subset \mathbb{R}^1$ is measurable, then $f(A) = \{y : y = f(x) \text{ for some } x \in A\}$.

The proof of Theorem (1.2) uses a capacity argument due to Frostman (see ADLER [A]) and relies on the following which has been discovered independently and at the same time by CSÁKI ET AL. [CsCsFR2]:

---

[1] Research supported in part by NSF grant DMS 91–00244 and AMS Centennial Research Fellowship

(1.3) Proposition. *There exists an almost surely jointly continuous family of "local times", $\{\ell_t^a; t \geq 0, a \in \mathbf{R}^1\}$, such that for all Borel measurable integrable functions, $f : \mathbf{R}^1 \mapsto \mathbf{R}^1$ and all $t \geq 0$,*

$$\int_0^t f(Z(s))ds = \int_{-\infty}^{\infty} f(a)\ell_t^a da.$$

Acknowledgements. We wish to thank T.M. Lewis for many enjoyable discussions. Also many thanks are due to J. Bertoin as well as E. Csáki and M. Csörgő for sending us the articles, [B] and [CsCsFR1,CsCsFR2].

§2. Local Times. If B is any Brownian motion, its process of local times will be denoted by $L_t^a(B)$. These satisfy the following occupation density formula: for any Borel measurable $f : \mathbf{R}^1 \mapsto \mathbf{R}^1$ and all $t \geq 0$,

$$\int_0^t f(B(r))dr = \int_{-\infty}^{\infty} f(a)L_t^a(B)da. \tag{2.1}$$

For a stochastic calculus description as well as many deep properties of local times see Revuz and Yor [RY]. Proposition (1.3) is a consequence of the following real variable fact:

(2.2) Proposition. *Let $K(b, da)$ be the kernel defined by*

$$K(b, da) = L_{da}^b(X^+) + L_{da}^b(X^-).$$

Then the local times ℓ are given by

$$\ell_t^b = \int_{Y([0,t])} L_t^a(Y)K(b, da).$$

Proof. Let $f : \mathbf{R}^1 \mapsto \mathbf{R}^1$ be as in Proposition (1.3). Viewing $f(Z)$ as $(f \circ X)(Y)$, we see from (2.1) that for all $t \geq 0$, a.s.,

$$\begin{aligned} \int_0^t f(Z(s))ds &= \int_{-\infty}^{\infty}(f \circ X)(a)L_t^a(Y)da \\ &= \int_0^{\infty}(f \circ X^+)(a)L_t^a(Y)da + \int_{-\infty}^0(f \circ X^-)(a)L_t^a(Y)da. \end{aligned} \tag{2.3}$$

By (2.1) and a monotone class argument, for any jointly measurable $F : \mathbf{R}^1 \times \mathbf{R}_+^1 \mapsto \mathbf{R}^1$,

$$\int_0^{\infty} F(X^{\pm}(s), s)ds = \int_{-\infty}^{\infty} \int_0^{\infty} F(a,t)L_{dt}^a(X^{\pm})da.$$

Applying (2.3),

$$\begin{aligned} \int_0^t f(Z(s))ds = &\int_{-\infty}^{\infty} f(b)db \int_0^{\infty} L_t^a(Y)L_{da}^b(X^+) \\ &+ \int_{-\infty}^{\infty} f(b)db \int_0^{\infty} L_t^{-a}(Y)L_{da}^b(X^-). \end{aligned}$$

Since $a \mapsto L_t^a(Y)$ is a.s. supported on $Y([0,t])$,

$$\int_0^t f(Z(s))ds = \int_{-\infty}^{\infty} f(b)db \int_{\mathbf{R}_+^1 \cap Y([0,t])} L_t^a(Y) L_{da}^b(X^+)$$
$$+ \int_{-\infty}^{\infty} f(b)db \int_{\mathbf{R}_+^1 \cap Y([0,t])} L_t^a(Y) L_{da}^b(X^-).$$

The proposition follows from a change of variables. The joint continuity of ℓ_t^a follows from that of $L_t^a(B)$ for any Brownian motion, B; see REVUZ AND YOR [RY]. ///

(2.4) Proposition. *For any $T > 0$, almost surely,*

$$\limsup_{\varepsilon \to 0} \sup_{a \in \mathbf{R}^1} \sup_{0 \le t \le T} \frac{\ell_{t+\varepsilon}^a - \ell_t^a}{\varepsilon^{3/4}\left(\ln(1/\varepsilon)\right)^{5/4}} \le 2^{3/2}, \qquad \text{a.s.}\ .$$

With more work, one can improve the upper bound of $2^{3/2}$ in the above. However, we do not even know whether the power of the logarithm is the correct one. Therefore, we will be satisfied with our simple proof of the upper bound.

Proof. By Proposition (2.2),

$$\ell_{t+\varepsilon}^a - \ell_t^a = \int_{Y([0,t])} \left(L_{t+\varepsilon}^r(Y) - L_t^r(Y)\right) K(a, dr)$$
$$(2.5) \qquad + \int_{Y([0,t+\varepsilon]) \setminus Y([0,t])} L_{t+\varepsilon}^r(Y) K(a, dr)$$
$$= \mathrm{I}_{(2.5)} + \mathrm{II}_{(2.5)}.$$

For all $r \in Y([0,t]) \setminus Y([t,t+\varepsilon])$, $L_{t+\varepsilon}^r(Y) = L_t^r(Y)$. Hence, as $\varepsilon \to 0^+$,

$$\mathrm{I}_{(2.5)} = \int_{Y([t,t+\varepsilon]) \cap Y([0,t])} \left(L_{t+\varepsilon}^r(Y) - L_t^r(Y)\right) K(a, dr)$$
$$\le \sup_r \left(L_{t+\varepsilon}^r(Y) - L_t^r(Y)\right) K\left(a, Y([t,t+\varepsilon])\right)$$
$$\le (1+o(1))\sqrt{2\varepsilon \ln(1/\varepsilon)}\, K\left(a, Y([t,t+\varepsilon])\right),$$

uniformly over all $a \in \mathbf{R}^1$ and all $0 \le t \le T$. We have used the uniform modulus of continuity of local times in time; see Lemma 5(c) of PERKINS [P]. By Lévy's modulus of continuity ([RY]), as $\varepsilon \to 0^+$,

$$\left|Y([t,t+\varepsilon])\right| \le (1+o(1))\sqrt{2\varepsilon \ln(1/\varepsilon)},$$

uniformly over all $0 \le t \le T$. Hence, by MCKEAN [Mc] and the independence of X^+ and X^-, as $\varepsilon \to 0^+$, uniformly over all $a \in \mathbf{R}^1$ and $0 \le t \le T$,

$$K\left(a, Y([t,t+\varepsilon])\right) \le (1+o(1))\sqrt{2\sqrt{2\varepsilon \ln(1/\varepsilon)}\left|\ln\sqrt{2\varepsilon \ln(1/\varepsilon)}\right|}$$
$$= (1+o(1))2^{1/4}\varepsilon^{1/4}\left(\ln(1/\varepsilon)\right)^{3/4}.$$

This implies that as $\varepsilon \to 0^+$,

$$\text{I}_{(2.5)} \leq (1+o(1))2^{3/4}\varepsilon^{3/4}\big(\ln(1/\varepsilon)\big)^{5/4}. \tag{2.6}$$

The bound for $\text{II}_{(2.5)}$ is very similar. Note that for all $r \in Y([0,t+\varepsilon]) \setminus Y([0,t])$, $L_t^r(Y) = 0$. Hence, making similar arguments as above, we see that as $\varepsilon \to 0^+$, uniformly over all $a \in \mathbb{R}^1$ and $0 \leq t \leq T$,

$$\begin{aligned}
\text{II}_{(2.5)} &= \int_{Y([0,t+\varepsilon])\setminus Y([0,t])} \big(L_{t+\varepsilon}^r(Y) - L_t^r(Y)\big) K(a,dr) \\
&\leq \sup_r \big(L_{t+\varepsilon}^r(Y) - L_t^r(Y)\big) K\big(a, Y([t,t+\varepsilon])\big) \\
&\leq (1+o(1))\sqrt{2\varepsilon \ln(1/\varepsilon)}\sqrt{2|Y([t,t+\varepsilon])| \cdot |\ln|Y([t,t+\varepsilon])||} \\
&= (1+o(1))2^{3/4}\varepsilon^{3/4}\big(\ln(1/\varepsilon)\big)^{5/4}.
\end{aligned}$$

Together with (2.5) and (2.6), the above implies the result. ///

§3. The proof of Theorem (1.2). Once there is a modulus of continuity of local times (in t), we proceed by Frostman's capacity method as outlined in ADLER [A], for example. Recall from KHOSHNEVISAN AND LEWIS [KL] that for any $T > 0$, almost surely,

$$\lim_{\varepsilon \to 0} \sup_{0 \leq t \leq T} \frac{|Z(t+\varepsilon) - Z(t)|}{\varepsilon^{1/4}\big(\ln(1/\varepsilon)\big)^{3/4}} = 1, \qquad \text{a.s.}\,. \tag{3.1}$$

In particular, we see that Z is Hölder continuous of order $\gamma < 1/4$. By Proposition (1.3) and Lemma 7 of [A], simultaneously over all $x \in \mathbb{R}^1$ $\dim_H \mathcal{L}_x(t) \leq 3/4$. Moreover, by Proposition (2.4), $t \mapsto \ell_t^a$ is Hölder continuous of order $\gamma < 3/4$, uniformly in $a \in \mathbb{R}^1$. By Frostman's lemma, (see the proof of Lemma 6 of [A]), simultaneously over all x in the interior of $Z([0,t])$, $\dim_H \mathcal{L}_x(t) \geq 3/4$. This proves the result. ///

We conclude this section with some open problem.

Problem 1. Define $Z^+(t) = \sup_{0 \leq s \leq t} Z(s)$. BERTOIN [B] proves that for all $T > 0$, almost surely,

$$\lim_{\varepsilon \to 0} \sup_{0 \leq t \leq T} \frac{|Z^+(t+\varepsilon) - Z^+(t)|}{\varepsilon^{1/4}\big(\ln(1/\varepsilon)\big)^{3/4}} = \frac{1}{2^{1/4} \cdot 3^{3/4}}.$$

In light of (3.1), this says that Z^+ is smoother than Z. Is there a probabilistic explanation for this, in terms of (say) path decompositions?

Problem 2. Define

$$\mathcal{S}^+(t) \triangleq \{t \in [0,1] : Z(t) > Z(s) \text{ for all } s < t\}.$$

According to BERTOIN [B], the Hausdorff dimension of $\mathcal{S}^+(t)$ is almost surely $1/4$. This is in sharp contrast with Theorem (1.2) and the analogous result for Brownian motion which is a consequence of Lévy's characterization of Brownian motion. Is there a probabilistic explanation for this apparent difference?

Problem 3. By being more careful, it is possible to show that φ–Hausdorff measure of $\mathcal{L}_0(t)$ is a.s. (strictly) positive, if $\varphi(\varepsilon) = \varepsilon^{3/4}\big(\ln(1/\varepsilon)\big)^{5/4}$. Is this sharp? For the corresponding problem for $\mathcal{S}^+(t)$, see BERTOIN [B].

References.

[A] R.J. ADLER (1978). The uniform dimension of the level sets of a Brownian sheet, *Ann. Prob.* **6** 509–515.

[B] J. BERTOIN (1995). Iterated Brownian motion and Stable (1/4) subordinator, to appear in *Prob. and Stat. Lett.*

[B1] K. BURDZY (1993). Some path properties of iterated Brownian motion. *Sem. Stoch. Proc.* 1992, 67–87 (Ed. K.L. Chung, E. Çinlar and M.J. Sharpe) Birkhäuser, Boston.

[B2] K. BURDZY (1994). Variation of iterated Brownian motion. *Measure-valued Processes, Stochastic Partial Differential Equations and Interacting Systems,* (Ed. D.A. Dawson) CRM Proceedings and Lecture Notes, **5** 35–53.

[CsCsFR1] E. CSÁKI, M. CSÖRGŐ, A. FÖLDES AND P. RÉVÉSZ (1995). Global Strassen type theorems for iterated Brownian motion, to appear in *Stoch. Proc. Their Appl.*

[CsCsFR2] E. CSÁKI, M. CSÖRGŐ, A. FÖLDES AND P. RÉVÉSZ (1995). The local time of iterated Brownian motion, Preprint.

[DM] P. DEHEUVELS AND D.M. MASON (1992). A functional LIL approach to pointwise Bahadur–Kiefer theorems, *Prob. in Banach Spaces,* **8**, 255–266 (eds.: R.M. Dudley, M.G. Hahn and J. Kuelbs)

[F] T. FUNAKI (1979). A probabilistic construction of the solution of some higher order parabolic differential equations, *Proc. Japan Acad.* **55**, 176–179.

[HPS] Y. HU, D. PIERRE LOTTI VIAUD AND Z. SHI (1994). Laws of the iterated logarithm for iterated Wiener processes, to appear in *J. Theor. Prob.*

[HS] Y. HU AND Z. SHI (1994). The Csörgő–Révész modulus of non–differentiability of iterated Brownian motion, to appear in *Stoch. Proc. Their Appl.*.

[IM] K. ITÔ AND H.P. MCKEAN (1965). *Diffusion Processes and Their Sample Paths*, Springer, Berlin, Heidelberg.

[KL1] D. KHOSHNEVISAN AND T.M. LEWIS (1995). Chung's law of the iterated logarithm for iterated Brownian motion, to appear in *Ann. Inst. Hen. Poinc.: Prob. et Stat.*

[KL2] D. KHOSHNEVISAN AND T.M. LEWIS (1995). The modulus of continuity for iterated Brownian motion, to appear in *J. Theoretical Prob.*

[Mc] H.P. MCKEAN (1962). A Hölder condition for Brownian local time, *J. Math. Kyoto Univ.*, **1–2**, 195–201.

[P] E.A. PERKINS (1981). The exact Hausdorff measure of the level sets of Brownian motion, *Z. Wahr. verw. Geb.* **58**, 373–388.

[RY] D. REVUZ AND M. YOR (1991). *Continuous Martingales and Brownian Motion*, Springer, New York.

[S] Z. SHI (1994). Lower limits of iterated Wiener processes, to appear in *Stat. Prob. Lett.*

K. BURDZY
Department of Mathematics
University of Washington
Seattle, WA. 98195
burdzy@math.washington.edu

D. KHOSHNEVISAN
Department of Mathematics
University of Utah
Salt Lake City, UT. 84112
davar@math.utah.edu

On the Spitzer and Chung laws of the iterated logarithm for Brownian motion

by

J.-C. Gruet[(1)] and Z. Shi[(2)]

(1) *Laboratoire de Probabilités, Université Paris VI, 4 Place Jussieu, 75252 Paris Cedex 05, France & Université de Reims, UFR Sciences, 51062 Reims Cedex, France*

(2) *L.S.T.A., Université Paris VI, 4 Place Jussieu, 75252 Paris Cedex 05, France*

1. Introduction.

Let $\{W(t);\ t \geq 0\}$ be a two-dimensional Brownian motion. It is well-known that the Brownian path is almost surely dense in the plane, but never hits a given point at positive time. A natural question is thus to study the rate with which the small values of $\|W - x\|$ (the symbol "$\|\cdot\|$" denoting the usual Euclidean modulus) approach 0 for any $x \in \mathbb{R}^2$. Without loss of generality, we assume $W(0) = (1,0)$ and $x = (0,0)$. Let

$$X(t) = \inf_{0 \leq s \leq t} \|W(s)\|, \qquad t > 0.$$

The following celebrated Spitzer (1958) integral test characterizes the lower functions of X:

Theorem A (Spitzer 1958). *For any non-decreasing function $f > 1$,*

$$\mathbb{P}\Big[\, X(t) < \frac{t^{1/2}}{f(t)},\ \text{i.o.}\,\Big] = \begin{cases} 0 \\ 1 \end{cases} \iff \int^{\infty} \frac{dt}{t \log f(t)} \begin{cases} < \infty \\ = \infty \end{cases}.$$

Here and in the sequel, "i.o." stands for "infinitely often" as t tends to infinity.

Spitzer's Theorem A answers the how-small-are-the-small-values-of-$\|W\|$ question. We propose to study the corresponding "how big" problem for the small values. Our Theorem 1, stated as follows, provides a characterization of the upper functions of X.

Theorem 1. *If $g > 1$ is non-decreasing, then*

$$\mathbb{P}\Big[\, X(t) > \exp\Big(-\frac{\log t}{g(t)}\Big),\ \text{i.o.}\Big] = \begin{cases} 0 \\ 1 \end{cases} \iff \int^{\infty} \frac{dt}{t\,g(t)\log t} dt \begin{cases} < \infty \\ = \infty \end{cases}.$$

Theorems A and 1 together give an accurate description of the almost sure asymptotic behaviours of X. For example, it is immediately seen from the aboves theorems that

$$\mathbb{P}\Big[\, X(t) > \exp\Big(-\frac{\log t}{(\log\log t)^a}\Big),\ \text{i.o.}\Big] = \begin{cases} 0 & \text{if } a > 1; \\ 1 & \text{otherwise.} \end{cases}$$

$$\mathbb{P}\Big[\, X(t) < \exp\Big(-(\log t)(\log\log t)^a\Big),\ \text{i.o.}\Big] = \begin{cases} 0 & \text{if } a > 1; \\ 1 & \text{otherwise.} \end{cases}$$

What about the lower functions of the big values of Brownian motion? Let us recall the classical Chung (1948) integral test for linear Brownian motion.

Theorem B (Chung 1948). *Let B be a real-valued Brownian motion. For every non-decreasing function $h > 0$ such that $t^{-1/2}h(t)$ is non-increasing, we have*

$$\mathbb{P}\Big[\sup_{0\le s\le t} |B(s)| < \frac{t^{1/2}}{h(t)},\ \text{i.o.}\Big] = \begin{cases} 0 \\ 1 \end{cases} \iff \int^{\infty} \frac{dt}{t} h^2(t)\exp\Big(-\frac{\pi^2}{8}h^2(t)\Big) \begin{cases} < \infty \\ = \infty \end{cases}.$$

Chung's Theorem B was obtained for linear Brownian motion. The following natural question was raised by Révész (1990, p.195): for the planar Brownian motion W, what can be said on the liminf behaviour of $\sup_{0\le s\le t} \|W(s)\|$?

The same question can be asked for a Brownian motion of any dimension. Let $\{V(t);\, t \geq 0\}$ denote a d-dimensional Brownian motion, and let

$$Y(t) = \sup_{0 \leq s \leq t} \|V(s)\|, \quad t > 0.$$

Our answer to the problem is the following

Theorem 2. *Let $d \geq 1$ and let $h > 0$ be a non-decreasing function. Then*

$$\mathbb{P}\Big[Y(t) < \frac{t^{1/2}}{h(t)}, \text{ i.o.} \Big] = \begin{cases} 0 \\ 1 \end{cases} \iff \int^{\infty} dt\, \frac{h^2(t)}{t} \exp\Big(-\frac{j_\nu^2}{2} h^2(t) \Big) \begin{cases} < \infty \\ = \infty \end{cases},$$

where j_ν denotes the smallest positive zero of the Bessel function $J_\nu(x)$ of index $\nu \equiv (d-2)/2$.

Remarks. (i) Since $j_{-1/2} = \pi/2$, Theorem B is a special case of the above result.

(ii) An interesting feature in Theorem 2 is that we do not suppose $t^2/h(t)$ to be non-decreasing. Thus the latter condition can be removed from Chung's Theorem B.

(iii) As usual, Theorem 2 has a "local" version for t tending to 0, of which the statement and proof are omitted.

Corollary 1. *We have, for $d \geq 1$,*

$$\liminf_{t \to \infty} \Big(\frac{\log \log t}{t} \Big)^{1/2} Y(t) = \frac{j_\nu}{2^{1/2}} \quad \text{a.s.}, \tag{1.1}$$

with rate of convergence

$$\liminf_{t \to \infty} \frac{(\log \log t)^{3/2}}{t^{1/2} \log \log \log t} \Big(Y(t) - \frac{j_\nu}{2^{1/2}} \frac{t^{1/2}}{(\log \log t)^{1/2}} \Big) = -\frac{j_\nu}{2^{1/2}}, \quad \text{a.s.}$$

Remark. The LIL (1.1) was previously obtained by Lévy (1953) for $d = 2$ and by Ciesielski & Taylor (1962) for any dimension d.

Theorem 1 is proved in Section 2, and Theorem 2 in Section 3.

2. The proof of Theorem 1.

Let W be as before a Brownian motion in the plane, starting from $(1,0)$. Define

$$H(x) = \inf\{ t > 0 : \|W(t)\| = x \}, \quad 0 < x \le 1,$$

the first hitting time of $\|W\|$ at level x. Obviously the process $y \mapsto H(1/y)$ (for $y \ge 1$) is increasing, and it has independent increments by using the strong Markov property of $\|W\|$. Since

$$H(2^{-n}) = \sum_{k=1}^{n} \Big(H(2^{-k}) - H(2^{-(k-1)}) \Big),$$

using Brownian scaling gives

$$H(2^{-n}) = \sum_{k=1}^{n} 2^{-2(k-1)} \xi_k,$$

where $(\xi_k)_{k\ge1}$ is an i.i.d. sequence of random variables having the same law as $H(1/2)$. Consequently,

$$2^{-2(n-1)} \sum_{k=1}^{n} \xi_k \le H(2^{-n}) \le \sum_{k=1}^{n} \xi_k. \tag{2.1}$$

Let us first establish a preliminary result for the partial sum of (ξ_k):

Lemma 1. *Let $\{ \Lambda(t); t \ge 0 \}$ be a subordinator, and assume that $\Lambda(1)$ has the same law as $H(1/2)$. Then for any function $f > 1$ such that $f(t)/t$ is non-decreasing, we have*

$$\limsup_{t\to\infty} \frac{\Lambda(t)}{f(t)} = \begin{cases} 0 \\ \infty \end{cases}, \quad \text{a.s.} \quad \Longleftrightarrow \quad \int^{\infty} \frac{dt}{\log f(t)} \begin{cases} < \infty \\ = \infty \end{cases}.$$

Proof of Lemma 1. The Laplace transform of $H(1/2)$ is well-known (see Kent (1978 Theorem 3.1)):

$$\mathbb{E} \exp(-\lambda H(1/2)) = \frac{K_0(\sqrt{2\lambda})}{K_0(\sqrt{\lambda/2})}, \quad \forall \lambda > 0,$$

where K_0 is the modified Bessel function. Recall that $\Lambda(1)$ has the same law as $H(1/2)$. Write

$$\mathbb{E} \exp(-\lambda\Lambda(1)) = \exp(-\Psi(\lambda)).$$

Thus

$$\Psi(\lambda) = \log K_0(\sqrt{\lambda/2}) - \log K_0(\sqrt{2\lambda}).$$

Using elementary asymptotics of K_0, we immediately arrive at the following estimate

$$\Psi(\lambda) - \lambda\Psi'(\lambda) \sim \frac{2\log 2}{\log(1/\lambda)}, \qquad \lambda \to 0,$$

(the usual symbol "$a(x) \sim b(x)$" ($x \to x_0$) means $\lim_{x\to x_0} a(x)/b(x) = 1$). Now the statement of Lemma 1 follows by applying a general result for subordinators (see for example Fristedt (1974 Theorem 6.1)) which tells that $\limsup_{t\to\infty} \Lambda(t)/f(t) = 0$ or ∞ (almost surely) according as

$$\int^{\infty} \Big(\Psi(\frac{1}{f(t)}) - \frac{1}{f(t)} \Psi'(\frac{1}{f(t)}) \Big) dt$$

converges or diverges. □

Lemma 2. *If $h > 1$ is a non-decreasing function with $\int^{\infty} dt/h(t) = \infty$, then*

$$\int^{\infty} \frac{dt}{t + h(t)} = \infty.$$

Proof of Lemma 2. The proof is briefly sketched, since it involves only elementary computations. Set $\mathcal{A} = \{ t : h(t) \le t \}$ and $\mathcal{B} = \{ t : h(t) > t \}$. Obviously, we have

$$\frac{1}{t + h(t)} \ge \frac{1}{2} \Big(\frac{1}{t} \mathbb{1}_{\mathcal{A}}(t) + \frac{1}{h(t)} \mathbb{1}_{\mathcal{B}}(t) \Big).$$

Assume $\int^{\infty} \mathbb{1}_{\mathcal{A}}(t)(dt/t) < \infty$. We only have to show $\int^{\infty} \mathbb{1}_{\mathcal{B}}(t)(dt/h(t)) = \infty$. Write $F_{\mathcal{A}}(t) \equiv \int_1^t \mathbb{1}_{\mathcal{A}}(s)ds$. Using integration by parts for $\int \mathbb{1}_{\mathcal{A}}(s)(ds/s)$, it is seen that $t \mapsto F_{\mathcal{A}}(t)/t$ is a Cauchy family for $t > 1$. Thus $\lim_{t\to\infty} F_{\mathcal{A}}(t)/t$ exists. If $\lim_{t\to\infty} F_{\mathcal{A}}(t)/t > 0$, then $\int_1^t F_{\mathcal{A}}(s)(ds/s^2)$ would diverge, which contredicts the convergence of $\int_1^{\infty} \mathbb{1}_{\mathcal{A}}(s)(ds/s)$ (the latter is obviously greater than $\int_1^t F_{\mathcal{A}}(s)(ds/s^2)$ by integration by parts). Consequently, $\lim_{t\to\infty} F_{\mathcal{A}}(t)/t = 0$. Thus $\int_1^t \mathbb{1}_{\mathcal{B}}(s)ds \ge t/2$ for sufficiently large t. Again using integration by parts, we obtain

$$\int_1^t \mathbb{1}_{\mathcal{B}}(t) \frac{ds}{h(s)} \ge \frac{1}{2} \int_1^t \frac{ds}{h(s)} + \text{a finite term},$$

which diverges as t tends to infinity. Lemma 2 is proved. □

Proof of Theorem 1. Pick $0 < x < 1$, and let us write $2^{-n} \le x \le 2^{-(n-1)}$ (which means $(n-1)\log 2 \le \log(1/x) \le n \log 2$). Then $H(2^{-(n-1)}) \le H(x) \le H(2^{-n})$. Using (2.1), we have

$$x^2 \Lambda\Big(\log(1/x)/\log 2 - 1\Big) \le H(x) \le \Lambda\Big(\log(1/x)/\log 2 + 1\Big). \tag{2.2}$$

First, we show the following integral test for H:

$$\limsup_{x\to 0+} \frac{H(x)}{f(x)} = \begin{cases} 0 \\ \infty \end{cases}, \quad \text{a.s.} \iff \int_{0+} \frac{dx}{x \log f(x)} \begin{cases} < \infty \\ = \infty \end{cases}, \tag{2.3}$$

for any function $f > 1$ such that $\log f(x)/\log(1/x)$ is non-increasing. Indeed, assume that $\int_{0+} dx/(x \log f(x))$ converges. Define $\hat{f}(t) = f(2^{-t/2})$. Then $\hat{f}(t)/t$ is non-decreasing, with $\int^\infty dt/\log \hat{f}(t) < \infty$. By Lemma 1, we have $\limsup_{t\to\infty} \Lambda(t)/\hat{f}(t) = 0$, with probability 1. Thus $\limsup_{t\to\infty} \Lambda(t)/f(2^{-(t-1)}) = 0$. Using the second part of (2.2), this implies $\limsup_{x\to 0+} H(x)/f(x) = 0$. It remains to verify the divergent half of (2.3). Suppose $\int_{0+} dx/(x \log f(x)) = \infty$. Then $\int^\infty dt/\log f(2^{-2(t+1)})$ diverges as well. According to Lemma 2, this implies

$$\int^\infty \frac{dt}{\log \tilde{f}(t)} = \infty,$$

for $\tilde{f}(t) \equiv 2^{2(t+1)} f(2^{-2(t+1)})$. Applying Lemma 1 gives $\limsup_{t\to\infty} \Lambda(t)/\tilde{f}(t) = \infty$, which, by means of the first part of (2.2), yields $\limsup_{x\to 0+} H(x)/f(x) = \infty$. Hence (2.3) is proved. By noting $[\,H(x) > t\,] = [\,X(t) > x\,]$ (for any $0 < x < 1$ and $t > 0$), several lines of standard calculation readily confirm that the integral test (2.3) is equivalent to that in Theorem 1. □

3. The proof of Theorem 2.

In this section, V denotes a d-dimensional Brownian motion, which, without loss of generality, is assumed to start from 0. Let $H(x) = \inf\{\, t > 0 : \|V(t)\| = x \,\}$ (for $x > 0$). The proof of Theorem 2 is essentially based on the following exact density function of $H(1)$ due to Ciesielski & Taylor (1962):

$$\mathbb{P}\,[\,H(1) \in dt\,]/dt = \frac{1}{2^\nu \Gamma(\nu+1)} \sum_{n=1}^{\infty} \frac{j_{\nu,n}^{\nu+1}}{J_{\nu+1}(j_{\nu,n})} \exp\Big(-\frac{j_{\nu,n}^2}{2} t\Big), \quad t > 0,$$

where $\nu = (d-2)/2$, and $0 < j_{\nu,1} < j_{\nu,2} < \cdots$ are the positive zeros of the Bessel function J_ν (and of course $J_{\nu+1}$ denotes the Bessel function of index $\nu+1$). Let Y be as before the supremum process of $\|V\|$. By Brownian scaling, we have, for any $x > 0$,

$$\mathbb{P}\left[Y(1) < x\right] = \mathbb{P}\left[H(x) > 1\right] = \mathbb{P}\left[H(1) > 1/x^2\right]$$
$$= \frac{2^{1-\nu}}{\Gamma(\nu+1)} \sum_{n=1}^{\infty} \frac{1}{j_{\nu,n}^{1-\nu} J_{\nu+1}(j_{\nu,n})} \exp\left(-\frac{j_{\nu,n}^2}{2x^2}\right),$$

which implies

$$\mathbb{P}\left[Y(1) < x\right] \sim \frac{2^{1-\nu}}{\Gamma(\nu+1) j_\nu^{1-\nu} J_{\nu+1}(j_\nu)} \exp\left(-\frac{j_\nu^2}{2x^2}\right), \quad \text{as } x \to 0, \tag{3.1}$$

(recall that $j_\nu \equiv j_{\nu,1}$ is the smallest positive zero of J_ν). Write in the sequel $\rho \equiv j_\nu^2/2$.

Let $h > 0$ be a non-decreasing function. In the rest of the note, generic constants will be denoted by K_i $(1 \le i \le 9)$.

We begin with the convergent part of Theorem 2, which is an immediate consequence of the tail estimation (3.1). Indeed, pick a sufficiently large initial value t_0 and define the sequence $(t_k)_{k\ge1}$ by $t_{k+1} = (1 + h^{-2}(t_k))t_k$ for $k = 0, 1, 2, \cdots$, and write $h_k \equiv h(t_k)$ for notational convenience. Obviously t_k increases to infinity (as $k \to \infty$). Assume that $\int^\infty (dt/t) h^2(t) \exp(-\rho h^2(t))$ converges. This implies, by several lines of elementary calculation, the convergence of $\sum_k \exp(-\rho h_k^2)$. From (3.1) and scaling it follows that

$$\mathbb{P}\left[Y(t_k) < \frac{t_k^{1/2}}{h_k - 1/h_k}\right] = \mathbb{P}\left[Y(1) < \frac{1}{h_k - 1/h_k}\right]$$
$$\le K_1 \exp(-\rho(h_k - 1/h_k)^2)$$
$$\le K_2 \exp(-\rho h_k^2),$$

which sums. According to Borel-Cantelli lemma, (almost surely) for large k, $Y(t_k) \ge t_k^{1/2}/(h_k - 1/h_k)$. Let $t \in [t_k, t_{k+1}]$. Then by our construction of (t_k),

$$Y(t) \ge Y(t_k) \ge \frac{t_k^{1/2}}{h_k - 1/h_k} = \frac{t_{k+1}^{1/2}}{(1+1/h_k^2)(h_k - 1/h_k)} \ge \frac{t_{k+1}^{1/2}}{h_k} \ge \frac{t^{1/2}}{h(t)},$$

which yields the convergent part of Theorem 2.

It remains to show the divergent part. Let h be such that

$$\int^\infty \frac{dt}{t} h^2(t) \exp(-\rho h^2(t)) = \infty. \tag{3.2}$$

In view of (1.1), we assume without loss of generality that

$$\frac{(\log\log t)^{1/2}}{(2\rho)^{1/2}} \le h(t) \le \frac{(2\log\log t)^{1/2}}{\rho^{1/2}}. \tag{3.3}$$

Define $t_k = \exp(k/\log k)$ (for $k \ge k_0$) and write as before $h_k \equiv h(t_k)$. In what follows, we only deal with the index k tending ultimately to infinity. Therefore our results, sometimes without further mention, are to be understood for sufficiently large k's. Obviously (3.2) is equivalent to the following

$$\sum_k \exp(-\rho h_k^2) = \infty. \tag{3.4}$$

Using (3.3) gives

$$\frac{(\log k)^{1/2}}{(3\rho)^{1/2}} \le h_k \le \log k. \tag{3.5}$$

Fix an $\varepsilon > 0$, then

$$k^{-1/2} h_k^2 \le \frac{\varepsilon}{\rho}, \tag{3.6}$$

(for $k \ge k_0$). Consider the measurable event $A_k = \{Y(t_k) < t_k^{1/2}/h_k\}$. From (3.1) it follows that for $k \ge k_0$,

$$\mathbb{P}(A_k) = \mathbb{P}\Big[Y(1) < \frac{1}{h_k}\Big] \ge (1-\varepsilon)\frac{2^{1-\nu}}{\Gamma(\nu+1) j_\nu^{1-\nu} J_{\nu+1}(j_\nu)} \exp\Big(-\rho h_k^2\Big), \tag{3.7}$$

which, by means of (3.4), yields

$$\sum_k \mathbb{P}(A_k) = \infty. \tag{3.8}$$

Let $k < \ell$. Since V has independent and stationary increments, we have

$$\begin{aligned} \mathbb{P}(A_k A_\ell) &= \mathbb{P}\Big[\sup_{0\le t\le t_k} \|V(t)\| < \frac{t_k^{1/2}}{h_k},\ \sup_{0\le t\le t_\ell} \|V(t)\| < \frac{t_\ell^{1/2}}{h_\ell}\Big] \\ &\le \mathbb{P}(A_k) \sup_{\|x\|\le t_k^{1/2}/h_k} \mathbb{P}\Big[\sup_{0\le t\le t_\ell - t_k} \|V(t)+x\| < \frac{t_\ell^{1/2}}{h_\ell}\Big]. \end{aligned}$$

Using a general property of Gaussian measures (see for example Ledoux & Talagrand (1991 p.73)), it follows that

$$\begin{aligned} \mathbb{P}(A_k A_\ell) &\le \mathbb{P}(A_k)\mathbb{P}\Big[\sup_{0\le t\le t_\ell - t_k} \|V(t)\| < \frac{t_\ell^{1/2}}{h_\ell}\Big] \\ &= \mathbb{P}(A_k)\mathbb{P}\Big[Y(1) < \frac{t_\ell^{1/2}}{(t_\ell - t_k)^{1/2} h_\ell}\Big]. \end{aligned} \tag{3.9}$$

For every $n > k_0$, define

$$\mathcal{E}(n) = \{\, k_0 \le k < \ell \le n : \ell - k \le (\log k)^3 \,\},$$
$$\mathcal{F}(n) = \{\, k_0 \le k < \ell \le n : \ell - k > (\log k)^3 \,\}.$$

It is seen that when $k < \ell < k + (\log k)^3$,

$$\begin{aligned}\frac{\ell}{\log \ell} - \frac{k}{\log k} &= \frac{\ell \log k - k \log \ell}{\log k \, \log \ell} \\ &= \frac{(\ell - k)\log k - k \log(1 + (\ell - k)/k)}{\log k \, \log \ell} \\ &\sim \frac{\ell - k}{\log k}, \qquad (k \to \infty),\end{aligned}$$

which implies

$$\frac{t_k}{t_\ell} \le \exp\Big(-\frac{\ell - k}{2 \log k}\Big).$$

Let $(k, \ell) \in \mathcal{E}(n)$. From the above estimate it follows that

$$\frac{t_\ell^{1/2}}{(t_\ell - t_k)^{1/2}} \le \Big(1 - \exp(-\frac{\ell - k}{2 \log k})\Big)^{-1/2} \le \Big[\, K_3 \min\Big((\ell - k)/\log k, 1 \Big) \Big]^{-1/2},$$

which, by means of (3.5), yields

$$\frac{t_\ell^{1/2}}{(t_\ell - t_k)^{1/2} h_\ell} \le \frac{t_\ell^{1/2}}{(t_\ell - t_k)^{1/2} h_k} \le \Big[\, K_4 \min(\ell - k, \log k) \Big]^{-1/2}.$$

From (3.9) and (3.1) it follows that

$$\mathbb{P}(A_k A_\ell) \le K_5 \mathbb{P}(A_k) \exp(-K_6(\ell - k)) + K_5 \mathbb{P}(A_k) k^{-K_6}.$$

Obviously,

$$\sum_{\ell > k} \exp(-K_6(\ell - k)) \le K_7,$$
$$\sum_{k < \ell < k + (\log k)^3} k^{-K_6} \le k^{-K_6} (\log k)^3 \le K_8.$$

Therefore,

$$\sum\sum_{(k,\ell) \in \mathcal{E}(n)} \mathbb{P}(A_k A_\ell) \le K_9 \sum_{k=k_0}^{n} \mathbb{P}(A_k). \tag{3.10}$$

Now let $(k, \ell) \in \mathcal{F}(n)$. In this case, $\ell - (\log \ell)^2 \ge k + (\log k)^3 - (\log(k + (\log k)^3))^2 > k$, thus $\ell - k > (\log \ell)^2$. Since

$$\frac{\ell}{\log \ell} - \frac{k}{\log k} = \frac{(\ell - k)\log k - k \log(1 + (\ell - k)/k)}{\log k \, \log \ell} \sim \frac{\ell - k}{\log \ell},$$

we have

$$\frac{t_\ell^{1/2}}{(t_\ell - t_k)^{1/2} h_\ell} \le \frac{1}{\left[1 - \exp(-(\ell - k)/2\log \ell)\right]^{1/2} h_\ell} \le \frac{1}{\left(1 - \ell^{-1/2}\right)^{1/2} h_\ell}.$$

By means of (3.9), (3.1), (3.6) and (3.7), this implies

$$\begin{aligned} \mathbb{P}(A_k A_\ell) &\le \mathbb{P}(A_k)(1+\varepsilon)\exp(\rho \ell^{-1/2} h_\ell^2) \frac{2^{1-\nu}}{\Gamma(\nu+1) j_\nu^{1-\nu} J_{\nu+1}(j_\nu)} \exp\left(-\rho h_k^2\right) \\ &\le (1+3\varepsilon)e^{\varepsilon}\mathbb{P}(A_k)\mathbb{P}(A_\ell). \end{aligned}$$

Combining the above estimate together with (3.10) and (3.8) yields

$$\liminf_{n\to\infty} \sum_{k=k_0}^{n} \sum_{\ell=k_0}^{n} \mathbb{P}(A_k A_\ell) \Big/ \Big(\sum_{k=k_0}^{n} \mathbb{P}(A_k) \Big)^2 \le 1.$$

According to a well-known version of Borel-Cantelli lemma (see for example Révész (1990 p.28)), we have $\mathbb{P}(A_k, \text{i.o.}) = 1$. The proof of the divergent part of Theorem 2 is completed. □

Acknowledgement. We are grateful to Davar Khoshnevisan for helpful comments on the first draft of this work.

REFERENCES

Chung, K.L. (1948): On the maximum partial sums of sequences of independent random variables. *Trans. Amer. Math. Soc.* 64, 205-233.

Ciesielski, Z. & Taylor, S.J. (1962). First passage times and sojourn times for Brownian motion in space and the exact Hausdorff measure of the sample path. *Trans. Amer. Math. Soc.* 103 434-450.

Fristedt, B.E. (1974): Sample functions of stochastic processes with stationary independent increments. In: *Advances in Probability 3* (Eds.: P. Ney & S. Port) pp. 241-396. Dekker, New York.

Itô, K. & McKean, H.P. (1965): *Diffusion Processes and their Sample Paths.* Springer, Berlin.

Ledoux, M. & Talagrand, M. (1991): *Probability in Banach Spaces: Isoperimetry and Processes.* Springer, Berlin.

Lévy, P. (1953): La mesure de Hausdorff de la courbe du mouvement brownien. *Giornale dell'Istituto Italiano degli Attuari* 16 1-37.

Révész, P. (1990): *Random Walk in Random and Non-Random Environments.* World Scientific, Singapore.

Revuz, D. & Yor, M. (1994): *Continuous Martingales and Brownian Motion.* 2nd Edition. Springer, Berlin.

Spitzer, F. (1958): Some theorems concerning 2-dimensional Brownian motion. *Trans. Amer. Math. Soc.* 87 187-197.

ON THE EXISTENCE OF DISINTEGRATIONS

by
Lester E. Dubins and Karel Prikry

Abstract. **Whether, with respect to every partition of the unit square, S, consisting of Borel subsets of S, Lebesgue measure on S admits a countably additive disintegration, is undecidable with the usual axioms for set theory. Also reported herein: There are Borel partitions of S with respect to which, Lebesgue measure admits no proper, integrable disintegrations, not even one that is finitely additive.**

Section 1. Introduction and Summary.

One is often concerned with the conditional probability of an event, B, given (the occurrence of) an event h, $P(B/h)$, where B ranges over a collection $\mathcal{B}$ of events and h ranges over a collection, π, of exhaustive and pairwise incompatible events. The condition,

$$P(h/h) = 1 \text{ for all } h \text{ in } \pi, \tag{1}$$

is sometimes, as in the theory of regular conditional distributions, not required. Presumably, the forfeiture of (1), an intuitively necessary condition, has been made in order to accommodate certain requirements of measurability and countable additivity.

In the present paper, which joins earlier ones in the study of the existence of proper disintegrations, Condition (1) is required.

If (1) holds, and if the integral, or expectation, of $P(B/h)$ with respect to a probability measure, Q, on π, has the unconditional probability, $P(B)$, for its value, for all B in a collection $\mathcal{B}$, then P on $\mathcal{B}$ has a *proper disintegration*, a notion more formally defined below. Since the present paper concerns no disintegrations other than those that are proper, "disintegration" will mean "proper disintegration".

Principal interest herein is the case in which P is countably additive and nonatomic, and defined on the sigma-field, $\mathcal{B}$, of all Borel subsets of some uncountable Borel subset, S, of a complete, separable metric space. Since all such

---

1991 Mathematics Subject Classification: Primary 28A50, 60A05, 60A10, 03E35, 03E75; Secondary 03E50, 04A30, 28E15.

Key words and phrases. Sigma-field, countably additive probability, Borel partition, disintegration: proper, integrable, countably additive, π-kernel.

are Borel isomorphic (see [P, 1967, Theorem 2.12]), it may be better to fix S to be the usual coin-tossing space, the unit square, or the unit interval, and P to be the usual fair-coin distribution, or Lebesgue measure.

If every element of a partition of S is a Borel subset of S, the partition itself is, in this paper, called *Borel.* A disintegration, D, is ***conventional*** if the partition π is Borel, and if these two additional usual conditions obtain:

(2) The marginal of P on π can serve as the integrating measure, Q;

(3) D is countably additive.

(Some formal definitions are postponed until the next section.)

It is known that there are Borel π for which Lebesgue measure possesses no proper conventional disintegrations. The set of atoms of the tail sigma-field for ordinary coin-tossing measure provides such an example; see [DH, 1983].

Therefore, interest arises in the existence of disintegrations if Condition (2) or (3) (or both) is not required, and the purpose of this paper is to report these two findings:

(4a) There is a Borel partition of the square, S, with respect to which Lebesgue measure possesses no proper disintegration that satisfies (2), not even one that is finitely additive.

(4b) The question whether, for every Borel partition of the square, S, Lebesgue measure possesses a disintegration that satisfies (3), is not decidable with the usual axioms for set theory.

The question whether, when neither (2) nor (3) is required, there exists, for each Borel π, a disintegration of Lebesgue measure, we do not see how to settle.

Introduce the notation, $\mathfrak{P}$, for the set of P that are countably additive and nonatomic. Of course, since all such P are Borel isomorphic, a fact corresponding to (4a) holds for each such P. As it turns out, however, this stronger fact obtains:

(4c) There is a Borel partition of S with respect to which no P in $\mathfrak{P}$, possesses an integrable disintegration.

Some references containing material related to the present study are: [De, 1930, 1972, 1974], [BR, 1963], [Bo, 1969, p. 39, Proposition 13], [BD, 1975, Theorem 2], [D, 1977], [SV, 1979, Theorem 1.1.8], [DH, 1983] and [MR, 1988].

Section 2. Definitions and Notations.

π-Measurability. Always, π designates a partition of S, sometimes Borel, and functions are real-valued, defined on S, and usually bounded. Let π^* designate the set of functions whose restriction to each member h of π is constant, and call such functions *π-measurable.* Plainly, each π-measurable function can be identified with a unique function whose domain is π, and vice versa. The useful convention of identifying a set with its indicator, that is, with that function that is 1 on the set and 0 off the set, is borrowed from de Finetti, and is used herein. Plainly, the π-measurable sets then constitute a sigma-field, indeed a complete Boolean algebra. Equally plainly, a function, f, is π-measurable if, and only if, the inverse image under f of every set is π-measurable. Calling an element of π a *π-fiber*, it is evident that a set is π-measurable if, and only if, it is a union of π-fibers.

Proper. A mapping, κ, of a collection of functions into π^* is proper, at a π-fiber, h, if the value of κf at h depends only on the values of f on h, that is, if f and f' are members of the collection that agree on h, then κf and $\kappa f'$ agree at h. If κ is proper at each h in π, then κ is *proper*.

π-Kernels. A π-proper, κ, defined on a linear space, F, of functions, that includes the constant functions, c, is a *π-kernel* if κ is linear and order-preserving, and normalized by the condition $\kappa(c) = c$ for constants c.

For later reference, it is noted here that if an f in F is π-measurable, then κf equals f. Moreover, even if not in F, if f is π-measurable, it is natural to define κf to be f. There is some convenience in enlarging the scope of κ to the linear space, $F + \pi^*$, of all $f + g$ for f in F and g in π^*, by

$$\begin{aligned} \kappa(f+g) &= \kappa f + \kappa g. \\ &= \kappa f + g. \end{aligned} \tag{5}$$

As is easily verified, this is a valid definition, and the enlarged κ, too, is a π-kernel. Thus enlarged, κ is idempotent, that is, $\kappa(\kappa)$ is κ, or more fully, $\kappa(\kappa(f))$ is $\kappa(f)$.

Expectations. Each π-kernel determines a family of conditional probabilities, or expectations, one for each π-fiber, h, where the expectation corresponding to h is supported by h. If not otherwise stated, an expectation is not required to be countably additive, so an expectation here is a linear functional, Q, defined on a linear space of bounded functions, that satisfies for each f, Qf is at most the least upper bound of f. Here, the expectation corresponding to h has as its domain the linear space, Fh, of fh for f in F, where fh agrees with f on h and is 0 off h, and the expectation assigns to fh the value of κf at h. Because κ is proper, this is indeed a well-defined expectation.

Following de Finetti again, [De, 1972, p. 117], probability measures and their corresponding expectations are designated by the same letter, herein usually by P or Q.

Disintegration. Consider the following two conditions that a pair κ and P may satisfy:

(6a) For all f, if κf is everywhere nonnegative then, for all positive numbers, ε, the P probability that f is less than $-\varepsilon$ is 0.

(6b) There is an expectation, Q, defined on the range of κ such that

$$Pf = Q(\kappa f), \text{ all } f \text{ in } F; \tag{7a}$$

or, more briefly,

(7b) $$P = Q\kappa.$$

Proofs, when straightforward, as for the following lemma, are often omitted.

Lemma 1. *(6a) implies (6b).*

If a π-kernel, κ, satisfies (6b), then P has a *π-disintegration*, and the pair $[\kappa, Q]$ constitutes a disintegration of P, or, more fully, a *π-disintegration* of P. Plainly, for any κ and P, there is at most one Q defined on the range of κ such that $[\kappa, Q]$ is a disintegration of P. So, when Q does exist, it is appropriate, too, to say that κ is a disintegration of P.

Integrable Disintegrations. If (6b) holds and κf is P-integrable, then P and Q agree on κf, for then:

(8) $$P(\kappa f) = Q[\kappa(\kappa(f))] = Q(\kappa f).$$

If, for all f in F, κf is P-integrable, κ is *P-integrable.* In this case, Q and P agree on the range of κ, and, letting P designate also its restriction to that range, $[\kappa, P]$ is a disintegration of P. Such a disintegration is *integrable.* Recapitulating, if κ is a P-integrable π-kernel, then

(7*a) $$Pf = P(\kappa f), \text{ all } f \text{ in } F;$$

or, more briefly,

(7*b) $$P = P\kappa.$$

Say that a partition π is *P-integrable* if, for some π-kernel, κ, (7*b) holds.

Proposition 0. *For P to have a π-disintegration, it is necessary and sufficient that some extension of P possess an integrable π-disintegration.*

Proof. Suppose that P has a π-disintegration, that is, suppose that (6b) holds. Then, if f and f' are in F, and if $\kappa f = f'$, then $Pf = Pf'$, as is evident by this calculation.

$$Pf = Q(\kappa f) = Q[\kappa(\kappa f)] = P(\kappa f) = Pf'.$$

As is now easily verified, if $\tilde{P}$ assigns to each bounded function of the form $f' + \kappa f$ the value $Pf' + Pf$, then $\tilde{P}$ is a well-defined expectation that extends P. And if $\tilde{\kappa}$ assigns to $f' + \kappa f$ the function $\kappa f' + \kappa f$, then $\tilde{\kappa}$ is a well-defined π-kernel that extends κ. It is then easily verified that for $g = f' + \kappa f$, $\tilde{P}g = \tilde{P}(\tilde{\kappa}g)$, that is, $\tilde{P}$ possesses an integrable disintegration. The converse is immediate from this easily verified fact.

Fact. *If $\tilde{P}$ is an extension of P, and $\tilde{P}$ has a π-disintegration, then so does P.*

Proof. Express $\tilde{P}$ as $\tilde{Q}\tilde{\kappa}$. Then let κ be the restriction of $\tilde{\kappa}$ to the domain of P, and let Q be the restriction of $\tilde{Q}$ to the range of κ. It is then a triviality to verify that $P = Q\kappa$.

Section 3. Universally Nonintegrable Partitions.

Theorem 1. *There exist Borel partitions with respect to which no non-atomic, countably additive probability, P, possesses an integrable disintegration.*

The proof of Theorem 1 requires some preliminaries.

Call π *connective* if, for any two disjoint uncountable Borel sets, there is a π-fiber that intersects each of them. Call π *binary* if each of its fibers has at most two elements. A significant step towards the proof of Theorem 1 is this Proposition.

Proposition 1. *There exist binary π that are connective.*

The proof of Proposition 1 is of a type that has frequently been used, at least since Felix Bernstein, [B, 1907].

Proof of Proposition 1. Index the pairs p of uncountable Borel sets by the ordinals less than the minimal ordinal whose cardinality is the continuum, and let α and β designate such ordinals. Let D be an enumerably infinite subset of S. Suppose that for some β, and for each $\alpha < \beta$, there is an h_α consisting of two elements, that satisfies: h_α is a subset of the union of the pair, p_α; h_α intersects each element of p_α, and, for the $\alpha < \beta$, the h_α are disjoint, and also disjoint from D. As is well-known, and as follows from [P, 1967, Theorem 2.8], each uncountable Borel set has the continuum as its cardinality, the process can continue, and the h_α become defined for all α less than the continuum. The subset of S not covered by the h_α, say, V is infinite, for it includes D. Partition V arbitrarily into sets, each of which has two elements, and let π consist of this partition together with the h_α. Plainly, such a π satisfies the lemma. □

[As is evident, the proof of Proposition 1 made use of the axiom of choice. The question arises whether it may be possible to find a proof that does not rely on that axiom. We believe that the answer is negative, and that that can be seen to follow from certain work of Solovay [S, 1970]. Also, from what Dellacherie has kindly told us, a negative answer perhaps follows from certain work of Sierpinski published in Fundamenta, but we do not know of a precise reference.]

In the interests of brevity of exposition introduce a definition. If no non-atomic, countably additive, probability, P, on S, possesses an integrable π-disintegration, call π universally nonintegrable. So, Theorem 1 asserts the existence of universally nonintegrable partitions.

Proposition 2. *Binary connective π are universally nonintegrable.*

Plainly, binary π are Borel, so Theorem 1 follows immediately from Propositions 1 and 2.

Preliminary to the proof of Proposition 2, call π ***ubiquitous*** if every uncountable Borel set includes a π-fiber. Of course, Proposition 2 follows from these two facts:

Proposition 2.1. *Binary connective π are ubiquitous.*

Proposition 2.2. *If π is both connective and ubiquitous, then it is universally nonintegrable.*

Proof of Proposition 2.1. Let B be an uncountable Borel set. Let U and V be disjoint, uncountable Borel subsets of B, which of course exist. Since π is connective, there is a π-fiber, h, that intersects U and V. Since every fiber has at most two elements, this h has two elements, and is a subset of the union of U and V, and hence of B. □

The next goal, a proof of Proposition 2.2, requires some preparation. First, two definitions: A subset of S is *thin*, if each of its Borel subsets is countable. A real-valued function, g, defined on S is *unwavering* if, there is a constant, c, such that, for each countably additive, nonatomic, P, for which g is integrable, g assumes the value c with P-probability 1.

Lemma 2. *For a set to be thin it is necessary and sufficient that it have measure zero for every countably additive, nonatomic P for which it is measurable.*

Proof. Each countably additive probability, P, on a complete separable metric space, is tight, see [P, 1967, Theorem 3.2]. Consequently, each P-measurable set A of positive P probability includes a compact and, therefore, Borel, K, of positive probability. Since P is nonatomic, K is uncountable. So K, and hence A, is not thin. Summarizing, the condition is necessary. To establish sufficiency, suppose that A is not thin, that is, that it includes an uncountable Borel B. As is well known, see, for example, [P, 1967, Theorem 2.8], such a B includes a subset, K, homeomorphic to the usual coin-tossing space. Therefore, some countably additive probability, P, is supported by K, and hence by B and by A. □

Lemma 3. *Each of the following conditions on π implies its successors.*

(a) *π is connective.*

(b) *For every π-measurable set, either it, or its complement, is thin.*

(c) *Suppose that V is a π-measurable set. Then, either V is thin and (therefore) a P-null set for all P in $\mathfrak{P}$ for which it is measurable, or its complement is.*

(d) *Every real-valued, π-measurable function is unwavering.*

Proof. The arguments that (a) implies (b) and that (b) implies (c) are straightforward and omitted. So suppose (c), and let g be real-valued and π-measurable. It may be supposed that, for some P in $\mathfrak{P}$, g is P-integrable, for otherwise, g is clearly unwavering. Fix a positive number ε, and partition the real line into half-closed intervals, I, of length ε. The inverse image of I under g, say V, is both π-measurable and P-measurable. By (c), either it is thin and $PV = 0$, or its complement is thin and $PV = 1$. Since the I are disjoint, so are the V. Therefore $PV = 1$ for at most one V. Since P is countably additive, PV cannot be 0 for all V. So the complement of V is thin for precisely one V, and, for that V, $PV = 1$. Consider the I corresponding to that V and label it $I(\varepsilon)$. It is now routine to let ε be of the form $1/k$ where k is a power of 2, and to verify that the corresponding sequence of I have as their intersection a single number c. So g is unwavering, and (d) holds. □

Remark 1. Were it useful for the main purposes of this paper, Lemma 3 could have been stated in a stronger form. For each of the four conditions are equivalent to the others. To verify this, it suffices to verify that if (a) does not hold, neither does (d). So, suppose that (a) does not hold. Then there exist disjoint, uncountable, Borel sets B and C with the property that each π-fiber that intersects B is disjoint from C. So the complement, N, of the smallest π-measurable set, M, that includes B, includes C. There is a P in $\mathfrak{P}$ for which PB and PC are 1/2. For such a P, M and N are P-measurable, and the π-measurable function, g, that is 1 on M and -1 off M is P-measurable. Plainly, g is not unwavering. So, (d) does not hold.

Lemma 4. *If π is connective, then, for all π-kernels, κ, and bounded, Borel f, κf is unwavering.*

Proof. Apply Lemma 3.

Turn now to the property of being ubiquitous.

Let the ***π-interior*** of a set, C, be the largest π-measurable set included in it, and designate it by πiC. It is obvious that $C \backslash \pi iC$ includes no π fibers.

Lemma 5. *Suppose π is ubiquitous, C is a Borel set, and P is in $\mathfrak{P}$. Then, if the π-interior of C is a P-null set, so is C.*

Proof. Express C as the union of two sets, its π-interior and the remainder of C, say, D. Since, both C and its π-interior are P-measurable, so is D. Since D includes no π-fibers and π is ubiquitous, D includes no uncountable Borel set. Therefore, D is a P-null set. Since C is the union of two P-null sets, it, too, is P-null. □

Lemma 6. *Suppose π is connective and ubiquitous. Suppose, too, κ is a π-kernel, and f is a bounded, real-valued, Borel function. Then, there is a constant c such that, for all P in $\mathfrak{P}$ for which κf is P-integrable, $f = c$ with P-probability 1.*

Proof. By Lemma 4, there is a c such that, for all described P, $\kappa f = c$ on a set of P-probability 1. Let ε be a positive real number, and let C be the event that f is at least $c + \varepsilon$. Then, on any π-fiber included in C, and, therefore, on the π-interior of C, κf is at least $c + \varepsilon$. So, the π-interior of C is a P-null set. Then, by Lemma 5, C, too, is a P-null event. Likewise, so is the event that f is at most $c - \varepsilon$. Plainly, since P is countably additive, $f = c$, with P-probability 1. □

Since Proposition 2.2 is an immediate consequence of Lemma 6, Proposition 2 has been proven. Therefore, the proof of Theorem 1 is complete.

This special consequence of Theorem 1 is recorded here.

Corollary 1. *There is a Borel partition of the unit interval with respect to which Lebesgue measure possesses no integrable disintegration, not even one that is finitely additive.*

Section 4. Countably Additive Disintegrations.

There is more than one notion for an expectation, E, on a linear space, F, to be countably additive; see [DH, 1984]. Herein, F includes only Borel functionss, and attention is restricted to the usual, and strongest notion in which E is the restriction to F of the usual L_1-space of some countably additive probability measure on $\mathcal{B}$. A kernel, κ, is *countably additive* if, for each h, the corresponding expectation is countably additive. A disintegration is *countably additive* if both κ and Q are countably additive.

Notice that, as defined, countably additive disintegrations need not be conventional. For a disintegration of P to be *conventional*, in addition to being countably additive, it is required to be P-integrable.

The purpose of this section is to prove:

Theorem 2. *The assertion that, with respect to every Borel partition, π, of the unit interval, Lebesgue measure possesses a countably additive disintegration, is undecidable with the usual axioms for set theory.*

The usual set of axioms of set theory, designated by ZFC, are the Zermelo-Fraenkel axioms, together with the axiom of choice.

Of course, the undecidability of an assertion is equivalent to the consistency of it, as well as of its negation, with ZFC. So Theorem 1 is equivalent to the conjunction of two propositions, the first of which is:

Proposition 3. *It is consistent with the usual axioms of set theory that Lebesgue measure on the unit interval possesses a countably additive π-disintegration for every Borel π.*

Recall two notions: If K is a collection of disjoint non-empty sets, then a *K-selection* is a set included in their union that has a single point in common with each of the members of K. In particular, a π-selection is a subset, V, of S that contains one, and only one, point of each π-fiber. And the *π-saturation* of a set, C, is the smallest π-measurable set that includes C.

Lemma 7. *Each of the following conditions on P and π implies its successors.*

(i) *If a set of π-fibers has cardinality less than the continuum, then each Borel subset of its union is a P-null set.*

(ii) *There is a π-selection, V, of outer P-probability 1.*

(iii) *There exists a countably additive, π-disintegration of P.*

Proof. (i) $\rightarrow$ (ii). Assign to each ordinal α of cardinality less than the continuum a Borel set, B_α, of positive P-probability so that all Borel sets of positive P-probability are listed. Fix α, and suppose that for each $\beta < \alpha$, there is assigned a point x_β in B_β, and let h_β be the π-fiber containing x_β. Plainly, the set of these fibers for $\beta < \alpha$ has cardinality less than the continuum. So, by (i), B_α contains a point, x_α, not in any fiber, h_β, for $\beta < \alpha$. Let V be the union of the set of these x_α with any selection from the set of fibers complementary to the set of h_α. Plainly, V satisfies (ii).

(ii) $\rightarrow$ (iii). Let κ be the π-kernel that assigns to the Borel set, B, the indicator of the π-saturation of VB, the intersection of V with B. Equivalently, κ associates to the π-fiber, h, the one-point dirac delta measure at the singleton Vh. Plainly, κ satisfies (6a) for V has outer measure 1. Hence, by Lemma 1, (6b) holds. Verify that the set of π-saturations of VB, B Borel, is a sigma-field, say $\mathcal{V}$, and each κf is measurable with respect to that sigma-field. There remains only to verify that Q is countably additive on $\mathcal{V}$. For this purpose, verify that $\mathcal{V}$ is isomorphic to the sigma-field, $\mathcal{W}$, of subsets of V of the form VB (which holds for any selection V). To conclude that Q is countably additive, one need only observe that $Q(\kappa(B)) = PB = P^*(VB)$ (where P^* denotes P-outer measure), for V has outer measure 1. For this shows that Q is isomorphic to the restriction of the outer measure P^* to $\mathcal{W}$, which, as is well-known, is countably additive.

Proof of Proposition 3. As is easily verified, it suffices to consider π's each of whose elements is a Borel set of Lebesgue measure zero. The continuum hypothesis then implies (i) of Lemma 7. For the union of countably many sets of measure zero has measure zero. Therefore, (i) is consistent with the usual axioms of set theory, and, in view of Lemma 7, so is (iii). □

Remark 2. Proposition 3 has wider validity than asserted. For the argument works for all countably additive P. Furthermore, all restrictions on the nature of π can be removed, at the expense of some complication of the proof.

Remark 3. The role played by the continuum hypothesis in the proof of Proposition 3 could have been played by a weaker axiom, an axiom known as Martin's axiom. For this axiom, too, is strong enough to imply (i) of Lemma 7. One formulation of Martin's axiom is: If a compact Hausdorff space, H, admits no disjoint collection of nonempty open sets, other than countable collections, then the union of fewer than a continuum of closed subsets of H, each of which has no interior, has no interior. (See [K, 1983, Theorem 3.4, p. 65].)

Proposition 4. *It is consistent with the usual axioms of set theory that there be a Borel partition of the unit interval with respect to which Lebesgue measure has no countably additive disintegration.*

Several lemmas are needed.

Lemma 8. *For any π and any countably additive π-kernel, κ, the sigma-field generated by the functions κB for B in $\mathcal{B}$, say $\mathcal{M}$ ($\mathcal{M}$ for marginal sigma-field) is countably generated. Moreover, π is the set of atoms of $\mathcal{M}$.*

Proof. By the monotone class argument, the image under κ of a countable boolean algebra that generates $\mathcal{B}$ generates $\mathcal{M}$, so $\mathcal{M}$ is countably generated. To see that π is the set of atoms of $\mathcal{M}$, notice first that π is a subset of $\mathcal{M}$, for $\kappa h = h$ for all h in π. Each π-fiber is an $\mathcal{M}$-atom, for none of its proper subsets is a member of $\mathcal{M}$. There are no other $\mathcal{M}$-atoms, for the union of these $\mathcal{M}$-atoms is the entire space, S. □

Recall certain terminology and facts: An *atom* of a sigma-field, $\mathcal{U}$, is a minimal element of $\mathcal{U}$; an atom of a probability, P, on $\mathcal{U}$, is a set of positive measure that has no subsets of smaller positive measure. Let α be the sum of the measures of the atoms. If α is 1, the measure is atomic, and, otherwise, P is a unique convex combination of an atomic and nonatomic probability. The next lemma is well-known.

Lemma 9. *Suppose that A is an atom for a countably additive probability measure, P, defined on a sigma-field, $\mathcal{U}$. Then, if $\mathcal{U}$ is countably generated, A is represented by an atom of $\mathcal{U}$.*

For the convenience of the reader, a less well-known fact in the literature [GP, 1984, Theorem 9.2] is reformulated as the next lemma, and a proof is provided. A definition facilitates the formulation.

A cardinal number is *small* if every set of reals of that cardinality has Lebesgue measure 0, or, as is equivalent, no set of positive outer Lebesgue measure is of that cardinality.

Lemma 10. *Every countably additive, finite measure, Q, defined on a sigma-field of subsets of a set, H, of small cardinality, is atomic. Consequently, only the Q that vanishes identically has no atoms.*

Proof. What must be seen is that the nonatomic part of Q vanishes. If it did not, then, by renorming that part, and changing notation, it may be assumed that Q itself is a nonatomic probability measure. Then there exist a sequence of independent events of probability 1/2. The indicators of these events provide a sequence of independent zero-one valued functions, which determine a mapping, φ, of the probability space, H, into fair coin-tossing space. Let C be any Borel subset of coin-tossing space that covers the range of φ. Plainly, its inverse image under φ, being H, certainly has measure 1. Since φ is measure-preserving, C has probability 1 (for the fair coin-tossing measure). So the range of φ has outer probability 1. Therefore, the range is not of small cardinality, so H certainly is not of small cardinality. This contradicts the hypothesis. □

Lemma 11. *Suppose that the cardinality of a partition, π, is small, and P is countably additive, with or without atoms, and has a countably additive π-disintegration. Then the sum of the P-probabilities of the π-fibers is 1.*

Proof. For any κ, $\kappa h = h$. So $Qh = Q\kappa h = Ph$, and it suffices to verify that the sum of the Qh is 1. By Lemmas 8 and 9, this sum is the same as the sum of QA over all Q-atoms A. Since Q is, in effect, defined on a sigma-field of subsets of π, and π is of small cardinality by hypothesis, Lemma 10 applies to complete the proof.

Lemma 11 has a corollary.

Corollary 2. *If π is a partition of small cardinality all of whose fibers are of P-probability 0, then P possesses no countably additive π-disintegrations.*

Proof Proposition 4. In the Cohen model for the negation of the continuum hypothesis, there is a set π, of small cardinality, whose members are disjoint Borel null subsets of the unit interval, I, that covers I, as is proven in [K, 1984]. Therefore, the existence of such π is indeed consistent with the usual axioms for set theory. This, together with Corollary 2, completes the proof.

Since Theorem 2 is nothing other than the conjunction of Propositions 3 and 4, its proof is complete.

REFERENCES

[B, 1907] F. Bernstein, *Zur Theorie der trigonometrischen Reihe*, Berichte über die Verhandlungen der Königlich Sächsischen Gesellschaft der Wissenschaften zu Leipzig, Math.-Phys. Klasse 59 (1907), 325–338.

[Bo, 1969] N. Bourbaki, *Éléments de mathématique*, Intégration, Chap. IX, Hermann, Paris, 1969.

[BD, 1975] D. Blackwell and L. Dubins, *On existence and non-existence of proper, regular, conditional distributions*, Ann. Probab. 3 (1975), 741–752.

[BR, 1963] D. Blackwell and C. Ryll-Nardzewski, *Non-existence of everywhere proper conditional distributions*, Ann. Math. Statist. 34 (1963), 223–225.

[D, 1975] L. Dubins, *Finitely additive conditional probabilities, conglomerability and disintegrations*, Ann. Probab. 3 (1975), 89–99.

[D, 1976] ———, *On disintegrations and conditional probabilities*, in Measure Theory, Lecture Notes in Math. 541 (1976), 53–59, Springer - Verlag, New York, (A. Bellow and D. Kölzow, eds.).

[D, 1977] ———, *Measurable, tail disintegrations of the Haar integral are purely finitely additive*, Proc. Amer. Math. Soc. 62 (1977), 34–36.

[De, 1930] Bruno de Finetti, *Sulla proprieta conglomerativa delle probabilita subordinata*, Rend. R. Inst. Lombardo (Milano) 63 (1930), 414–418.

[De, 1972] ———, *Probability, Induction and Statistics*, J. Wiley, New York, 1972.

[De, 1974] ———, *Theory of Probability, Vol. I*, J. Wiley, New York, 1974.

[DH, 1983] L. Dubins and D. Heath, *With respect to tail sigma-fields, standard measures possess measurable disintegrations*, Proc. Amer. Math. Soc. 88 (1983), 416–418.

[DH, 1984] ———, *On means with countably additive discontinuities*, Proc. Amer. Math. Soc. 91 (1984), 270–274.

[GP, 1984] R. Gardner, W. Pfeffer, *Borel measures*, in Handbook of set-theoretic topology, North- Holland Publ. Co., Amsterdam, 1984, pp. 957–1039, (K. Kunen and J. Vaughan, eds.).

[K, 1983] K. Kunen, *Set theory*, North-Holland Publ. Co., Amsterdam, 1983.

[K, 1984] ———, *Random and Cohen reals*, in Handbook of set-theoretic topology, North-Holland Publ. Co., Amsterdam, 1984, pp. 887–911, (K. Kunen and J. Vaughan).

[MR, 1988] A. Maitra and S. Ramakrishnan, *Factorization of measures and normal conditional distributions*, Proc. Amer. Math. Soc. 103 (1988), 1259–1267.

[P, 1967] K.R. Parthasarathy, *Probability measures on metric spaces*, Academic Press, New york 1967.

[S. 1970] R.M. Solovay, *A model of set-theory in which every set of reals is Lebesgue measurable*, Ann. of Math. 92 (1970), 1-56.

[SV, 1979] D. Stroock and S. Varadhan, *Multidimensional diffusion processes*, Springer-Verlag, New York 1979.

University of California, Berkeley, California 94720
University of Minnesota, Minneapolis, Minnesota 55455

A Counterexample for the Markov Property of Local Time for Diffusions on Graphs

by

Nathalie Eisenbaum* and Haya Kaspi**

*Laboratoire de Probabilités
Universite Pierre et Marie Curie
Tour 56, 4 Place Jussieu
75252 Paris, Cedex 05
FRANCE

**Faculty of Industrial Engineering and Management
Technion—Israel Institute of Technology
Haifa 32000
ISRAEL

1 Introduction

Let X be a transient Markov process taking values in an interval $E \subset \mathbb{R}$, admitting a local time at each point, and such that points communicate (i.e. every point may be reached from any other point). Let L^x be the total accumulated local time at x. Then $(L^x)_{x \in E}$ is a Markov process (indexed by the states of E) if, and only if X has continuous sample paths and fixed birth and death points. The sufficiency is the famous Ray-Knight theorem (see [R], [K], [W], [S] and [E]); the necessity was proved in [E.K].

In the symmetric case, the proof that Sheppard [S] and Eisenbaum [E] gave to the Ray-Knight theorem uses a centered Gaussian field $\{\phi_x : x \in E\}$ with covariance the Green function of X that was introduced in Dynkin's Isomorphism theorem [D]. They have shown that the Markov property of the local time process can be derived from the Markov property of the Gaussian field. Dynkin [D] and Atkinson [A] have shown that when $E \subset \mathbb{R}^d$ the field $(\phi_x)_{x \in E}$ is a Markov field if, and only if, X has continuous paths.

This research was supported by the Fund for Promotion of Research at the Technion.

In view of the above, it is natural to ask whether the conditions for the Markov property of the local time, for processes on $E \subset \mathbb{R}$, are also sufficient for processes taking values in $E \subset \mathbb{R}^d$, and admitting a local time at each point of E.

The following definition of the Markov property for processes indexed by $E \subset \mathbb{R}^d$ was used by Dynkin [D] and Atkinson [A] in the above-mentioned result.

Definition *A process $\{L^x : x \in E\}$ where $E \subset \mathbb{R}^d$ has the Markov property provided for every open, relatively compact set A, contained in E, $\{L^x : x \in \bar{A}\}$ and $\{L^x : x \in A^c\}$ are conditionally independent given $\{L^x : x \in \partial A\}$.*

With this definition of the Markov property, the answer to our question is that continuity and fixed birth and death points are not sufficient for $\{L^x : x \in E\}$ to be Markov. To see this, we consider the Brownian motion X on the unit circle S^1 born at $(1,0)$ and killed when the local time at $(-1,0)$ exceeds an exponential variable that is independent of X. This process has continuous sample paths and fixed birth and death points. Denoting by 1, 2, 3 and 4, respectively, the points $(1,0)$, $(0,1)$ $(-1,0)$ and $(0,-1)$ we shall show:

Proposition *Given (L^1, L^3), L^2 and L^4 are not conditionally independent.*

Remark Leuridan [L] has considered the local time process of a Brownian motion on a circle $(L^x_{\tau_r})_{x \in X^1}$, killed when the local time at $(-1,0)$ exceeds $r > 0$. He has obtained a Ray-Knight type theorem that describes the law of the process in terms of Bessel processes laws. This may be used to prove our result. It seems to us, however, that the direct computations presented here are easier.

2 Proof of the Proposition

We consider the process X restricted to the four points $\{1,2,3,4\}$. That is, we let $L_t = L^1_t + L^2_t + L^3_t + L^4_t$, where, for $i = 1, \ldots, 4$, L^i_t is the local time at i up to time t, and (τ_t) is the right continuous inverse of (L_t). The process (X_{τ_t}) is a pure jump process on $S = \{1,2,3,4\}$, and for $i \in S$, L^i is also the total time (X_{τ_t}) spends at i. For $i \in S$, let $T_i = \inf\{t \geq 0 : X_{\tau_t} = i\}$, and $E_i = E(\ |X_0 = i)$. For $\theta = (\theta_1, \theta_2, \theta_3, \theta_4)$, set

$$\varphi_i(\theta) = E_i\left(e^{-\theta_1 L^1 - \theta_2 L^2 - \theta_3 L^3 - \theta_4 L^4}\right).$$

Then

$$\varphi_1(\theta) = E_1\left(e^{-\theta_1 L^1_{T_2 \wedge T_4}} ; T_2 < T_4\right)\varphi_2(\theta) + E_1\left(e^{-\theta_1 L^1_{T_2 \wedge T_4}} ; T_4 < T_2\right)\varphi_4(\theta)$$

which by the symmetry is equal to

$$\frac{1}{2}\, E_1\left(e^{-\theta_1 L^1_{T_2 \wedge T_4}}\right)(\varphi_2(\theta) + \varphi_4(\theta)).$$

Under P_1, $L^1_{T_2\wedge T_4}$ has an exponential distribution with a parameter which we denote by λ, and by symmetry again

$$\frac{1}{\lambda} = E_1(L^1_{T_2\wedge T_4}) = E_2(L^2_{T_1\wedge T_3}) = E_4(L^4_{T_1\wedge T_3}) \, .$$

Further, let σ be the independent exponentially distributed random variable, which, when exceeded by the local time of 3 the process is killed. Then

$$P_3(T_2 \wedge T_4 < \zeta) = P_3(L^3_{T_2\wedge T_4} < \sigma) \, .$$

We take the parameter of σ to be λ as well. We thus obtain the following set of equations:

$$\begin{aligned}
\varphi_1(\theta) &= \frac{\lambda}{\lambda+\theta_1}\left\{\frac{1}{2}\,\varphi_2(\theta)+\frac{1}{2}\,\varphi_4(\theta)\right\}\\
\varphi_2(\theta) &= \frac{\lambda}{\lambda+\theta_2}\left\{\frac{1}{2}\,\varphi_1(\theta)+\frac{1}{2}\,\varphi_3(\theta)\right\}\\
\varphi_4(\theta) &= \frac{\lambda}{\lambda+\theta_4}\left\{\frac{1}{2}\,\varphi_1(\theta)+\frac{1}{2}\,\varphi_3(\theta)\right\}\\
\varphi_3(\theta) &= \frac{\lambda}{2\lambda+\theta_3}+\frac{\lambda}{2\lambda+\theta_3}\left\{\frac{1}{2}\,\varphi_2(\theta)+\frac{1}{2}\,\varphi_4(\theta)\right\} \, .
\end{aligned}$$

We may, without loss of generality, assume that $\lambda = 1$, and solve the system. The result

$$\varphi_1(\theta) = \frac{2+\theta_2+\theta_4}{4(1+\theta_1)(2+\theta_3)(1+\theta_2)(1+\theta_4)-(2+\theta_2+\theta_4)(3+\theta_1+\theta_3)} \, .$$

First, note that

$$\varphi_1(\theta_1,0,0,0) = \frac{1}{1+3\theta_1} \, ,$$

from which it follows that L^1 has an exponential distribution with parameter 1/3, and

$$\text{(1)} \qquad \varphi_1(\theta) = \int_0^\infty e^{-\theta_1\ell} E_1\left(e^{-\theta_2 L^2-\theta_3 L^3-\theta_4 L^4} \middle| L^1=\ell\right) P_1(L^1\in d\ell)$$

and also

$$\varphi_1(\theta) = \frac{2+\theta_2+\theta_4}{4(2+\theta_3)(1+\theta_2)(1+\theta_4)-(2+\theta_2+\theta_4)(3+\theta_3)} \times \frac{1}{1+\theta_1\left(\frac{4(2+\theta_3)(1+\theta_2)(1+\theta_4)-(2+\theta_2+\theta_4)}{4(2+\theta_3)(1+\theta_2)(1+\theta_4)-(2+\theta_2+\theta_4)(3+\theta_3)}\right)}$$

Set,

$$\lambda(\theta_2,\theta_3,\theta_4) = \frac{4(2+\theta_3)(1+\theta_2)(1+\theta_4)-(2+\theta_2+\theta_4)(3+\theta_3)}{4(2+\theta_3)(1+\theta_2)(1+\theta_4)-(2+\theta_2+\theta_4)} \, .$$

Then it can be easily checked that $\lambda(\theta_2,\theta_3,\theta_4) > 0$ for all $(\theta_2,\theta_3,\theta_4)$ and that

$$\varphi_1(\theta) = \frac{2+\theta_2+\theta_4}{4(2+\theta_3)(1+\theta_2)(1+\theta_4)-(2+\theta_2+\theta_4)(3+\theta_3)} \int_0^\infty e^{-\theta_1\ell}\lambda(\theta_2,\theta_3,\theta_4)e^{-\lambda(\theta_2,\theta_3,\theta_4)\ell}d\ell$$

which, when compared with (1) gives

$$E_1\left(e^{-\theta_2 L^2-\theta_3 L^3-\theta_4 L^4}|L^1=\ell\right)$$
$$=\frac{3(2+\theta_2+\theta_4)}{4(2+\theta_3)(1+\theta_2)(1+\theta_4)-(2+\theta_2+\theta_4)}\ e^{-\{\lambda(\theta_2,\theta_3,\theta_4)-\frac{1}{3}\}\ell}\ .$$

In particular, for $\theta_2=\theta_4=0$ we obtain

$$E_1\left(e^{-\theta_3 L^3}|L^1=\ell\right)=\frac{3/2}{3/2+\theta_3}\ e^{-\frac{1}{6}\ell\left(\frac{\theta_3}{3/2+\theta_3}\right)}\ .$$

But

$$e^{-\frac{1}{6}\ell\left(\frac{\theta_3}{3/2+\theta_3}\right)}=e^{-\frac{1}{6}\ell\left(1-\frac{3/2}{3/2+\theta_3}\right)}$$

which is the Laplace transform of a compound Poisson process with rate 1/6 and jump distribution exponential (3/2) at time ℓ. Therefore, given $L^1=\ell$ the law of L^3 is the convolution of an exponential law with the above compound Poisson at time ℓ.

Set

$$\beta(\theta_2,\theta_4)=\frac{6+7(\theta_2+\theta_4)+8\theta_2\theta_4}{4(1+\theta_2)(1+\theta_4)}$$
$$\lambda_\ell(\theta_2,\theta_4)=\frac{\ell}{6}\ \frac{2+5(\theta_2+\theta_4)+8\theta_2\theta_4}{2(1+\theta_2)(1+\theta_4)}\quad ,$$

and for λ and β positive, let $f_{(\beta,\lambda)}(m)$ be the density of a convolution of an exponential law with parameter β and of a compound Poisson law with jump distribution exponential with parameter β, and Poisson parameter λ. Then

(2) $$E_1\left(e^{-\theta_2 L^2-\theta_3 L^3-\theta_4 L^4}|L^1=\ell\right)$$
$$=\int_0^\infty E_1\ \left(e^{-\theta_2 L^2-\theta_4 L^4}|\,L^1=\ell,\ L^3=m\right)\ e^{-\theta_3 m}P_1(L^3\in dm|\,L^1=\ell)$$
$$=\int_0^\infty E_1\ \left(e^{-\theta_2 L^2-\theta_4 L^4}|\,L^1=\ell,\ L^3=m\right)\ f_{\beta(0,0),\lambda_\ell(0,0)}(m)\ e^{-\theta_3 m}dm$$

On the other hand, by our above computation

(3) $$E_1\left(e^{-\theta_2 L^2-\theta_3 L^3-\theta_4 L^4}|L^1=\ell\right)$$
$$=\frac{3(2+\theta_2+\theta_4)}{6+7(\theta_2+\theta_4)+8\theta_2\theta_4}\ \exp\left\{-\frac{\ell}{3}\ \frac{8(\theta_2+\theta_4)+16\theta_2\theta_4}{6+7(\theta_2+\theta_4)+8\theta_2\theta_4}\right\}$$
$$\times\ \frac{1}{1+\frac{\theta_3}{\beta(\theta_2,\theta_4)}}\ \exp\left\{-\lambda_\ell(\theta_2,\theta_4)\left(1-\frac{1}{1+\frac{\theta_3}{\beta(\theta_2,\theta_4)}}\right)\right\}$$
$$=\frac{3(2+\theta_2+\theta_4)}{6+7(\theta_2+\theta_4)+8\theta_2\theta_4}\ \exp\left\{-\frac{\ell}{3}\ \frac{8(\theta_2+\theta_4)+16\theta_2\theta_4}{6+7(\theta_2+\theta_4)+8\theta_2\theta_4}\right\}$$
$$\times\int_0^\infty e^{-\theta_3 m}\ f_{\beta(\theta_2,\theta_4),\lambda_\ell(\theta_2,\theta_4)}(m)dm\ .$$

Comparing (2) and (3) we obtain

$$E\left(e^{-\theta_2 L^2-\theta_4 L^4}\,|\,L^1=\ell,\ L^3=m\right)=$$

$$\frac{3(2+\theta_2+\theta_4)}{6+7(\theta_2+\theta_4)+8\theta_2\theta_4}\exp\left\{-\frac{\ell}{3}\frac{8(\theta_2+\theta_4)+16\theta_2\theta_4}{6+7(\theta_2+\theta_4)+8\theta_2\theta_4}\right\}\frac{f_{\beta(\theta_2,\theta_4),\lambda_\ell(\theta_2,\theta_4)}(m)}{f_{\beta(0,0),\lambda_\ell(0,0)}(m)}\,.$$

Suppose now that conditioned on (L^1, L^3), L^2 and L^4 are independent, that is for almost every $(m,\ell)\in\mathbb{R}^2$

$$E_1\left(e^{-\theta_2 L^2-\theta_4 L^4}\,|\,L^1=\ell,\ L^3=m\right)=E_1\left(e^{-\theta_2 L^2}\,|\,L^1=\ell,\ L^3=m\right)E_1\left(e^{-\theta_4 L^4}\,|\,L^1=\ell,\ L^3=\right.$$

Using the above computation, this is equivalent to

$$\frac{3(2+\theta_2+\theta_4)}{6+7(\theta_2+\theta_4)+8\theta_2\theta_4}\exp\left\{-\frac{\ell}{3}\frac{8(\theta_2+\theta_4)+16\theta_2\theta_4}{6+7(\theta_2+\theta_4)+8\theta_2\theta_4}\right\}f_{\beta(\theta_2,\theta_4),\lambda_\ell(\theta_2,\theta_4)}(m)\cdot f_{\beta(0,0),\lambda_\ell(0,0)}(m$$

$$=\frac{3(2+\theta_2)}{6+7\theta_2}\cdot\frac{3(2+\theta_4)}{6+7\theta_4}\exp\left\{-\frac{\ell}{3}\left[\frac{8\theta_2}{6+7\theta_2}+\frac{8\theta_4}{6+7\theta_4}\right]\right\}f_{\beta(\theta_2,0),\lambda_\ell(\theta_2,0)}f_{\beta(0,\theta_4),\lambda_\ell(0,\theta_4)}\,. \tag{4}$$

Recall that

$$f_{\beta,\lambda}(m)=\beta e^{-(\lambda+\beta m)}\sum_{n=0}^{\infty}\frac{(\lambda\beta m)^n}{(n!)^2}\,.$$

Thus, since (4) has to hold for every (m,ℓ), the constant terms have to be equal, which amounts to

$$\frac{3(2+\theta_2+\theta_4)}{6+7(\theta_2+\theta_4)+8\theta_2\theta_4}\cdot\frac{6+7(\theta_2+\theta_4)+8\theta_2\theta_4}{4(1+\theta_2)(1+\theta_4)}\cdot\frac{3}{2}$$

$$=\frac{9(2+\theta_2)(2+\theta_4)}{(6+7\theta_2)(6+7\theta_4)}\cdot\frac{6+7\theta_4}{4(1+\theta_4)}\cdot\frac{6+7\theta_2}{4(1+\theta_2)}\,.$$

This is equivalent to

$$2(2+\theta_2+\theta_4)=(2+\theta_2)(2+\theta_4)$$

which is absurd. □

References

[A] Atkinson, B. (1983). "On Dynkin's Markov property of random fields associated with symmetric processes". *Stoch. Proc. Appl.* **15**, 193–201.

[D] Dynkin, E.B. (1984). "Local times and quantum fields". *Sem. on Stoch. Proc.*, **1983**, Birkhauser, 69–84.

[E] Eisenbaum, N. (1994). "Dynkin's isomorphism theorem and the Ray-Knight theorems". *Probability Theory and Related Fields* **99**, No. 2, 321–335.

[EK] Eisenbaum, N. and Kaspi, H. (1993). "A necessary and sufficient condition for the Markov property of the local time process". *Ann. of Probab.* **21**, No. 3, 1591–1598.

[K] Knight, F.B. (1963). "Random walks and a sojourn density process of Brownian motion". *Trans. Amer. Math. Soc.* **109**, 56–86.

[L] Leuridan, C. "Problèmes liés aux temps locaux du mouvement Brownien: Estimation de norme H^p, théorèmes de Ray-Knight sur le tore, point le plus visité". Thèse de doctorat de mathématiques de l'Université Joseph Fourier.

[R] Ray, D.B. (1963). "Sojourn time of a diffusion process". *Ill. J. Maths.* **7**, 615–630.

[S] Sheppard, P. (1985). "On the Ray-Knight property of local times". *J. London Math. Soc.* **31**, 377–384.

[W] Walsh, J.B. (1978). "Excursions and local time". *Temps Locaux, Asterisque* **52–53**, 159–192.

UNE VERSION SANS CONDITIONNEMENT DU THEOREME D'ISOMORPHISME DE DYNKIN

Nathalie Eisenbaum

Laboratoire de Probabilités - Université Paris VI - 4, place Jussieu - Tour 56 - 3 ème étage - 75252 Paris Cedex 05

Abstract We establish here an unconditioned version of Dynkin's isomorphism theorem and use it to give short proofs and extensions of known results on local times.

Introduction Considérons un processus de Markov symétrique admettant une fonction de Green finie et un deuxième processus gaussien centré ayant pour covariance cette fonction de Green et indépendant du premier. Le théorème d'isomorphisme de Dynkin [D] est une identité en loi mettant en relation ce processus gaussien avec la famille des temps locaux du processus de Markov conditionné à mourir en un état fixe. Plusieurs auteurs ont utilisé ce théorème pour, entre autres, transférer certaines propriétés du processus gaussien au processus des temps locaux. Ainsi, Marcus et Rosen se sont intéressés au transfert de la propriété de continuité [M-R 1], Eisenbaum [E] et Sheppard [S] à celui de la propriété de Markov.

Nous établissons dans la partie I une version sans conditionnement du théorème d'isomorphisme de Dynkin. Cette version nous permet de donner dans la partie II des preuves rapides des théorèmes de Ray-Knight usuels ([Ra],[K 1]) et de retrouver, dans la partie III , des théorèmes limites dus à Yor [Y 1] pour le cas du mouvement brownien et à Rosen [R] pour les processus stables en général. Dans la partie IV , nous étendons des résultats de Marcus et Rosen [M-R 2] sur les transformées de Laplace des accroissements du temps local.

I - Le théorème d'isomorphisme de Dynkin sans conditionnement

Rappelons le théorème d'isomorphisme de Dynkin [D] tel qu'il a été énoncé par Marcus et Rosen dans [M-R 1].

Théorème 1.1 : *Soit* $(\ell^i)_{i\in\mathbb{N}}$ *une suite de variables aléatoires réelles définies sur un espace probabilisé* (Ω, Q). *Soit* $\left\{G_\alpha, G_\beta, (G_i)_{i\in\mathbb{N}}\right\}$ *une famille gaussienne centrée, indépendante de* ℓ, *définie sur un espace* Ω' *indépendant de* Ω. *Sur* Ω' *l'espérance est notée* $<\ >$.

Supposons que : $\forall\ j_1,\dots,j_n \in \mathbb{N}^*$

$$(1)\qquad Q\left[\prod_{i=1}^{n} \ell^{j_i}\right] = \sum_{\pi\in\mathcal{S}_n} <G_\alpha G_{j_{\pi(1)}}><G_{j_{\pi(1)}} G_{j_{\pi(2)}}>\dots<G_{j_{\pi(n)}} G_\beta>$$

$\mathcal{S}_n$ *désignant l'ensemble des permutations de* $\{1,2,\dots,n\}$

alors :

$$(2)\qquad Q<F\left(\ell+\frac{G^2}{2}\right)> = <\frac{G_\alpha G_\beta}{<G_\alpha G_\beta>} F\left(\frac{G^2}{2}\right)>$$

pour toute fonctionnelle F *mesurable, positive.*

L'exemple suivant est détaillé par Marcus et Rosen dans [M-R 1] p.1635.

Exemple 1.2 : Soit X un processus de Markov symétrique à valeurs dans $\mathbb{R}$ admettant une fonction de Green finie $\left(g(x,y)\ ;\ (x,y)\in\mathbb{R}^2\right)$. Soit $(G_x,\ x\in\mathbb{R})$ un processus gaussien centré de covariance g. On note $(L^x_\xi,\ x\in\mathbb{R})$ le processus des temps locaux de X pris en son temps de vie ξ. Soit $\tilde{P}_{ab}$ la loi de X démarrant en a et tué au dernier temps de passage en b. Alors les processus L_ξ et G vérifient la relation suivante :

$\forall\ x_1, x_2,\dots,x_n \in \mathbb{R}$

$$\tilde{P}_{ab}\left(\prod_{i=1}^{n} L_\xi^{x_i}\right) = \sum_{\pi\in\mathcal{S}_n} <G_a, G_{x_{\pi(1)}}><G_{x_{\pi(1)}} G_{x_{\pi(2)}}>\dots\frac{<G_{x_{\pi(n)}}, G_b>}{<G_a, G_b>}.$$

On deduit du Théorème 1.1 que pour toute fonction F mesurable positive:

$$\tilde{P}_{ab}<F\left(L_\xi+\frac{G^2}{2}\right)> = <\frac{G_a G_b}{<G_a G_b>} F\left(\frac{G^2}{2}\right)>.$$

Nous proposons maintenant un théorème analogue au Théorème 1.1.

<u>**Théorème 1.3**</u> : *Soit* $(\ell^{i})_{i\in\mathbb{N}}$ *une suite de variables aléatoires réellesdéfinies sur un espace probabilisé* (Ω, Q). *Soit* $\left\{G_\alpha, (G_i)_{i\in\mathbb{N}}\right\}$ *une famille gaussienne centrée, indépendante de* ℓ, *définie sur un espace* Ω' *indépendant de* Ω. *Sur* Ω' *l'espérance est notée* $< \ >$.

Les relations (3) *et* (4) *sont équivalentes :*

$$(3)\quad Q\left[\prod_{i=1}^{n} \ell^{j_i}\right] = \sum_{\pi\in\mathcal{S}_n} \langle G_\alpha G_{j_{\pi(1)}}\rangle \langle G_{j_{\pi(1)}} G_{j_{\pi(2)}}\rangle \dots \langle G_{j_{\pi(n-1)}} G_{j_{\pi(n)}}\rangle$$

$\forall\, j_1,\dots,j_n \in \mathbb{N}^*$

$\mathcal{S}_n$ *désigne l'ensemble des permutations de* $\{1,2,\dots,n\}$.

$$(4)\qquad Q\left\langle F\left(\ell + \frac{(G+\varepsilon)^2}{2}\right)\right\rangle = \frac{1}{\varepsilon}\left\langle (G_\alpha + \varepsilon)\, F\left(\frac{(G+\varepsilon)^2}{2}\right)\right\rangle$$

pour tout ε *de* $\mathbb{R}$ *et toute fonctionnelle mesurable* F.

Ce théorème présente une équivalence entre deux relations. Nous verrons que les arguments employés pour l'établir sont utilisables pour prouver également l'équivalence entre les relations (1) et (2) du Théorème 1.1.

Reprenons l'exemple 1.2. Cette fois, on note P_a la loi de X issu de a. Les processus L_ξ et G vérifient la formule suivante :

$\forall\, x_1, x_2, \dots, x_n \in \mathbb{R}$,

$$P_a\left(\prod_{i=1}^{n} L_\xi^{x_i}\right) = \sum_{\pi\in\mathcal{S}_n} \langle G_a, G_{x_{\pi(1)}}\rangle \dots \langle G_{x_{\pi(n-1)}}, G_{x_{\pi(n)}}\rangle.$$

Grâce au Théorème 1.3, nous avons :

$$P_a\left\langle F\left(L_\xi + \frac{(G+\varepsilon)^2}{2}\right)\right\rangle = \frac{1}{\varepsilon}\left\langle (G_a+\varepsilon)\, F\left(\frac{(G+\varepsilon)^2}{2}\right)\right\rangle \qquad \forall \varepsilon \in \mathbb{R}^*.$$

L'intérêt de cette formule réside en ce qu'elle ne fait pas intervenir de conditionnement sur X. Nous verrons que dans le cadre des parties II, III et IV cela en rend l'utilisation beaucoup plus rapide que celle du théorème d'isomorphisme .

<u>Démonstration du Théorème 1.3</u> :

Supposons que (3) soit vérifiée, établissons (4). Nous nous inspirons pour cela de la démonstration du Théorème 1.1 proposée par Marcus et Rosen dans [M-R 1].On commence par établir (4) pour des fonctionnelles F du type suivant : $F(Y) = F\left((Y_{x_i})_{1\le i\le n}\right) = \prod_{i=1}^{n} Y_{x_i}$, où Y est un processus à valeurs dans $\mathbb{R}$,indexé sur $\mathbb{R}$.

$$\frac{1}{\varepsilon} \left\langle (G_a + \varepsilon)\, F\left(\frac{(G+\varepsilon)^2}{2} \right) \right\rangle = \frac{1}{\varepsilon \cdot 2^n} \left\langle \prod_{i=0}^{2n} \left(G_{x_i} + \varepsilon\right) \right\rangle$$

où l'on a noté : $x_0 = a$ et $(x_i, x_{n+i}) = (x_i, x_i)$ pour $1 \le i \le n$

$$= \frac{1}{\varepsilon \cdot 2^n} \sum_{\substack{H_1 \cup H_2 = \{0,1,..,2n\} \\ H_1 \cap H_2 = \emptyset}} \varepsilon^{|H_1|} \left\langle \prod_{i\in H_2} G_{x_i} \right\rangle$$

G étant centré, seuls les termes tels que $|H_2|$ soit pair, sont non nuls

$$= \frac{1}{\varepsilon \cdot 2^n} \sum_{\substack{H_1 \cup H_2 = \{1,..,2n\} \\ H_1 \cap H_2 = \emptyset}} \varepsilon^{|H_1|+1} \left\langle \prod_{i\in H_2} G_{x_i} \right\rangle$$

$$+ \frac{1}{\varepsilon \cdot 2^n} \sum_{\substack{H_1 \cup H_2 = \{1,..,2n\} \\ H_1 \cap H_2 = \emptyset \\ |H_2| \text{ impair}}} \varepsilon^{|H_1|} \left\langle \prod_{i\in H_2 \cup \{0\}} G_{x_i} \right\rangle$$

$$= I + II.$$

On utilise alors un lemme connu sur les processus gaussiens centrés (voir par exemple la démonstration dans [M-R 1]). H désigne un sous-ensemble fini de $\mathbb{N}$, de cardinal pair .

<u>**Lemme 1.4**</u> : *Soit* $(G_i)_{i\in\mathbb{N}}$ *un processus gaussien centré .Pour tout* H *sous-ensemble fini de* $\mathbb{N}$, *on a:*

$$\langle \prod_{i\in H} G_i \rangle = \sum_{\substack{D = D_1 \cup \ldots \cup D_{|H|/2} \\ D \in \mathcal{D}(H)}} \prod_{j=1}^{|H|/2} \langle \prod_{i\in D_j} G_i \rangle.$$

$\mathcal{D}(H)$ *représentant l'ensemble des partitions de* H *en sous-ensembles à deux éléments.*

Grâce au Lemme 1.4 ,

$$\langle \prod_{i\in H_2 \cup \{0\}} G_{x_i} \rangle = \sum_{\substack{D = D_1 \cup \ldots \cup D_{\frac{|H_2 \cup \{0\}|}{2}} \\ D \in \mathcal{D}(H_2 \cup \{0\})}} \prod_{j=1}^{\frac{|H_2 \cup \{0\}|}{2}} \langle \prod_{i\in D_j} G_{x_i} \rangle.$$

A chaque D élément de $\mathcal{D}(H_2 \cup \{0\})$, on peut associer une chaîne partant de 0, de la façon suivante :

- il existe un, et un seul, élément D^1 de D tel que $0 \in D^1$. D^1 est de la forme $\{0, i\}$,
- il existe alors un, et un seul, élément D^2 de D tel que $i+n$ (ou $i-n$) $\in D^2$. $D^2 = \{i\pm n, j\}$
- puis il existe un seul D^3 tel que : $j\pm n \in D^3$. etc...

Cette suite s'arrête à un rang inférieur ou égal à $\dfrac{|H_2 \cup \{0\}|}{2}$.

On note J_1 l'ensemble des éléments de H_2 ne faisant pas partie de cette chaîne issue de 0.

$H_2 \setminus J_1$ est de la forme $\{i, i+n \,;\, i \in C\} \setminus \{k\}$ où C est un sous-ensemble de $\{1,2,\ldots,n\}$ et k un élément de $\{i, i+n \,;\, i \in C\}$. (En fait : $k \in H_1$).

On en déduit que :

$$II = \frac{1}{\varepsilon . 2^n} \sum_{\substack{H_1 \cup H_2 = \{1,..,2n\} \\ H_1 \cap H_2 = \emptyset}} \varepsilon^{|H_1|} \sum_{k\in H_1} \sum_{\substack{\tilde{C} \cup J_1 = H_2 \cup \{k\} \\ \tilde{C} \cap J_1 = \emptyset}} \langle \prod_{i\in J_1} G_{x_i} \rangle \langle G_{x_0} \prod_{\substack{i \in \tilde{C} \\ i \neq k}} G_{x_i} \rangle$$

avec $\tilde{C}$ de la forme $\{i, i+n \,;\, i \in C\}$ tel que $k \in \tilde{C}$ et $J_1 = H_2 \cup \{k\} \setminus \tilde{C}$.

On pose alors : $\tilde{B} = (H_1 \setminus \tilde{C}) \cup J_1$

$$H'_1 = H_1 \setminus \tilde{C} \;;$$

$\tilde{B}$ est donc l'union de toutes les paires de la forme $\{i, i+n\}$ de $\{1, 2, \dots, 2n\} \setminus \tilde{C}$.

Puis, on pose : $B = \tilde{B} \cap \{1, \dots, n\}$, $C = \tilde{C} \cap \{1, \dots, n\}$.
On obtient alors :

$$II = \frac{1}{2^n \varepsilon} \sum_{\substack{B \cup C = \{1,..,n\} \\ B \cap C = \emptyset}} \sum_{k \in \tilde{C}} \langle G_{x_0} \prod_{\substack{i \in \tilde{C} \\ i \neq k}} G_{x_i} \rangle \sum_{H'_1 \cup J_1 = \tilde{B}} \varepsilon^{|H'_1|+1} \langle \prod_{i \in J_1} G_{x_i} \rangle$$

Maintenant, on remarque, en utilisant l'hypothèse (3) et le Lemme 1.4, que pour C sous-ensemble de $\{1, 2, \dots, n\}$:

$$Q\Big(\prod_{i \in C} \ell^{x_i}\Big) = \sum_{\pi \in \mathcal{G}(|C|)} \langle G_a G_{x_{\pi(1)}} \rangle \langle G_{x_{\pi(1)}} G_{x_{\pi(2)}} \rangle \dots \langle G_{x_{\pi(|C|-1)}} G_{x_{\pi(|C|)}} \rangle$$

$$= \frac{1}{2^{|C|}} \sum_{k \in \tilde{C}} \langle G_{x_0} \prod_{\substack{i \in \tilde{C} \\ i \neq k}} G_{x_i} \rangle$$

avec $\tilde{C} = \{i, i+n \;;\; i \in C\}$
D'où :

$$II = \frac{1}{2^{|B|}} \sum_{\substack{B \cap C = \emptyset \\ B \cup C = \{1,..,n\} \\ |C| \geq 1}} Q\Big(\prod_{i \in C} \ell^{x_i}\Big) \langle \prod_{i \in B} (G_{x_i} + \varepsilon)^2 \rangle.$$

Par ailleurs, on constate que :

$$I = \frac{1}{2^n} \sum_{\substack{H_1 \cup H_2 = \{1,..,2n\} \\ H_1 \cap H_2 = \emptyset}} \varepsilon^{|H_1|} \langle \prod_{i \in H_2} G_{x_i} \rangle = \langle \prod_{i=1}^{n} \frac{\big(G_{x_i} + \varepsilon\big)^2}{2} \rangle.$$

Donc I correspond au terme manquant dans l'expression II, soit : $C = \emptyset$ et $B = \{1, 2, \dots, n\}$. On en déduit que :

$$\frac{1}{\varepsilon} \langle (G_a + \varepsilon)\, F\left(\frac{(G+\varepsilon)^2}{2}\right) \rangle$$

$$= \frac{1}{2^{|B|}} \sum_{\substack{B \cap C = \emptyset \\ B \cup C = \{1,..,n\}}} Q\left(\prod_{i \in C} \ell^{x_i}\right) \langle \prod_{i \in B} (G_{x_i} + \varepsilon)^2 \rangle$$

$$= Q \langle \prod_{i=1}^{n} \left(\ell^{x_i} + \frac{\left(G_{x_i} + \varepsilon\right)^2}{2} \right) \rangle.$$

Par des vérifications analogues à celles de Marcus et Rosen dans ([M-R 1] p.1639) on s'assure qu'il suffit de vérifier la relation (4) pour de telles fonctionnelles F .

Supposons maintenant que la relation (4) soit vérifiée; nous établissons à présent la relation (3).
En reprenant le développement de la démonstration précédente, nous savons que :

$$\frac{1}{\varepsilon} \langle (G_a + \varepsilon) \prod_{i=1}^{n} \frac{\left(G_{x_i} + \varepsilon\right)^2}{2} \rangle$$

$$= \frac{1}{2^{|C|}} \sum_{\substack{B \cap C = \emptyset \\ B \cup C = \{1,2,\dots,n\}}} \langle \prod_{i \in B} \frac{\left(G_{x_i} + \varepsilon\right)^2}{2} \rangle \sum_{k \in \tilde{C}} \langle G_{x_0} \prod_{\substack{i \in \tilde{C} \\ i \neq k}} G_{x_i} \rangle$$

où : $\tilde{C} = \{i,\ i + n\ ;\ i \in C\}$.
Or, on a :

$$\frac{1}{2^{|C|}} \sum_{k \in \tilde{C}} \langle G_{x_0} \prod_{\substack{i \in \tilde{C} \\ i \neq k}} G_{x_i} \rangle = \sum_{\pi \in \mathcal{S}(|C|)} \langle G_a G_{x_{\pi(1)}} \rangle \dots \langle G_{x_{\pi(|C|-1)}} G_{x_{\pi(|C|)}} \rangle$$

d'où :

$$\frac{1}{\varepsilon}\left\langle (G_a+\varepsilon)\prod_{i=1}^{n}\frac{\left(G_{x_i}+\varepsilon\right)^2}{2}\right\rangle$$

$$= \sum_{\substack{B\cup C=\{1,2,\dots,n\}\\ B\cap C=\emptyset}} \left\langle \prod_{i\in B}\frac{\left(G_{x_i}+\varepsilon\right)^2}{2}\right\rangle$$

$$\times \sum_{\pi\in\mathcal{S}(|C|)} \langle G_aG_{x_{\pi(1)}}\rangle\langle G_{x_{\pi(1)}}G_{x_{\pi(2)}}\rangle\dots\langle G_{x_{\pi(|C|-1)}}G_{x_{\pi(|C|)}}\rangle.$$

Par ailleurs en utilisant l'hypothèse (4):

$$\frac{1}{\varepsilon}\left\langle (G_a+\varepsilon)\prod_{i=1}^{n}\frac{\left(G_{x_i}+\varepsilon\right)^2}{2}\right\rangle = Q\left\langle \prod_{i=1}^{n}\left(\ell^{x_i}+\frac{\left(G_{x_i}+\varepsilon\right)^2}{2}\right)\right\rangle$$

$$= \sum_{\substack{B\cup C=\{1,2,\dots,n\}\\ B\cap C=\emptyset}} \left\langle \prod_{i\in B}\frac{\left(G_{x_i}+\varepsilon\right)^2}{2}\right\rangle Q\left\langle \prod_{i\in C}\ell^{x_i}\right\rangle.$$

On en déduit que :

$$(*)\qquad 0 = \sum_{\substack{B\cup C=\{1,2,\dots,n\}\\ B\cap C=\emptyset}} \left\langle \prod_{i\in B}\frac{\left(G_{x_i}+\varepsilon\right)^2}{2}\right\rangle \left\{Q\left(\prod_{i\in C}\ell^{x_i}\right) - \sum_{\pi\in\mathcal{S}(|C|)} \langle G_aG_{x_{\pi(1)}}\rangle\dots\langle G_{x_{\pi(|C|-1)}}G_{x_{\pi(|C|)}}\rangle\right\}.$$

Nous allons établir (3) par récurrence :

$$Q(\ell^{x_1}) = \frac{1}{\varepsilon}\left\langle (G_a+\varepsilon)\frac{\left(G_{x_1}+\varepsilon\right)^2}{2}\right\rangle - \left\langle\frac{\left(G_{x_1}+\varepsilon\right)^2}{2}\right\rangle$$

$$= \frac{1}{2\varepsilon}\langle G_aG_{x_1}^2\rangle + \frac{\varepsilon}{2}\langle G_a\rangle + \langle G_aG_{x_1}\rangle$$

$$= \langle G_aG_{x_1}\rangle.$$

D'où (3) est vérifiée au rang 1.

Supposons que (3) soit vérifiée au rang (n-1).

Le seul terme non nul dans la sommation intervenant dans l'équation (*) est :

$$Q\left(\prod_{i \in \{1,\dots,n\}} \ell^{x_i}\right) - \sum_{\pi \in \mathcal{S}(n)} \langle G_a G_{x_{\pi(1)}} \rangle \dots \langle G_{x_{\pi(n-1)}} G_{x_{\pi(n)}} \rangle$$

(3) est donc vérifiée en rang n.

Remarque : Soit X un processus de Markov symétrique à valeurs dans $\mathbb{R}^d$ admettant des densités de transition $(p_t(x,y),\ t>0;\ (x,y) \in \mathbb{R}^d \times \mathbb{R}^d)$, de fonction de Green $g(x,y)$. Dans le cas où l'on a seulement : $\int_s^{+\infty} p_t(x,x)dt < +\infty$, on peut écrire, en reprenant les définitions de Dynkin [D], une généralisation de la formule (4) du Théorème 1.3 : pour tout ε de $\mathbb{R}^*$, pour toute probabilité μ sur $\mathbb{R}^d$ telle que $\int \mu(dx)\mu(dy)g(x,y) < +\infty$, on a :

$$P_\mu \langle F\left(L^\lambda + \xi_\lambda + \varepsilon\phi_\lambda\ ;\ \lambda \in \Lambda^2\right)\rangle = \frac{1}{\varepsilon} \langle (\phi_\mu + \varepsilon)\, F\left(\xi_\lambda + \varepsilon\phi_\lambda\ ;\ \lambda \in \Lambda^2\right)\rangle$$

où $\Lambda^2 = \left\{\lambda \text{ mesure positive sur } \mathbb{R}^d : \int \lambda(dx)\ \lambda(dy)\ g^2(x,y) < +\infty\right\}$ et

$$\xi_\lambda = \lim_{s \downarrow 0} \int \lambda(dx) \left\{\phi^2_{\mu(s,x)} - \langle\phi^2\rangle_{\mu(s,x)}\right\}, \quad \text{avec } \mu(s,x)\ (dy) = p_s(x,y)dy.$$

II - Applications aux Théorèmes de Ray-Knight

Nous retrouvons ici les deux principaux théorèmes de Ray-Knight [Ra] [K 1] comme illustration immédiate du Théorème 1.3.

1) Cas d'un mouvement brownien B tué en son premier temps d'atteinte de zéro.

La fonction de Green d'un tel processus est : $g(x,y) = 2(x \wedge y)$. Soit $(\beta_x, x\geq 0)$ mouvement brownien issu de 0, alors $\sqrt{2}\,\beta$ est un processus gaussien centré de covariance $g(x,y)$. On note $\left(L^x_{T(0)},\ x \in \mathbb{R}^*_+\right)$ le processus des temps locaux du mouvement brownien tué en $T(0) = \inf\{t \geq 0 : B_t = 0\}$.

On a , grâce au Théorème 1.3 : $\forall\alpha \in \mathbb{R}^*$, $\forall a \in \mathbb{R}_+^*$

(1) $$P_a \langle F\left(L_{T(0)} + (\beta + \alpha)^2\right)\rangle = \langle \frac{(\beta_a + \alpha)}{\alpha} F((\beta + \alpha)^2)\rangle.$$

Pour identifier le terme de droite, on commence par montrer que :

$$\langle \frac{(\beta_a + \alpha)}{\alpha} F((\beta + \alpha)^2)\rangle = \langle \frac{|\beta_a + \alpha|}{|\alpha|} 1_{\{\forall x \in (0,a) \ ; \ |\beta_x + \alpha| > 0\}} F((\beta + \alpha)^2)\rangle.$$

Cette propriété peut se démontrer en utilisant la propriété de Markov forte au temps $\tilde{T}(0) = \inf\{s \geq 0 : \alpha + \beta_s = 0\}$. En effet, elle peut se traduire par :

$$\langle(\beta_a + \alpha) \mid |\beta_x + \alpha| \ ; \ 0 \leq x \leq a\rangle = |\beta_a + \alpha| \, 1_{\{\tilde{T}(0) > a\}} \times \operatorname{sgn}(\alpha).$$

C'est aussi une utilisation immédiate de (1) avec pour fonctionnelle F :

$$F(X) = G(X) \times 1_{\{\exists x \in (0,a) \ ; \ X_x = 0\}}$$

où G fonctionnelle positive mesurable. Pour cela, il suffit de remarquer que sous P_a, notre mouvement brownien B tué en T(0), atteint toutes les valeurs de (0,a].

On utilise alors les notations canoniques pour réécrire (1) : $(X_x,\ x \geq 0)$ est le processus des coordonnées. $\mathbb{Q}_a^d$ désigne la loi d'un carré de processus de Bessel issu de a, de dimension d. ${}^a\mathbb{L}$ désigne la loi de $(L_{T(0)}^x,\ x \geq 0)$ sous P_a .L'opération de convolution de deux lois est notée $*$. En notant : $\tau(0) = \inf\{x \geq 0 : X_x = 0\}$, (1) se transcrit par:

$${}^a\mathbb{L} * \mathbb{Q}^1_{\alpha^2} = \left(\frac{X_{a\wedge\tau(0)}}{\alpha^2}\right)^{1/2} \mathbb{Q}^1_{\alpha^2}.$$

On utilise alors la relation d'absolue continuité suivante (voir Knight [K 2] p.124 ou Pitman-Yor [P-Y]) :

$$\mathbb{Q}^3_{\alpha^2}\Big|_{\mathcal{F}_a} = \left(\frac{X_{a\wedge\tau(0)}}{\alpha^2}\right)^{1/2} \mathbb{Q}^1_{\alpha^2}\Big|_{\mathcal{F}_a}$$

ainsi que la propriété d'additivité des carrés de Bessel : $\mathbb{Q}_0^2 * \mathbb{Q}^1_{\alpha^2} = \mathbb{Q}^3_{\alpha^2}$

pour obtenir :

$$ {}^{a}\mathbb{L}\Big|_{\mathcal{F}_a} = \mathbb{Q}_0^2\Big|_{\mathcal{F}_a} . $$

Cette identité s'écrit également :

$$ P_a\left(L^x_{T(0)} \; ; \; 0 \le x \le a\right) \overset{(d)}{=} \mathbb{Q}_0^2 \,(X_x \; ; \; 0 \le x \le a) . $$

2) <u>Cas d'un mouvement brownien tué en l'inverse de son temps local en zéro</u>.

Soit $(L^x_t \,,\, x \in \mathbb{R},\, t \ge 0)$ la famille de temps locaux d'un mouvement brownien B issu de zéro .On note : $\tau_t = \inf\{s \ge 0 : L^0_s > t\}$.On considère alors le processus B tué en τ_t. La fonction de Green de ce processus est :

$$ g(x,y) = \begin{cases} 2(|x| \wedge |y|) + t & \text{si } x \text{ et } y \text{ sont de même signe} \\ 0 & \text{sinon} \end{cases} $$

Soit $(\beta_x,\, x \ge 0)$ un mouvement brownien issu de 0, indépendant de B, alors

$\left(\sqrt{2}\,\beta_{x+\frac{t}{2}} \; ; \; x \ge 0\right)$ est un processus gaussien centré de covariance

$(g(x,y)),(x,y)\in\mathbb{R}^2_+)$. On a pour tout $\alpha > 0$:

$$ (2) \quad P_0\left\langle F\left[L^x_{\tau_t} + (\beta_{x+t/2} + \alpha)^2 \; ; \; x \ge 0\right]\right\rangle = \left\langle \frac{(\beta_{t/2} + \alpha)}{\alpha} F\left((\beta_{x+t/2} + \alpha)^2 \; ; \; x \ge 0\right)\right\rangle. $$

Cette fois, l'identification du terme de droite est immédiate :

$$ \left\langle \frac{(\beta_{t/2} + \alpha)}{\alpha} F((\beta_{\cdot+t/2} + \alpha)^2)\right\rangle = \mathbb{Q}^1_\mu(F) $$

où μ est la loi de $(\beta_{t/2} + 2)^2$ sous $\left\langle \frac{\beta_{t/2} + \alpha}{\alpha} \,,\, \cdot \right\rangle$.
En fait, grâce à (2), on sait aussi que : $\mu = \delta_t * \nu$, où ν est la loi de $(\beta_{t/2} + \alpha)^2$.

En utilisant la propriété d'additivité des carrés de Bessel, on obtient :

$$P_0\left(L^x_{\tau_t}\ ;\ x \geq 0\right) \overset{(d)}{=} Q^0_t\ (X_x\ ;\ x \geq 0).$$

Remarquons que l'indépendance des processus $(L^x_{\tau_t},\ x \geq 0)$ et $((L^{-x}_{\tau_t},\ x \geq 0)$ se lit également dans la formule (2).

III - Application aux théorèmes limites des temps locaux des processus stables.

Pour tout γ de $(0,1]$, il existe un processus $(B^{(\gamma)}_t(x),\ x\in\mathbb{R}\ ;\ t\geq 0)$ gaussien centré continu de covariance :

$$E\left(B^{(\gamma)}_s(x).B^{(\gamma)}_t(y)\right) = (s\wedge t).\Gamma^{(\gamma)}(x,y)$$

où $\Gamma^{(\gamma)}(x,y) = \frac{1}{2}\ (|x|^\gamma + |y|^\gamma - |x-y|^\gamma)$ (voir [Y 2]).

Nous appellerons un tel processus un drap brownien fractionnaire d'indice γ. Soit $(L^x_t,\ x \in R\ ;\ t \geq 0)$ le processus des temps locaux d'un processus stable X, issu de 0, à valeurs réelles, symétrique d'indice $\beta > 1$. Soit $(p_t(x), x\in\mathbb{R}; t\geq 0)$ les densités de transition de X. On pose :

$$c_\beta = \int_0^{+\infty} (p_t(0) - p_t(1))dt$$

Dans le cas où $\beta=2$, c'est à dire le cas d'un mouvement brownien, Yor [Y 1] a établi le résultat suivant :

$$\left(\frac{1}{\varepsilon^{\frac{\beta-1}{2}}}\left(L^{\varepsilon x}_t - L^0_t\right)\ ;\ x \in \mathbb{R};\ t\geq 0\right) \xrightarrow[\varepsilon \to 0]{(d)} \left(\sqrt{c_\beta}\ \ B^{(1)}_{2L^0_t}(x)\ ;\ x \in \mathbb{R}\ ;\ t\geq 0\right)$$

avec $(B^{(1)}_t(x), x\in\mathbb{R}\ ;\ t\geq 0)$ indépendant de X.

Rosen [R] a ensuite prouvé ce résultat pour $1<\beta<2$ mais en établissant l'indépendance de $(B^{(\beta-1)}_t(x),\ x\in\mathbb{R}\ ;\ t\geq 0)$ par rapport à L^0_t seulement. De plus, il a directement prouvé pour certains temps d'arrêt T de X, notamment pour un temps exponentiel indépendant, que :

$$(1)\qquad \left(\frac{1}{\varepsilon^{\frac{\beta-1}{2}}}\left(L^{\varepsilon x}_T - L^0_T\right)\ ;\ x \in \mathbb{R}\right) \xrightarrow[\varepsilon \to 0]{(d)} \left(\sqrt{c_\beta}\ \ B^{(\beta-1)}_{2L^0_T}(x)\ ;\ x \in \mathbb{R}\right)$$

avec $(B^{(\beta-1)}_t(x), x\in\mathbb{R}\ ;\ t\geq 0)$ indépendant de L^0_T.

Nous présentons ici une extension de leurs théorèmes consistant à introduire un paramètre supplémentaire. Notre démonstration utilise le Théorème 1.3. Par souci de clarté, nous considerons le paramètre β fixé et omettons de le faire apparaitre dans l'écriture de $B^{(\beta-1)}$. Par ailleurs, nous marquerons la dépendance de $B^{(\beta-1)}$ en un paramétre y , en le notant : $B^{[y]}$.

Théorème 3.1 : *Pour* $y_1, y_2, \ldots, y_n$ *n réels distincts, on a:*

$$\left(\frac{1}{\varepsilon^{\frac{\beta-1}{2}}}\left(L_t^{\varepsilon x+y_i} - L_t^{y_i}\right) \; ; \; x\in\mathbb{R}; 1\le i\le n; t\ge 0\right)$$

$$\xrightarrow[\varepsilon \to 0]{(d)} \left(\sqrt{c_\beta}\, B^{[y_i]}_{2L_t^{y_i}}(x) \; ; \; x\in\mathbb{R}; 1\le i\le n; t\ge 0\right)$$

où $\{ B_t^{[y_i]}(x),\ x\in\mathbb{R}; 1\le i\le n; t\ge 0 \}$ *est un système gaussien indépendant de X, composé de draps browniens fractionnaires d'indice* $(\beta-1)$ *tous indépendants.*

Remarque : Le résultat de convergence fini-dimensionnelle en y du Théorème 3.1, ne peut être étendu à une convergence en loi. Une des façons de s'en rendre compte consiste à supposer que l'on ait pour t>0 et x réel fixé :

$$\left(\frac{1}{\varepsilon^{\frac{\beta-1}{2}}}\left(L_t^{\varepsilon x+y} - L_t^{y}\right) ; y\in\mathbb{R}\right) \xrightarrow[\varepsilon \to 0]{(d)} \left(\sqrt{c_\beta}\, B^{[y]}_{2L_t^{y}}(x) \; ; y\in\mathbb{R}\right)$$

et à constater que le processus de droite ne peut être mesurable. Pour cela , on se ramène au cas d'un processus $(Z_y, y\in\mathbb{R})$ mesurable, borné dans L^1 sur tout compact et tel que les variables Z_y soient toutes indépendantes . On a :

$\mathrm{var}\left(n \int_x^{x+1/n} Z_y dy\right) = 0$, donc pour tout n>0 , $n \int_x^{x+1/n} Z_y dy$ est une variable constante. Par ailleurs $(n \int_x^{x+1/n} Z_y dy;\ n>0)$ tend vers Z_x p.s. quand n tend vers l'infini. Z est donc un processus déterministe.

Démonstration :

1) Soit T un temps exponentiel de paramètre λ , indépendant de X. On pose : $\mathbb{A} = \{y_1, y_2, \ldots y_n\}$ .Nous commençons par établir à l'aide du Théorème 1.3 , le résultat suivant :

$$\left(\frac{1}{\varepsilon^{\frac{\beta-1}{2}}}\left(L_T^{\varepsilon x+y} - L_T^{y} \quad ; \quad (x,y) \in \mathbb{R}\times\mathbb{A}\right)\right) \xrightarrow[\varepsilon \to 0]{(d)} \left(\sqrt{c_\beta}\; B^{[y]}_{2L_T^y}(x) \quad ; \quad (x,y) \in \mathbb{R}\times\mathbb{A}\right).$$

La fonction de Green de X tué en T est :

$$u^\lambda(x,z) = u^\lambda(|x-z|) = \int_0^{+\infty} e^{-\lambda t}\, p_t(x-z)dt$$

Soit ϕ le processus gaussien centré de covariance $u^\lambda(x,z)$. Grâce au Théorème 1.3, nous avons pour tout $\alpha > 0$:

$$P_0 \left\langle F\left(\frac{1}{\varepsilon^{\frac{\beta-1}{2}}}\left(L_T^{\varepsilon x+y} - L_T^{y}\right) + \frac{\tilde{\phi}^2_{\varepsilon x+y} - \tilde{\phi}^2_y}{2\,\varepsilon^{\frac{\beta-1}{2}}} \; ; \; (x,y) \in \mathbb{R}^2\right)\right\rangle$$

$$= \left\langle \frac{\tilde{\phi}_0}{\alpha} F\left(\frac{\tilde{\phi}^2_{\varepsilon x+y} - \tilde{\phi}^2_y}{2\,\varepsilon^{\frac{\beta-1}{2}}} \; ; \; (x,y) \in \mathbb{R}^2\right)\right\rangle$$

où $\quad \tilde{\phi} = \phi + \alpha.$

On remarque que : $\quad \dfrac{\tilde{\phi}^2_{\varepsilon x+y} - \tilde{\phi}^2_y}{2\,\varepsilon^{\frac{\beta-1}{2}}} = \left(\dfrac{\phi_{\varepsilon x+y} - \phi_y}{\varepsilon^{\frac{\beta-1}{2}}}\right) \cdot \left(\dfrac{\phi_{\varepsilon x+y} + \phi_y + 2\alpha}{2}\right).$

On a les convergences suivantes :

(a) $$\left\langle\left(\frac{\phi_{\varepsilon x+y} - \phi_y}{\varepsilon^{\frac{\beta-1}{2}}}\right)\left(\frac{\phi_{\varepsilon z+y} - \phi_y}{\varepsilon^{\frac{\beta-1}{2}}}\right)\right\rangle \xrightarrow[\varepsilon \to 0]{} c_\beta\, \Gamma^{(\beta-1)}(x,z)$$

(b) $$\left\langle\left(\frac{\phi_{\varepsilon x+y} - \phi_y}{\varepsilon^{\frac{\beta-1}{2}}}\right)\phi_z\right\rangle \xrightarrow[\varepsilon \to 0]{} 0$$

(c) $$\left\langle\left(\frac{\phi_{\varepsilon x+y} - \phi_y}{\varepsilon^{\frac{\beta-1}{2}}}\right)\left(\frac{\phi_{\varepsilon x'+y'} - \phi_{y'}}{\varepsilon^{\frac{\beta-1}{2}}}\right)\right\rangle \xrightarrow[\varepsilon \to 0]{} 0$$

En effet :

(a) $$\left\langle\left(\frac{\phi_{\varepsilon x+y} - \phi_y}{\varepsilon^{\frac{\beta-1}{2}}}\right)\left(\frac{\phi_{\varepsilon z+y} - \phi_y}{\varepsilon^{\frac{\beta-1}{2}}}\right)\right\rangle = \frac{1}{\varepsilon^{\beta-1}}\{u^\lambda(\varepsilon x-\varepsilon z) - u^\lambda(\varepsilon x) - u^\lambda(\varepsilon z) + u^\lambda(0)\}$$

$$= \frac{1}{\varepsilon^{\beta-1}}\{\, (u^\lambda(\varepsilon x-\varepsilon z) - u^\lambda(0)) - (u^\lambda(\varepsilon x) - u^\lambda(0)) - (u^\lambda(\varepsilon z) - u^\lambda(0))\,\}$$

On calcule : $\frac{1}{\varepsilon^{\beta-1}} (u^\lambda(0) - u^\lambda(\varepsilon x))$.

Grâce aux propriétés de scaling de X, on a :

$$p_t(x) = \frac{1}{|x|} p_{t/_{|x|^\beta}}(1) \qquad \forall x \in \mathbb{R}^*, \ \forall t > 0 .$$
$$p_t(0) = \frac{1}{|x|} p_{t/_{|x|^\beta}}(0)$$

$$\frac{1}{\varepsilon^{\beta-1}} (u^\lambda(0) - u^\lambda(\varepsilon x)) = \frac{1}{\varepsilon^\beta} \frac{1}{|x|} \int_0^{+\infty} e^{-\lambda t} \left(p_{t/_{|\varepsilon x|^\beta}}(0) - p_{t/_{|\varepsilon x|^\beta}}(1) \right) dt$$
$$= |x|^{\beta-1} \int_0^{+\infty} e^{-\lambda t |\varepsilon x|^\beta} (p_t(0) - p_t(1)) dt.$$

On obtient : $\frac{1}{\varepsilon^{\beta-1}} (u^\lambda(0) - u^\lambda(\varepsilon x)) \xrightarrow[\varepsilon \to 0]{} c_\beta \, |x|^{\beta-1}$.

(b) Pour x et y fixés, on calcule : $\frac{1}{\varepsilon^{\beta-1}} (u^\lambda(y) - u^\lambda(\varepsilon x+y))$

$$\frac{1}{\varepsilon^{\beta-1}} (u^\lambda(y) - u^\lambda(\varepsilon x+y)) = \frac{1}{\varepsilon^{\beta-1}} (u^\lambda(y) - u^\lambda(0)) + \frac{1}{\varepsilon^{\beta-1}} (u^\lambda(0) - u^\lambda(\varepsilon x+y))$$

$$= \int_0^{+\infty} \frac{1}{\varepsilon^{\beta-1}} \left(|\varepsilon x+y|^{\beta-1} e^{-\lambda t |\varepsilon x+y|^\beta} - |y|^{\beta-1} e^{-\lambda t |y|^\beta} \right) (p_t(0) - p_t(1)) dt .$$

On suppose $y > 0$. Pour ε suffisamment petit : $|\varepsilon x+y| = \varepsilon x+y$

$$\frac{1}{\varepsilon^{\beta-1}} \left((\varepsilon x+y)^{\beta-1} e^{-\lambda t (\varepsilon x+y)^\beta} - y^{\beta-1} e^{-\lambda t y^\beta} \right)$$

$$= y^{\beta-2} e^{-\lambda t y^\beta} \left(\beta-1 - \lambda t \beta y^\beta + \frac{o(\varepsilon)}{\varepsilon} \right) x \, \varepsilon^{2-\beta}.$$

On obtient le lemme suivant :

<u>**Lemme 3.2**</u> : *Pour* $x \in \mathbb{R}$, $y \in \mathbb{R}^*$

- *si* $\beta < 2$ $\quad \frac{u^\lambda(y) - u^\lambda(\varepsilon x+y)}{\varepsilon^{\beta-1}} \xrightarrow[\varepsilon \to 0]{} 0$

- *si* $\beta = 2$ $\quad \frac{u^\lambda(y) - u^\lambda(\varepsilon x+y)}{\varepsilon^{\beta-1}} \xrightarrow[\varepsilon \to 0]{} x \, \mathrm{sgn}(y) \, f_\lambda(y^2)$

$$avec \quad f_\lambda(y^2) = \int_0^{+\infty} e^{-\lambda t y^2}(1-2\lambda t y^2)(p_t(0) - p_t(1))dt \; .$$

Au vu du lemme 3.2, (b) est immédiat .

$$(c) \quad < \left(\frac{\phi_{\varepsilon x+y} - \phi_y}{\varepsilon^{\frac{\beta-1}{2}}} \right) \left(\frac{\phi_{\varepsilon x'+y'} - \phi_{y'}}{\varepsilon^{\frac{\beta-1}{2}}} \right) >$$

$$= \frac{1}{\varepsilon^{\beta-1}} \{u^\lambda(\varepsilon(x-x') + y-y') - u^\lambda(\varepsilon x'+y'-y)) - u^\lambda(\varepsilon x+y-y')) + u^\lambda(y-y')\}$$

$$= \frac{1}{\varepsilon^{\beta-1}} \{u^\lambda(\varepsilon(x-x') + y-y') - u^\lambda(y-y')) + (u^\lambda(y-y') - u^\lambda(\varepsilon x'+y'-y))$$

$$+ (u^\lambda(y-y') - u^\lambda(\varepsilon x+y-y'))\}$$

Dans tous les cas, grâce au lemme 3.2, la limite de cette somme quand ε tend vers zéro, est nulle.

Grace à (a), (b), et (c) on a obtient la convergence en loi suivante :

$$\left(\left(\frac{\phi_{\varepsilon x+y} - \phi_y}{\varepsilon^{\frac{\beta-1}{2}}} \; ; \; (x,y) \in \mathbb{R}\times\mathbb{A} \right), \left(\frac{\phi_{\varepsilon x+y} + \phi_y + 2\alpha}{2} \; ; \; (x,y) \in \mathbb{R}\times\mathbb{A} \right), (\phi_x ; \; x \in \mathbb{R}) \right)$$

$$\Big\downarrow (d) \quad \text{quand} \quad \varepsilon \to 0$$

$$\left(\left(\sqrt{c_\beta} \; B_1^{[y]}(x), \; (x,y) \in \mathbb{R}\times\mathbb{A} \right), (\tilde{\phi}_y ; \; y \in \mathbb{A}) \; , (\phi_x ; \; x \in \mathbb{R}) \right)$$

où pour tout y, $B^{[y]}$ est un drap brownien fractionnaire d'indice β-1.

Les processus $B^{[y]}$,y variant dans $\mathbb{A}$, étant tous indépendant entre eux et indépendant de ϕ.

En particulier, on a :

$$\left(\frac{\tilde{\phi}^2_{\varepsilon x+y} - \tilde{\phi}^2_y}{2\,\varepsilon^{\frac{\beta-1}{2}}} \; ; \; (x,y) \in \mathbb{R}\times\mathbb{A} \right) \xrightarrow[\varepsilon \to 0]{(d)} \left(\sqrt{c_\beta} \; B^{[y]}(x) \; ; \; (x,y) \in \mathbb{R}\times\mathbb{A} \right)$$

$\tilde{\phi}^2_y$

Ce résultat reste vrai sous la mesure $< \frac{\tilde{\phi}_0}{\alpha} \, , \, \cdot >$.

Il s'agit maintenant de trouver un processus aléatoire Y indépendant de $\tilde{\phi}^2$ et tel que :

$$Y + \tilde{\phi}^2 \overset{(d)}{=} \tilde{\phi}^2 \text{ sous } \langle \frac{\tilde{\phi}_0}{\alpha}, \cdot \rangle.$$

La solution est donnée par le Théorème 1.3 ; on a :

$$2L_T^y + \tilde{\phi}_y^2 \overset{(d)}{=} \tilde{\phi}_y^2 \text{ sous } \langle \frac{\tilde{\phi}_0}{\alpha}, \cdot \rangle$$

On a donc :

$$\bar{B}^{[y]}_{2L_T^y} + B^{[y]}_{\tilde{\phi}_y^2} \overset{(d)}{=} B^{[y]}_{\tilde{\phi}_y^2} \text{ sous } \langle \frac{\tilde{\phi}_0}{\alpha}, \cdot \rangle$$

où $(\bar{B}^{[y]};y\in\mathbb{A})$ est une copie indépendante de $(B^{[y]};y\in\mathbb{A})$, et indépendante de L_T.

D'où :

$$\left(\frac{1}{\varepsilon^{\frac{\beta-1}{2}}}\left(L_T^{\varepsilon x+y} - L_T^y \;;\; (x,y)\in\mathbb{R}\times\mathbb{A}\right)\right) \xrightarrow[\varepsilon\to 0]{(d)} \left(\sqrt{c_\beta}\; B^{[y]}_{2L_T^y}(x) \;;\; (x,y)\in\mathbb{R}\times\mathbb{A}\right).$$

Remarquons que $(B^{[y]};y\in\mathbb{A})$ étant indépendante de L_T, elle est indépendante de $T = \int_{\mathbb{R}} L_T^x\, dx$. On en déduit que pour presque tout $t>0$, on a :

$$\left(\frac{1}{\varepsilon^{\frac{\beta-1}{2}}}\left(L_t^{\varepsilon x+y} - L_t^y \;;\; (x,y)\in\mathbb{R}\times\mathbb{A}\right)\right) \xrightarrow[\varepsilon\to 0]{(d)} \left(\sqrt{c_\beta}\; B^{[y]}_{2L_t^y}(x) \;;\; (x,y)\in\mathbb{R}\times\mathbb{A}\right)$$

avec $(B^{[y]};y\in\mathbb{A})$ indépendante de L_t.

Par scaling, ce résultat s'étend à tout $t>0$.

2) Montrons que :

$$\left(\frac{1}{\varepsilon^{\frac{\beta-1}{2}}}\left(L_t^{\varepsilon x+y} - L_t^y \;;(x,y)\in\mathbb{R}\times\mathbb{A};t>0\right)\right) \xrightarrow[\varepsilon\to 0]{(d)} \left(\sqrt{c_\beta}\; B^{[y]}_{2L_t^y}(x) \;;(x,y)\in\mathbb{R}\times\mathbb{A};t>0\right)$$

avec B indépendant de L .

Nous commençons par montrer la convergence fini-dimensionnelle en t.

Soient $t,s > 0$ et $y_1,y_2,\ldots,y_n$ n réels distincts fixés. On considère F_1 et F_2 fonctionnelles bornées sur $\mathcal{C}(\mathbb{R}^2,\mathbb{R})$ telles que pour $i = 1,2$:

$$F_i(X) = \exp\left(-\sum_{k=1}^{n}\int_{\mathbb{R}} X_{x,y_k}\, d\mu_{i_k}(x,y_k)\right)$$ où pour tout k ,μ_{i_k} est une

mesure σ-finie sur $\mathbb{R}^2$.

$$E_o\left(F_1\left(\frac{L_t^{\varepsilon x+y}-L_t^y}{\varepsilon^{\frac{\beta-1}{2}}}\ ;\ (x,y)\in\mathbb{R}^2\right) F_2\left(\frac{L_{t+s}^{\varepsilon x+y}-L_{t+s}^y}{\varepsilon^{\frac{\beta-1}{2}}}\ ;\ (x,y)\in\mathbb{R}^2\right)\right)$$

$$= E_o\left(F_1F_2\left(\frac{L_t^{\varepsilon x+y}-L_t^y}{\varepsilon^{\frac{\beta-1}{2}}}\ ;\ (x,y)\in\mathbb{R}^2\right) F_2\left(\frac{L_s^{\varepsilon x+y}-L_s^y}{\varepsilon^{\frac{\beta-1}{2}}}\ ;\ (x,y)\in\mathbb{R}^2\right) o\ \theta_t\right)$$

$$= \int_{\mathbb{R}} P_o(X_t\in da)\ E_o\left(F_1F_2\left(\frac{L_t^{\varepsilon x+y}-L_t^y}{\varepsilon^{\frac{\beta-1}{2}}}\ ;\ (x,y)\in\mathbb{R}^2\right)\middle| X_t=a\right)$$

$$E_a\left(F_2\left(\frac{L_s^{\varepsilon x+y}-L_s^y}{\varepsilon^{\frac{\beta-1}{2}}}\ ;\ (x,y)\in\mathbb{R}^2\right)\right).$$

Pour tout $a\in\mathbb{R}$, on a :

$$E_a\left(F_2\left(\frac{L_s^{\varepsilon x+y}-L_s^y}{\varepsilon^{\frac{\beta-1}{2}}}\ ;\ (x,y)\in\mathbb{R}^2\right)\right)\xrightarrow[\varepsilon\to 0]{} E_a\left(F_1F_2\left(\sqrt{c_\beta}\ B^{[y]}_{2L_s^y}(x)\ ;\ (x,y)\in\mathbb{R}^2\right)\right)$$

et

$$E_o\left(F_1F_2\left(\frac{L_t^{\varepsilon x+y}-L_t^y}{\varepsilon^{\frac{\beta-1}{2}}}\ ;\ (x,y)\in\mathbb{R}^2\right)\middle| X_t=a\right)$$

$$\xrightarrow[\varepsilon\to 0]{} E_o\left(F_1F_2\left(\sqrt{c_\beta}\ B^{[y]}_{2L_t^y}(x)\ ;\ (x,y)\in\mathbb{R}^2\right)\middle| X_t=a\right)$$

En utilisant le Théorème de convergence dominée de Lebesgue, on obtient

$$E_o\left(F_1\left(\frac{L_t^{\varepsilon x+y}-L_t^y}{\varepsilon^{\frac{\beta-1}{2}}}\ ;\ (x,y)\in\mathbb{R}^2\right) F_2\left(\frac{L_{t+s}^{\varepsilon x+y}-L_{t+s}^y}{\varepsilon^{\frac{\beta-1}{2}}};\ (x,y)\in\mathbb{R}^2\right)\right)$$

$$\Big\downarrow\ \varepsilon\to 0$$

$$\int_{\mathbb{R}} P_o(X_t\in da)\ E_o\left(F_1F_2\left(\sqrt{c_\beta}\ B^{[y]}_{2L_t^y},\ y\in\mathbb{R}\right)\middle| X_t=a\right).E_a\left(F_2\left(\sqrt{c_\beta}\ \overline{B}^{[y]}_{2L_t^y},\ y\in\mathbb{R}\right)\right)$$

(où $\overline{B}$ est une copie indépendante de B , indépendante de X)

La limite ci-dessus s'écrit aussi :

$$E_o\left[F_1\left(\sqrt{c_\beta}\; B^{[y]}_{2L^y_t}\;,\; y\in\mathbb{R}\right) F_2\left(\sqrt{c_\beta}\left(B^{[y]}_{2L^y_t} + \overline{B}^{[y]}_{2L^y_s o\theta_t}\right);\; y\in\mathbb{R}\right)\right]$$

et est donc égale à : $E_o\left[F_1\left(\sqrt{c_\beta}\; B^{[y]}_{2L^y_t}\;,\; y\in\mathbb{R}\right) F_2\left(\sqrt{c_\beta}\; B^{[y]}_{2L^y_{t+s}}\;;\; y\in\mathbb{R}\right)\right]$

avec $(B^{[y]}_t,\ y\in\mathbb{R};\ t>0)$ indépendant de (L_t, L_{t+s}).

On étend aisément la démonstration à une suite finie de temps $(t_i)_{1\le i\le n}$.

Il reste à prouver que pour tout y, la famille $\left(\dfrac{L^{\varepsilon x+y}_t - L^y_t}{\varepsilon^{\frac{\beta-1}{2}}}\;;\; x\in\mathbb{R}\;,\; t>0\right)$

est tendue. On utilise pour cela, les lemmes 1 et 2 établis par Rosen dans [R], pour obtenir immédiatement :

$$(*)\qquad E_o(L^x_t - L^y_t)^{2k} \le c_k\, |x-y|^{k(\beta-1)}\, t^{k(1-\frac{1}{\beta})}$$
$$\forall x,y\in\mathbb{R}\;,\ \forall t>0,\ \forall k\in\mathbb{N}^*.$$

Pour (x,t) et $(x',t+s)$ dans un compact de $\mathbb{R}\times\mathbb{R}_+$, on a :

$$E_o\left[\left(L^{\varepsilon x+y}_{t+s} - L^y_{t+s}\right) - \left(L^{\varepsilon x'+y}_t - L^y_t\right)\right]^{2k}$$

$$\le cE_o\left[\left(L^{\varepsilon x+y}_{t+s} - L^{\varepsilon x'+y}_{t+s}\right)^{2k}\right] + cE_o\left[\left(L^{\varepsilon x'+y}_s - L^y_s\right) o\ \theta_t\right]^{2k}$$

$$\le c\ \varepsilon^{k(\beta-1)}\left\{|x-x'|^{k(\beta-1)} + s^{k(1-\frac{1}{\beta})}\right\}\qquad\text{grâce à } (*)$$

C'est une condition suffisante pour affirmer que la famille des lois de probabilité de $\left(\dfrac{L^{\varepsilon x+y}_t - L^y_t}{\varepsilon^{\frac{\beta-1}{2}}}\;;\; x\in\mathbb{R}\;,\; t>0\right)$ est faiblement relativement compacte (voir par exemple [R-Y] p.492).

Sachant que B est indépendante de $(L^x_t; x\in\mathbb{R}; t>0)$, B est également indépendante de $(\int_{\mathbb{R}} x\, L^x_t\, dx\ ;\ t>0)$ et de $(\int_0^{+\infty} g(t)dt\int_{\mathbb{R}} x\, L^x_t\, dx\ ;\ g\in\mathcal{C})$. Pour toute fonction g continue à support compact, on a :

$$\int_0^{+\infty} g(t)dt \int_{\mathbb{R}} x\, L_t^x\, dx = \int_0^{+\infty} X_s ds \int_s^{+\infty} g(t)dt.$$

D'où B est indépendante de $\left(\int_0^{+\infty} G(s)X_s ds\ ;\ G\in C^1\right.$ et G à support compact).

On en déduit que B est indépendante de X.

IV - Loi d'un accroissement du temps local d'un processus de Markov symétrique.

Marcus et Rosen donnent dans [M-R 2] une version sans conditionnement du théorème d'isomorphisme de Dynkin, valable uniquement pour un processus de Markov tué en un temps exponentiel indépendant. Cette version leur permet de calculer la transformée de Laplace d'un accroissement du temps local en un temps exponentiel indépendant. Grâce au Théorème 1.3 , nous étendons leur résultat à tout temps terminal T ainsi qu'au dernier instant d'atteinte de zéro avant T .

On reprend les notations de l'exemple 1.2. De plus , on pose: $g_\xi = \sup\{t<\xi : X_t=0\}$ et $\ell^x = L_\xi^x - L_{g_\xi}^x$. ℓ représente donc le processus des temps locaux accumulés au cours de la dernière excursion à partir de 0. Pour x et y fixés dans $\mathbb{R}^2$, on notera :

$$\begin{cases} a^2 = g(x,x) + g(y,y) - 2g(x,y) \\ b^2 = g(x,x) + g(y,y) + 2g(x,y) \\ c\ = (g(0,x) - g(0,y))b^2 + (g(0,x) + g(0,y))a^2 \end{cases}$$

Théorème 4.1 :

(1) *Les processus* $(\ell^x,\ x\in\mathbb{R})$ *et* $(L_{g_\xi}^x\ ,\ x\in\mathbb{R})$ *sont indépendants.*

(2) *Pour tout* $x\in\mathbb{R}$:

$$E_0\left[e^{\theta L_\xi^x}\right] = \left(1 - \frac{g(0,x)}{g(x,x)}\right) + \frac{g(0,x)}{g(x,x)}\left(\frac{1}{1-\theta g(x,x)}\right)$$

$$E_0\left[e^{\theta L^x_{g_\xi}}\right] = \left(1 - \frac{g^2(0,x)}{g(x,x)\ g(0,0)}\right) + \frac{g^2(0,x)}{g(0,0)\ g(x,x)}\left(\frac{1}{1-\theta g(x,x)}\right).$$

D'où : $$E\left[e^{\theta \ell^x}\right] = \frac{1 + (g(x,x) - g(0,x))\theta}{1 + \left(g(x,x) - \dfrac{g^2(0,x)}{g(0,0)}\right)\theta}.$$

(3) *Pour tout* $(x,y) \in \mathbb{R}^2$:

$$E_0\left[e^{\theta(L^x_\xi - L^y_\xi)}\right] = 1 + \frac{\theta\left(g(0,x)-g(0,y) + \dfrac{\theta a^2}{2}\ (g(0,x)+g(0,y)\right)}{1 - \dfrac{\theta^2 a^2 b^2}{4}}$$

$$E_0\left[e^{\theta(L^x_{g_\xi} - L^y_{g_\xi})}\right] = 1 + \frac{\dfrac{\theta}{2}\left(g^2(0,x)-g^2(0,y) + \dfrac{\theta}{2}\ c\right)}{\left(1 - \dfrac{\theta^2 a^2 b^2}{4}\right) g(0,0)}.$$

D'où : $$E\left[e^{\theta(\ell^x-\ell^y)}\right] = \frac{1 + (g(0,x)-g(0,y))\theta + \dfrac{\theta^2 a^2}{2}\ (g(0,x)+g(0,y)) - \dfrac{b^2}{2}\Big)}{1 + \left(\dfrac{g^2(0,x)-g^2(0,y)}{2g(0,0)}\right)\theta + \dfrac{\theta^2}{4g(0,0)}\ (c - a^2 b^2)}.$$

<u>Démonstration</u> :

(1) C'est le résultat d'une utilisation simple des formules clés de la théorie des excursions. Soit $H_t = \int_{\mathbb{R}} L^x_t\ f(x)dx$; pour f fonction mesurable bornée positive. M désigne l'ensemble des extrémités gauches des excursions de X autour de 0, R le temps de vie de l'excursion et ${}^*P^0$ la mesure des excursions de X autour de 0.

$$\begin{aligned} E_0\left[e^{i\lambda H_{g_\xi} + i\mu(H_\xi - H_{g_\xi})}\right] &= E_0\left[\sum_{s\in M} e^{i\lambda H_s}\ e^{i\mu H_R \circ \theta_s}\ 1_{(R\circ\theta_s = \infty)}\right] \\ &= E_0\left[\int_0^\infty e^{i\lambda H_s}\ {}^*P^0\left(e^{i\mu H_R}\ 1_{(R=\infty)}\right) dL^0_s\right] \\ &= {}^*P^0\left(e^{i\mu H_R}\ 1_{(R=\infty)}\right) \times E_0\left(\int_0^\infty e^{i\lambda H_s}\ dL^0_s\right). \end{aligned}$$

(2) $$E_0\left[e^{\theta L^x_\xi}\right] = P_0(L^x_\xi = 0) + E_0\left(e^{\theta L^x_\xi} \mid L^x_\xi > 0\right) P_0(L^x_\xi > 0)$$

$$P_0(L^x_\xi > 0) = \frac{g(0,x)}{g(x,x)} \qquad ; \; E_0(L^x_\xi) = g(0,x)$$

On sait que conditionnellement à $L^x_\xi > 0$, L^x_ξ suit une loi exponentielle.

D'où : $$E_0\left[e^{\theta L^x_\xi}\right] = \left(1 - \frac{g(0,x)}{g(x,x)}\right) + \frac{g(0,x)}{g(x,x)}\left(\frac{1}{1 - \theta g(x,x)}\right).$$

Le raisonnement précédent n'est plus utilisable sous $\tilde{P}_{0,0}$, la loi de X n'étant plus homogène. Mais on remarque que :

$$\tilde{P}_{0,0}\left(e^{\theta L^x_\xi}\right) = E_0\left(e^{\theta L^x_{g_\xi}}\right).$$

On utilise donc le théorème d'isomorphisme :

$$\tilde{P}_{0,0}\left(e^{\theta L^x_\xi}\right) \times \langle e^{\theta \frac{G_x^2}{2}} \rangle = \frac{1}{\langle G_0^2 \rangle} \langle G_0^2 e^{\theta \frac{G_x^2}{2}} \rangle$$

Pour calculer le membre de droite dans l'équation ci-dessus, il suffit alors de définir ρ variable gaussienne centrée indépendante de G_x , en posant :

$$G_0 = \frac{\langle G_0, G_x \rangle}{\langle G_x^2 \rangle} G_x + \rho .$$

On obtient $E\left(e^{\theta \ell^x}\right)$ en remarquant que : $L^x_\xi = L^x_{g_\xi} + \ell^x$, et en utilisant (1)

(3) On pose $\tilde{G} = G + \varepsilon$. En utilisant le Théorème 1.3, on a :

$$E_0\left(e^{\theta(L^x_\xi - L^y_\xi)}\right) = \frac{\langle (G_0+\varepsilon)\, e^{\frac{\theta}{2}(\tilde{G}_x^2 - \tilde{G}_y^2)} \rangle}{\varepsilon \langle e^{\frac{\theta}{2}(\tilde{G}_x^2 - \tilde{G}_y^2)} \rangle}$$

On remarque que : $\tilde{G}_x^2 - \tilde{G}_y^2 = \left(\tilde{G}_x - \tilde{G}_y\right)\left(\tilde{G}_x + \tilde{G}_y\right) = \phi\ (\psi + 2\varepsilon)$.

ϕ et ψ sont deux gaussiennes centrées indépendantes.

On pose : Var $\phi = a^2 = g(x,x) + g(y,y) - 2g(x,y)$

Var $\psi = b^2 = g(x,x) + g(y,y) + 2g(x,y)$.

On calcule facilement : $\dfrac{\langle (G_0+\varepsilon)\, e^{\frac{\theta}{2}\phi(\psi+2\varepsilon)} \rangle}{\varepsilon \langle e^{\frac{\theta}{2}\phi(\psi+2\varepsilon)} \rangle}$,

en posant : $G_0 = \dfrac{\langle G_0\, \phi\rangle}{\langle \phi^2\rangle}\phi + \dfrac{\langle G_0\, \psi\rangle}{\langle \psi^2\rangle}\psi + \rho$

avec ρ gaussienne centrée indépendante de (ϕ,ψ).
De même, par des calculs élémentaires , on obtient la loi de $(L^x_{g_\xi} - L^y_{g_\xi})$

à partir de : $$\tilde{P}_{0,0}\left(e^{\theta(L^x_\xi - L^y_\xi)}\right) = \frac{\langle G_0^2\, e^{\frac{\theta}{2}(G_x^2-G_y^2)} \rangle}{\langle e^{\frac{\theta}{2}(G_x^2-G_y^2)}\rangle \langle G_0^2\rangle} .$$

(1) nous donne le dernier résultat.

Remerciements L'idée d'établir les théorèmes limites sur les processus stables à partir du théorème d'isomorphisme revient à Jay Rosen . Je le remercie pour nos nombreuses discussions lors de mon séjour au Technion.

REFERENCES

[D] **E.B. Dynkin** : Local times and quantum fields. *Sem. on Stoch. Processes, Birkhauser eds. P. Huber and M. Rosenblatt (69-89) 1983.*

[E] **N. Eisenbaum** : Dynkin's isomorphism theorem and the Ray-Knight theorems. *Probab.Theory Relat.Fields 99,321-335 (1994).*

[K 1] **F.B. Knight** : Random walks and a sofourn density process of Brownian motions.*Trans. Amer. Math. Soc.,109 (56-86) 1963.*

[K 2] **F.B. Knight** : Essentials of Brownian motion and diffusion. *Mathematical Surveys 18 Amer.Math.Soc.Providence,1981.*

[M-R 1] **M.B. Marcus** and **J. Rosen** : Sample path properties of the local times of strongly symmetric Markov processes via Gaussian processes. *Annals of Proba.,V.20,n°4,(1603-16684),1992.*

[M-R 2] **M.B. Marcus** and **J.Rosen** : Moment generating functions for local times of strongly symmetric Markov processes and random walks. *Proceedings of the conference Probability in Banach Spaces,8. Birkhauser.1992.*

[P-Y] **J. Pitman** and **M. Yor** : Bessel processes and infinitely divisible laws.*In "Stochastic Integrals" D. Williams ed. LNM 851. Springer.1981.*

[Ra] **D.B. Ray** : Sojourn times of a diffusion process.*Ill. J. Maths., 77 (615-630) 1963.*

[R] **J. Rosen** : Second order limit laws for the local times of stable processes. *Sém. de Probabilités XXV.LNM 1485 (407-425).Springer.1991*

[R-Y] **D. Revuz** and **M. Yor** : Continuous martingales and Brownian motion. *Springer-Verlag.1991.Second edition 1994.*

[S] **P. Sheppard** : On the Ray-Knight property of local times. *J. London Math. Soc. 31,(377-384),1985.*

[Y 1] **M. Yor** : Le drap brownien comme limite en loi des temps locaux d'un mouvement brownien linéaire.*Sém. de Probabilités XVII. LNM 986 (89-105).Springer. 1983.*

[Y 2] **M. Yor** : Remarques sur certaines constructions des mouvements browniens fractionnaires.*Sém. de Probabilités XXII.LNM 1321 (217-225).Springer.1988.*

Sur la représentation des $(\mathcal{F}_t^- = \sigma\{B_s^-, s \leq t\})$-martingales

Y. Hu

Laboratoire de Probabilités, Université Paris VI
Tour 56, 4 Place Jussieu, 75252 Paris Cedex 05, France

1. Introduction.

Soit (B_t) un mouvement brownien réel issu de 0 et notons $(\mathcal{F}_t)$ sa filtration naturelle. Soit $(\mathcal{F}_t^+)$ (resp: $(\mathcal{F}_t^-)$) la filtration engendrée par la partie positive (resp: négative) de (B_t), i.e: $\mathcal{F}_t^+ = \sigma(B_s^+ : s \leq t)$ (resp: $\mathcal{F}_t^- = \sigma(B_s^- : s \leq t)$) rendue continue à droite, et complétée. En étudiant l'équation de structure $d[X,X]_t = dt - X_{t\_}^+ dX_t$, Azéma & Rainer (1994) ont démontré que toute $(\mathcal{F}_t^-)$-martingale de carré intégrable peut s'écrire comme une intégrale stochastique par rapport à une martingale, qui est à un changement de temps près l'unique solution de cette équation de structure, notée encore par (X_t), dont la partie discontinue fait intervenir la "première" martingale d'Azéma. (Pour une référence sur les équations de structure, voir Emery (1989)). On introduit $g_t = \sup\{s \leq t : B_t = 0\}$, le dernier zéro de B avant t, et on pose

$$X_t = -B_t^- + \mathbb{1}_{(B_t>0)} \sqrt{\frac{\pi}{2}(t - g_t)}.$$

Théorème (Azéma & Rainer (1994)): *Toute $(\mathcal{F}_t^-)$-martingale de carré intégrable peut s'écrire comme une intégrale stochastique par rapport à la martingale (X_t).*

Plus généralement, soit f une fonction borélienne et de carré intégrable sur toute intervalle bornée de $\mathbb{R}^1$; On étudie la filtration engendrée par la martingale continue $(\int_0^t f(B_s)dB_s, t \geq 0)$. Un problème ouvert est de savoir quand cette filtration est brownienne. Ce problème difficile, que l'on ne traite pas ici, est partiellement résolu par D. Lane (1978) pour une fonction f continue dont l'ensemble des zéros est de mesure de Lebesgue nulle. Un autre problème lié à cette filtration est de savoir si la martingale ci-dessus $(\int_0^t f(B_s)dB_s)$ est pure. Une façon de l'aborder, (voir Knight (1987)), est de le transformer en celui de l'existence de solution forte d'une certaine équation associée. En particulier, prenons $f(x) = a1_{(x\leq 0)} + b1_{(x>0)}$ avec a et b deux constantes non nulles, et $a \neq b$, Knight (1987) a montré que pour que la martingale $(\int_0^t f(B_s)dB_s)$ soit pure, il faut et il suffit que a et b aient le même signe. (Voir aussi Stroock & Yor (1981), et Barlow (1988) pour une étude de solution forte de cette équation associée). Le cas critique où $b = 0$ est décrit par le théorème précédent. En effet, en appliquant la formule de Tanaka à (B_t^-), on montrera dans Section 2 que la filtration $(\mathcal{F}_t^-)$ est engendrée par la martingale $(\int_0^t \mathbb{1}_{(B_s<0)} dB_s)$. (Pour une référence générale, voir Chaleyat-Maurel & Yor (1978)). On en déduit en particulier que la martingale $(\int_0^t \mathbb{1}_{(B_s<0)} dB_s)$ n'est pas pure. En fait, elle n'est même pas extrêmale, puisque dans sa filtration propre figurent des martingales purement discontinues.

Dans les sections suivantes, on démontrera ce théorème plus directement et on explicitera les parties continue et discontinue de la martingale (X_t) à l'aide du calcul stochastique. On établit tout d'abord trois lemmes préliminaires dans Section 2, et la démonstration du théorème se trouve à Section 3.

Cette preuve est issue de discussions avec J. Azéma, C. Rainer et M. Yor que je tiens à remercier vivement.

2. L'indépendance entre $\mathcal{F}^+$ et $\mathcal{F}^-$.

On note κ la filtration engendrée par les signes de B définie par

$$\kappa_t = \sigma\{\mathrm{sgn}(B_s) : s \leq t\}, \qquad t \geq 0.$$

Rappelons que $\mathcal{F}_{g_t} \stackrel{def}{=} \sigma\{\zeta_{g_t} : \zeta \text{ est un processus prévisible pour } (\mathcal{F}_t)\}$. On a le lemme suivant:

Lemme 1: *Pour tout $t > 0$, les deux tribus κ_∞ et $\mathcal{F}_{g_t} \bigvee \sigma(\mathrm{sgn} B_t)$ sont conditionnellement indépendantes par rapport à la tribu κ_t.*

Preuve du Lemme 1: Il suffit de montrer que pour toute variable $\Phi \in L^2(\mathcal{F}_{g_t})$, on a

$$\mathbb{E}(\Phi \mid \kappa_\infty) = \mathbb{E}(\Phi \mid \kappa_t).$$

On note $d_t = \inf\{s \geq t : B_s = 0\}$. Alors

$$\mathbb{E}(\Phi \mid \kappa_\infty) = \mathbb{E}(\Phi \mid \kappa_t, d_t).$$

Prenons $\xi \in L^2(\kappa_t)$, f une fonction bornée borélienne, et posons $h(x) = \mathbb{E} f(t + T_x)$ où T_x est le premier temps d'atteinte de x par un mouvement brownien réel issu de 0, en particulier h est paire. On définit un noyau markovien Λ par $\Lambda f(x) = \mathbb{E}(f(x m_1))$, où (m_t) est le méandre brownien défini par $m_u = |B_{g_t+u(t-g_t)}|/\sqrt{t-g_t}$, pour $0 \leq u \leq 1$. Alors

$$\begin{aligned}
\mathbb{E}(\Phi\xi f(d_t)) &= \mathbb{E}(\Phi\xi\mathbb{E}(f(d_t) \mid \mathcal{F}_t)) \\
&= \mathbb{E}(\Phi\xi h(|B_t|)) \\
&= \mathbb{E}\Big(\Phi\xi\Lambda h(\sqrt{t-g_t})\Big) \\
&= \mathbb{E}\Big(\mathbb{E}(\Phi \mid \kappa_t)\xi\Lambda h(\sqrt{t-g_t})\Big) \\
&= \mathbb{E}\Big(\mathbb{E}(\Phi \mid \kappa_t)\xi f(d_t)\Big).
\end{aligned}$$

où la troisième égalité résulte de l'indépendance entre la tribu $\mathcal{F}_{g_t} \bigvee \sigma(\text{sgn}B_t)$ et le méandre $(m_u; 0 \leq u \leq 1)$, (voir par exemple Lemme 1 de Azéma & Yor (1989)). Donc, $\mathbb{E}(\Phi \mid \kappa_\infty) = \mathbb{E}(\Phi \mid \kappa_t)$. □

On note (L_t) le temps local en 0 de (B_t). Soient M^+ et M^- deux martingales continues définies par

$$M_t^+ = -\int_0^t \mathbb{1}_{(B_s>0)} dB_s, \qquad M_t^- = \int_0^t \mathbb{1}_{(B_s<0)} dB_s.$$

Rappelons que la filtration $(\mathcal{F}_t^+)$ (resp: $(\mathcal{F}_t^-)$) est définie dans l'introduction comme la filtration engendrée par la partie positive (resp: négative) de (B_t). On écrit la formule de Tanaka pour B_t^+, $B_t^+ = -M_t^+ + \frac{1}{2}L_t$. Il vient du lemme de Skorokhod pour les équations de réflexions (voir aussi El Karoui & Chaleyat-Maurel (1978) pour le problème de réflexion) que $\frac{1}{2}L_t = \sup_{0\leq s\leq t} M_s^+$. De même pour (B_t^-), on en déduit aisément que la filtration naturelle de la martingale (M_t^+) (resp: (M_t^-)) est égale à $(\mathcal{F}_t^+)$ (resp: $(\mathcal{F}_t^-)$).

Lemme 2: *Conditionnellement à κ_∞, les tribus $\mathcal{F}_\infty^+$ et $\mathcal{F}_\infty^-$ sont indépendantes.*

Preuve du Lemme 2: On note A^+ (resp: A^-) le processus croissant associé à la martingale continue M^+ (resp: M^-). L'inverse continue à droite de A^+ (resp: A^-) est notée par α^+ (resp: α^-). i.e. pour $t \geq 0$, $A_t^{+,-} = < M^{+,-} >_t = \int_0^t 1_{(B_s \in \mathbb{R}_{+,-})} ds$ et $\alpha_t^{+,-} = \inf\{s > 0 : A_s^{+,-} > t\}$. Il résulte du théorème de Knight (1970) sur les martingales orthogonales qu'il existe deux mouvements browniens réels indépendants δ^+et δ^- tels que

$$M^{+,-} = \delta^{+,-}_{A^{+,-}}.$$

Le lemme de Skorokhod implique (voir par exemple Karatzas & Shreve (1988)), que

$$\frac{1}{2} L_{\alpha_t^{+,-}} = \sup\{\delta_s^{+,-} : 0 \leq s \leq t\} \stackrel{def}{=} S_t^{+,-}.$$

Or

$$\alpha_t^+ = t + A^-_{\alpha_t^+} = t + A^-_{\tau(L_{\alpha_t^+})} = t + A^-_{\tau(2S_t^+)},$$

où τ est l'inverse continu à droite de L, et il est facile de voir que $S^{+,-}$ est l'inverse de $A_\tau^{+,-}$. Donc

$$\sigma(\alpha^+) \subset \sigma(S^+, S^-), \text{ et de même pour } \alpha^-$$

Par conséquent,

$$\sigma(S^+, S^-) \subset \kappa_\infty = \sigma(\text{sgn}(B_t) : t > 0) = \sigma(A^+, A^-) = \sigma(\alpha^+, \alpha^-) \subset \sigma(S^+, S^-).$$

On peut donc écrire

$$\mathcal{F}^+_\infty = \sigma(\delta^+, \kappa_\infty) = \sigma(\delta^+, S^-), \quad \mathcal{F}^-_\infty = \sigma(\delta^-, \kappa_\infty) = \sigma(\delta^-, S^+).$$

Pour montrer le lemme 2, il s'agit de prouver que pour toute variable aléatoire bornée Z de la forme $Z = \phi\chi$ avec $\phi \in L^\infty(\delta^+), \chi \in L^\infty(\kappa_\infty)$, on a

$$\mathbb{E}(Z \mid \mathcal{F}^-_\infty) = \mathbb{E}(Z \mid \kappa_\infty).$$

Or,

$$\begin{aligned}\mathbb{E}(Z \mid F^-_\infty) &= \chi\mathbb{E}(\phi \mid \sigma(\delta^-, S^+, S^-)) \\ &= \chi\mathbb{E}(\phi \mid \sigma(S^+)) \\ &= \chi\mathbb{E}(\phi \mid \kappa_\infty) \\ &= \mathbb{E}(Z \mid \kappa_\infty),\end{aligned}$$

où la seconde égalité découle de l'indépendance entre δ^+ et δ^-. □

Lemme 3 : *Pour tout $t \geq 0$, conditionnellement à κ_t, les deux tribus $\mathcal{F}^+_t$ et $\mathcal{F}^-_t$ sont indépendantes.*

Preuve du Lemme 3: Soient Φ, Ψ deux fonctionnelles continues bornées, $t > 0$; Il suffit de montrer que

$$\mathbb{E}\Big(\Phi(B^+_s, s \leq t)\Psi(B^-_s, s \leq t)\Big) = \mathbb{E}\Big(\mathbb{E}(\Phi(B^+_s, s \leq t) \mid \kappa_t)\Psi(B^-_s, s \leq t)\Big).$$

Pour simplifier les notations, on note $\Phi_t = \Phi(B^+_s, s \leq t)$ et $\Psi_t = \Psi(B^-_s, s \leq t)$. Alors,

$$\begin{aligned}\mathbb{E}\Big(\Phi_t\Psi_t\mathbb{1}_{(B_t>0)}\Big) &= \mathbb{E}\Big(\Phi_t\Psi_{g_t}\mathbb{1}_{(B_t>0)}\Big) \\ &= \mathbb{E}\Big(\mathbb{E}(\Phi_t \mid \kappa_\infty)\mathbb{E}(\mathbb{1}_{(B_t>0)}\Psi_{g_t} \mid \kappa_\infty)\Big) \\ &= \mathbb{E}\Big(\mathbb{E}(\Phi_t \mid \kappa_\infty)\mathbb{E}(\Psi_{g_t} \mid \kappa_t)\mathbb{1}_{(B_t>0)}\Big) \\ &= \mathbb{E}\Big(\mathbb{E}(\Phi_t \mid \kappa_t)\mathbb{E}(\Psi_{g_t} \mid \kappa_t)\mathbb{1}_{(B_t>0)}\Big) \\ &= \mathbb{E}\Big(\mathbb{E}(\Phi_t \mid \kappa_t)\mathbb{E}(\Psi_t \mid \kappa_t)\mathbb{1}_{(B_t>0)}\Big),\end{aligned}$$

où la seconde égalité provient du Lemme 2, et la troisième égalité du Lemme 1. De même pour la partie $(B_t < 0)$, on a

$$\mathbb{E}\Big(\Phi_t\Psi_t\mathbb{1}_{(B_t<0)}\Big) = \mathbb{E}\Big(\mathbb{E}(\Phi_t \mid \kappa_t)\mathbb{E}(\Psi_t \mid \kappa_t)\mathbb{1}_{(B_t<0)}\Big).$$

En sommant, on obtient le Lemme 3. □

3. Démonstration du Théorème.

Soit (μ_t) la première martingale d'Azéma relative à la filtration (κ_t), définie par $\mu_t = \operatorname{sgn}(B_t)(t-g_t)^{\frac{1}{2}}$. Soient de plus:

$$Y_t^+ = \int_0^t \mathbb{1}_{(\mu_{s-}>0)} d\mu_s, \qquad Y_t^- = \int_0^t \mathbb{1}_{(\mu_{s-}<0)} d\mu_s.$$

Rappelons que M^+ et M^- sont définies dans Section 2. La formule de Tanaka montre que $M_t^+ = \frac{1}{2}L_t - B_t^+$. Il déduit facilement des propriétés du méandre brownien (voir Lemme 1 de Azéma & Yor (1989)) que

$$\mathbb{E}\Big(M_t^+ \mid \kappa_t\Big) = \frac{1}{2}L_t - 1_{(B_t>0)}\sqrt{\pi(t-g_t)/2} = -\sqrt{\pi/2}Y_t^+.$$

Quelques lignes de calculs montrent que

$$X_t = M_t^- + \sqrt{\pi/2}Y_t^+.$$

On déduit du lemme 3 que la projection de M^+ sur la tribu κ_t est la même que sur $\mathcal{F}_t^-$. Donc $Y_t^+ = -\sqrt{2/\pi}\mathbb{E}\Big(M_t^+ \mid \mathcal{F}_t^-\Big)$, est une $(\mathcal{F}_t^-)$-martingale. D'une part, on sait (voir Azéma & Yor (1989) ou Emery (1989)), que (μ_t) est une martingale purement discontinue par rapport à la filtration (κ_t). Il en découle que (Y_t^+) en est aussi une. Donc on a, d'après la formule d'Itô-Meyer (Voir Meyer (1976)),

$${Y_t^+}^2 - \sum_{s\leq t}(\Delta Y_s^+)^2 = 2\int_0^t Y_{s-}^+ dY_s^+,$$

qui est aussi, d'après le calcul précédent, une $(\mathcal{F}_t^-)$-martingale. On en déduit que, par rapport à la filtration $(\mathcal{F}_t^-)$, $[Y^+,Y^+]_t = \sum_{s\leq t}(\Delta Y_s^+)^2$, et il s'en suit que Y^+ est une $(\mathcal{F}_t^-)$-martingale purement discontinue. De la même manière, on peut démontrer que Y^- est une $(\mathcal{F}_t^+)$-martingale purement discontinue. Rappelons que $X_t = M_t^- + \sqrt{\pi/2}Y_t^+$, on en déduit que X est bien une $\mathcal{F}^-$-martingale. Pour finir la démonstration du Théorème, il suffit de montrer qu'il existe une famille dense de $\mathcal{F}^-$-martingales qui peuvent s'écrire comme intégrales stochastiques par rapport à la martingale X. Soit $f \in L^2(\mathbb{R}_+, ds)$; On pose

$$N_t = \exp\left(\int_0^t f(s)dB_s - \frac{1}{2}\int_0^t f^2(s)ds\right)$$

et

$$N_t^{+,-} = \exp\left(\int_0^t f(s)\mathbb{1}_{(B_s\in\mathbb{R}_{+,-})}dB_s - \frac{1}{2}\int_0^t f^2(s)\mathbb{1}_{(B_s\in\mathbb{R}_{+,-})}ds\right),$$

avec les notations évidentes. Il suffit de montrer que la martingale $\mathbb{E}(N_t \mid \mathcal{F}_t^-)$ vérifie l'assertion du théorème. En fait,

$$\mathbb{E}(N_t \mid \mathcal{F}_t^-) = N_t^-\mathbb{E}(N_t^+ \mid \mathcal{F}_t^-),$$

et $N_t^- = 1 + \int_0^t N_s^- f(s) \mathbb{1}_{(B_s \in \mathbb{R}_-)} dB_s = 1 + \int_0^t N_s^- f(s) dM_s^-$. Remarquons que $M_t^- = \int_0^t \mathbb{1}_{(X_{s-}<0)} dX_s$, et $Y_t^+ = \sqrt{2/\pi} \int_0^t \mathbb{1}_{(X_{s-}\geq 0)} dX_s$. En s'appuyant sur la formule d'Itô-Meyer (voir Meyer (1976)), il ne reste qu'à prouver l'existence d'un processus (n_s) κ-prévisible tel que:

$$\mathbb{E}(N_t^+ \mid \mathcal{F}_t^-) = 1 + \int_0^t n_s dY_s^+.$$

Pour montrer ceci, utilisons le lemme 3:

$$\mathbb{E}(N_t^+ \mid \mathcal{F}_t^-) = \mathbb{E}(N_t^+ \mid \kappa_t) = 1 + \int_0^t z_s d\mu_s,$$

pour un processus (z_s) qui est κ-prévisible avec $\mathbb{E}(\int_0^\infty (z_s)^2 ds) < +\infty$, où on a utilisé la propriété de représentation prévisible de la martingale d'Azéma (μ_t) par rapport à la filtration (κ_t) (de plus, (μ_t) admet la propriété de représentation chaotique, voir Azéma & Yor (1989) ou Emery (1989)). Comme Y^- est une $(\mathcal{F}_t^+)$-martingale purement discontinue, et que N^+ est une $(\mathcal{F}_t^+)$-martingale continue, on a

$$\mathbb{E}\Big(N_t^+ \int_0^t z_s dY_s^-\Big) = 0.$$

Or,

$$\mathbb{E}\Big(N_t^+ \int_0^t z_s dY_s^-\Big) = \mathbb{E}\Big(\mathbb{E}(N_t^+ \mid \mathcal{F}_t^-) \int_0^t z_s dY_s^-\Big)$$
$$= \mathbb{E} \int_0^t z_s^2 \mathbb{1}_{(\mu_{s-}<0)} ds.$$

D'où $\mathbb{E} \int_0^t z_s^2 \mathbb{1}_{(\mu_{s-}<0)} ds = 0$. On en déduit que $\mathbb{E}(N_t^+ \mid \mathcal{F}_t^-) = 1 + \int_0^t n_s dY_s^+$, avec $n_s = z_s 1_{(\mu_{s-}>0)}$. D'où le résultat cherché. □

Bibliographie:

J. Azéma & C. Rainer. (1994). Sur l'équation de structure $d[X,X]_t = dt - X_{t-}^+ dX_t$. *Séminaire de Probabilités XXVIII* (Eds: J. Azéma, P.-A. Meyer & M. Yor) Lecture Notes in Maths. 1583 pp. 236-255. Springer Berlin.

J. Azéma & M. Yor. (1989). Etude d'une martingale remarquable. *Séminaire de Probabilités XXIII* (Eds: J. Azéma, P.-A. Meyer & M. Yor) Lecture Notes in Maths.1372 pp. 88-130. Springer Berlin.

M. T. Barlow. (1988). Skew Brownian motion and a one-dimensional stochastic differential equation. *Stochastics* 25 1-2.

M. Chaleyat-Maurel & M. Yor. (1978). Les filtrations de $|X|$ et X^+, lorsque X est une semimartingale continue. *Astérisque 52-53* pp.193-196.

N. El Karoui & M. Chayelat-Maurel. (1978). Un problème de réflexion et ses applications au temps local et aux équations différentielles stochastiques sur $\mathbb{R}$, cas continu. *Astérisque 52-53* pp. 117-144.

M. Emery. (1989). On the Azéma martingales. *Séminaire de Probabilités XXIII* (Eds: J. Azéma, P.-A. Meyer & M. Yor) Lecture Notes in Maths.1372 pp. 66-87. Springer Berlin.

I. Karatzas & S.E. Shreve. (1988). *Brownian Motion and stochastic calculus.* Springer, Berlin.

F.B. Knight. (1970). A reduction of continuous square-integrable martingales to Brownian motion. *Lecture notes in Mathematics, vol.190.* pp. 19-31. Springer, Berlin.

F.B. Knight. (1987). On the invertibility of martingale time changes. *Seminar on Stochastic Processes.* 1987. pp.193-221. Birkhäuser, Basel 1988.

D. Lane. (1978). On the fields of some Brownian martingales. *The Annals of Probability* Vol. 6, No. 3, pp. 499-506.

P.A. Meyer. (1976). Un cours sur les intégrales stochastiques. *Séminaire de Probabilités X.* Lecture Notes in Maths. 511. pp. 245-400. Springer, Berlin.

D. Stroock & M. Yor. (1981). Some remarkable martingales. *Séminaire de Probabilités XV.* Lecture Notes in Maths. 850. pp. 590-603. Springer, Berlin.

C-semigroups on Banach spaces and functional inequalities.

Shiqi SONG, Equipe d'Analyse et Probabilités, Université d'Evry – Val d'Essonne, Boulevard des Coquibus, 91025 EVRY Cedex, FRANCE. E-mail: song@dmi.univ-evry.fr

Abstract. We introduce the notion of C-semigroup on a Banach space. This notion is intimately relevent to classical Dirichlet forms on Banach spaces. We shall prove a sufficient condition for a semigroup on R^d to be a C-semigroup. Then, we prove that C-semigroups satisfy various functional inequalities such as Poincaré inequality, logarithmical Sobolev inequality and Stein-Meyer-Bakry inequalities (Riesz transform).

Key words. C-semigroup, classical Dirichlet form, well-admissible measure on Banach space, symmetric Markov process, Poincaré inequality, logarithmical Sobolev inequality, Stein-Meyer-Bakry inequalities (Riesz transform), stochastic flow.

AMS classification. 31C25, 44A15, 47B99, 60J35, 60J60.

Démontrer l'inégalité de Meyer à partir de la formule

$$\frac{1}{2}\tilde{\nabla}f(x) = E^{\infty}[\int_0^{\tau} \tilde{\nabla}Cf(X_s,B_s)\, e^{s}\, dB_s\, e^{-\tau} \,|\, X_\tau = x].$$

Introduction.

We consider a separable real Banach space B. We assume that B is <u>riggid</u>, i.e. there is a separable real Hilbert space H such that $B^* \subset H \subset B$ densely and continuously. We choose an orthonormal basis $(k_i)_{i\geq 1}$ of H, consisting of elements of B^*. We denote by $\langle \cdot,\cdot \rangle$ the scalar product in H and by $\|\cdot\|$ the associated norm on H.

Let μ be a positive measure on B, charging every open set in B. We suppose that μ is σ-finite on the cylindrical sets. Let $FC_f^2(B)$ denote the family of real cylindrical two times continuously differentiable bounded functions f on B such that $\mu(\{f \neq 0\}) < \infty$. For $u \in FC_f^2(B)$, we define

$$\frac{\partial u}{\partial k_i}(x) = \lim_{t\downarrow 0} \frac{1}{t}(u(x + tk_i) - u(x)),\ i = 1, 2, \ldots.$$

We in addition assume that the measure μ is <u>well-admissible</u>, i.e., for any $i \geq 1$, there exists a function $\eta_i \in L^2(B,\mu)$ such that

$$\int \frac{\partial u}{\partial k_i}(x)\, \mu(dx) = -\int u(x)\, \eta_i(x)\, \mu(dx),\ \forall\, u \in FC_f^2(B).$$

For a well-admissible measure μ, we can define a differential operator in the following way. For a function $f \in L^2(B,\mu)$, we shall say that f is <u>differentiable</u> (in directions k_i), if there exists $g \in L^2(B,H,\mu)$ such that, for any $i \geq 1$, for any $u \in FC_f^2(B)$,

$$\int \left[\frac{\partial u}{\partial k_i}(x) + u(x)\, \eta_i(x)\right] f(x)\, \mu(dx) = -\int u(x)\, \langle g(x), k_i\rangle\, \mu(dx).$$

The function g, which is uniquely determined by f (cf. Song [29]), will be denoted by $\tilde{\nabla} f$. Clearly, $\tilde{\nabla} u$ coincides with $\nabla u = (\frac{\partial u}{\partial k_i}, i \geq 1)$ if $u \in FC_f^2(B)$.

The <u>gradient operator</u> $\tilde{\nabla}$ being defined, we can now introduce the notion of C-semigroup. In the following definition, $\Lambda(H)$ denotes the space of bounded linear transformations in H equipped with the usual operator norm $\|\cdot\|$ (there will be no confusion with the norm on H).

Definition 1. Let (Q_t) be a semigroup of symmetric Markov operators on $L^2(B,\mu)$. We shall say that (Q_t) is a <u>C-semigroup</u> (with respect to the measure μ), if

i. for any $f \in L^2(B,\mu)$, for any $t > 0$, $Q_t f$ is differentiable;

ii. there exists a B-valued Hunt process X such that Q_t is the transition semigroup of X;

iii. there exists a $\Lambda(H)$-valued càdlàg process C_t (which will be called a <u>C-process</u>), adapted with respect to the natural filtration of X, with bounded variation, such that $\tilde{\nabla} Q_t f(x) = E_x[\tilde{\nabla} f(X_t)\, C_t]$, μ-a.s. x, for any differentiable function $f \in L^2(B,\mu)$, for any $t > 0$;

iv. $\alpha(Q_\cdot) = \text{esssup}\ \sup_{t>0} \frac{1}{t} \log \|C_t\| < \infty$.

v. C_t^{-1} exists and has bounded variation.

vi. for any $T > 0$, the identity $C_{T-s} \circ \iota_T = (C_s')^{-1}\, C_T'$ holds, where C_s' denotes the adjoint operator of C_s, and ι_T denotes the time inversion operator of the Hunt process X.

vi. for any $s \geq 0$, $t \geq 0$, $C_{t+s} = (C_s \circ \theta_t) C_t$, where θ_t, $t \geq 0$, denotes the translation operator associated with X.

Relatively to a C-semigroup (Q_t), we introduce the operators $\tilde{Q}_t$, $t \geq 0$, on the space $L^2(B,H,\mu)$: $\tilde{Q}_t(F)(x) = E_x[F\, C_t]$, $F \in L^2(B,H,\mu)$. We have the following lemma:

Lemma 2. The operators $(\tilde{Q}_t)$ are bounded, symmetric and form a semigroup.

This lemma will be proved in §5. We shall call the semigroup $(\tilde{Q}_t)$ the <u>tengent semigroup</u> of (Q_t).

A priori, C-semigroup property can concern any semigroup on $L^2(B,\mu)$. But, the far intimately relevent semigroup is the semigroup associated with the Dirichlet form defined as follows: We define the space $W(\mu)$ to be the family of differentiable functions f in $L^2(B,\mu)$. We introduce the form:

$$\mathcal{E}_\mu(f,h) = \int \langle \tilde{\nabla} f, \tilde{\nabla} h\rangle(x)\, \mu(dx), \quad f, h \in W(\mu).$$

It is known that the form $(\mathcal{E}_\mu, W(\mu))$ is a Dirichlet form on $L^2(B,\mu)$ (see Albeverio-Röckner [1]). This Dirichlet form will be called the <u>classical Dirichlet form</u> associated with the (well-admissible) measure μ.

Occasionally, we also need the next definition. As it will be seen below, this notion concerns especially the C-semigroups associated with a classical Dirichlet form.

Definition 3. Let γ be a measurable mapping from B into $\Lambda(H)$. Let (Q_t, X_t, C_t) be a C-semigroup with its associated Hunt process X and C-process (C_t). We shall say that the C-process is <u>logarithmically differentiable with log-derivative</u> γ, if $\int_0^t \|\gamma(X_s)\|\, ds < \infty$ almost surely, for any $t > 0$, and C_t satisfies the stochastic integral equation

$$C_t = I + \int_0^t \gamma(X_s)\, C_s\, ds, \ \forall\, t > 0,$$

where I denotes the identity transform in H.

We shall see that the constant $\alpha(Q_\bullet)$ in Definition 1.iv is a functional of the log-derivative of C_t, if it exists.

The introduction of the notion of C-semigroup has been stimulated by our experiences on the studies of classical Dirichlet spaces on Banach spaces. We have noticed that, for a symmetric Markovian semigroup Q_t on $L^2(B,\mu)$, many problems will have very simple solutions, if we can say about $\tilde{\nabla}Q_t$. The notion of C-semigroup synthesizes what about $\tilde{\nabla}Q_t$ would be useful.

Examples of interventions of C-semigroups are numerous. The Brownian semigroups, the Ornstein-Uhlenbeck semigroups, the Bessel semigroups (cf. Song [31]), the symmetric convolution semigroups, etc., are C-semigroups. Also, the transition semigroup of a stochastic differentiable flow on R^d of the form

$$dX_t = d\beta_t + \nabla b(X_t)dt,$$

where b is a $C_b^2(R^d)$-function, gives rise to a C-semigroup. Notice that, in this last example, a C-process is simply given by $C_t = (\partial_j X^i(t),\ 1 \le i,j \le d)$, which is obviously logarithmically differentiable with log-derivative $(\partial_{ij} b)$. We shall see later (Theorem 1.1, Part I) that this will remain true for a larger class of functions b on R^d.

The notion of C-semigroup already had been introduced in [31], where we proved that, for a Markovian semigroup Q_t on R^1 to be the transition semigroup of a two parameter continuous symmetric Markov process in the sense of Hirsch-Song [13], [14], Q_t must be a C-semigroup with $\alpha(Q_\bullet) \le 0$. More recently in [30], we show that C-semigroup property can be used to prove the Markovian uniqueness for Dirichlet operators on Banach spaces. In the present paper we shall show that C-semigroups satisfy various functional inequalities.

The paper is organised in two parts. In the first part, we shall give a sufficient condition for a semigroup on R^d to be a C-semigroup.

In the second part, we shall study the Poincaré inequality, the logarithmical Sobolev inequality, the Stein-Meyer-Bakry inequalities, for a C-semigroup associated with a classical Dirichlet form.

Let us say two words on the hypothesis of our paper. First, we have limited ourselves to consider only Banach spaces. But, our knowledge on Bakry's paper [3], in which Bakry has already used the tengent semigroup $\tilde{Q}_t$, convinces us of the possibility to apply our method to the studies of diffusions on Riemmanian manifolds. Secondly, the notion of C-semigroup is introduced by use of the gradient operator $\tilde{\nabla}$. This deprives us of considering that semigroups which possess an "opérateur carré du champs" $\Gamma(f,f)$ other than $\langle\tilde{\nabla}f,\tilde{\nabla}f\rangle$.

Part I. Existence of C-semigroups on R^d.

§1. Hypothesis.

In this section, we describe the semigroups which will be proved to be C-semigroups. Our description uses the notion of Dirichlet form, for which we refer to the book of Fukushima [9]. For the special case of Dirichlet forms on Banach spaces, we also can refer to Ma-Röckner [19], Song [29], [28], and the references therein.

We work on R^d equipped with the Lebesgue measure dx. We consider the classical Sobolev space $H^{1,2}(R^d)$ and the Dirichlet form

$$\mathcal{E}(u,v) = \int \sum_{1\le i\le d} \partial_i u(x)\, \partial_i v(x)\, dx = \int \langle \nabla u(x), \nabla v(x)\rangle\, dx,\ u, v \in H^{1,2}(R^d),$$

where $\nabla u(x) = (\partial_1 u(x),\dots,\partial_d u(x))$, and ∂_i is in the distribution sense. Let P_t be the semigroup associated with $\mathcal{E}$. This semigroup has a nice representation: $P_t f(x) = E[f(x + \sqrt{2}\,\beta_t)]$, where β denotes the standard d-dimensional Brownian motion started from zero.

Let Ω denote $C(R_+,R^d)$. The points in Ω will be denoted by ω, while the coordinate process will be denoted by $(\omega_t, t \ge 0)$. We shall denote the law of $x + \sqrt{2}\beta$ on Ω by P_x. If ξ denotes an R^d-valued random variable whose law is the Lebesgue measure, independent of β, the law of $\xi + \sqrt{2}\beta$ on Ω will be denoted by P.

We choose now a function b in $L^2(R^d,dx)$ and put $b_n = P_{1/n} b$ for any integer $n \ge 1$. We assume the following hypotheses on the function b.

Hy1: The fonction b is an $\mathcal{E}$-quasi-continuous function in the space $H^{1,2}$ and there are constants $C > 0$, $\upsilon > 0$, $2 > \rho > 0$ such that $|b(x)| \le C \exp\{\upsilon |x|^\rho\}$.

Under this hypothesis, the functions b_n belong to $C^\infty(R^d)$.

Hy2: $e^b \in L^2(R^d)$.

Let us show then that $e^{b_n} \in L^2(R^d)$ for any $n \ge 1$. Indeed, for any $t > 0$, we have

$$P_t[e^{2b}] \ge e^{2P_t b}.$$

This implies

$$\int e^{2b_n(x)}\,dx \le \int P_{1/n}[e^{2b}](x)\,dx = \int e^{2b(x)}\,dx < \infty.$$

Hy3: The functions e^b and e^{b_n} belong to $H^{1,2}(R^d)$.

Hy4: For any $1 \le i, j \le d$, $\partial_{ij}b$ are signed measures whose positive and negative parts are of finite energy with respect to $\mathcal{E}$.

Hy5: There exists a constant $0 < C < \infty$ such that, for any $n \ge 1$, for dx-a.s. $x \in R^d$, the matrix $\partial^2 b_n(x) - CI_d$ is definite negative.

The measures $\partial_{ij}b$ are decomposed in positive parts and negative parts: $\partial_{ij}b = \partial_{ij}b^+ - \partial_{ij}b^-$. We denote $\partial_{ij}b^+ + \partial_{ij}b^-$ by $|\partial_{ij}b|$. By definition, $|\partial_{ij}b|$ is a finite $\mathcal{E}$-energy measure. Let $B_t(\omega) = (B_{ij}(\omega)(t))$ be the matrix-valued process on Ω such that each $B_{ij}(\omega)(t)$ is, under P, the additive functional associated with the measures $\partial_{ij}b$. We shall denote the matrix $(\partial_{ij}b)$ by $\partial^2 b$ and $(\partial_{ij}b_n)$ by $\partial^2 b_n$. Remark that, when $\partial_{ij}b$ are functions, $B_{ij}(\omega)(dt)$ is just $\partial_{ij}b(X_t(\omega))\,dt$.

By hypothesis Hy2 and its consequence, we can define the bounded measures $\mu(dx) = e^{2b(x)}dx$ and $\mu_n(dx) = e^{2b_n(x)}dx$. With respect to the measure μ, we have an integration by parts formula:

$$\int \partial_i u(x)\,\mu(dx) = -\int u(x)\,2\partial_i b(x)\,\mu(dx),\quad u \in C_c^1(R^d).$$

Notice that by Hy3, $\partial_i b \in L^2(R^d,\mu)$. Therefore, the measure μ is well-admissible and the operator $\tilde{\nabla}$ introduced in Introduction is well define on the space $L^2(R^d,\mu)$. We then consider the classical Dirichlet form $(\mathcal{E}_\mu, W(\mu))$ associated with μ.

Let Q_t denote the Markov semigroup on $L^2(R^d,\mu)$ associated with $\mathcal{E}_\mu$ (cf. Fukushima [9]). The main object of Part I is to prove the following theorem:

Theorem 1. Under the hypotheses Hy1 to Hy5, (Q_t) is a C-semigroup. Moreover, if $\partial_{ij}b$, $1 \le i, j \le d$, are functions on R^d, a C-process can be associated with (Q_t), which is logarithmically differentiable with logarithmic derivative $\partial^2 b = (\partial_{ij}b)$.

§2. Estimate on the function b.

Recall that there is a conservative μ-symmetric diffusion process X in R^d whose Dirichlet form coincides with $\mathcal{E}_\mu$ (see Takeda [34]). The process X satisfies the stochastic equation:

$$X_t = X_0 + \sqrt{2}\,\beta_t + \int_0^t 2\nabla b(X_s)\, ds.$$

The law of X on Ω will be denoted by Q_x when $X_0 = x$, and by Q when X_0 has the law of μ. We know how to describe the semigroup Q_t of X: According to Takeda [34], there is a multiplicative functional N defined on Ω such that

$$N_t(\omega) = \exp\{\int_0^t \langle 2\nabla b(\omega_s), d\omega_s\rangle - 4\int_0^t \langle \nabla b, \nabla b\rangle(\omega_s)\, ds\},$$

for P_x-a.s. $\omega \in \Omega$, for any $x \in R^d$. Then, $Q_t f(x) = P_x[N_t(\omega) f(\omega_t)]$, or more generally, $Q_x[F(\omega^t)] = P_x[N_t F(\omega^t)]$, where $\omega^t = (\omega_s,\ 0 \le s \le t)$. It is known that N is a P_x-martingale for any $x \in R^d$.

Notice that the above results hold again when the function b is replaced by the function b_n. We shall denote the corresponding objects by $(X^n, \mu_n, Q^n_x, Q^n, Q^n_t, N^n_t)$.

Lemma 1. We have the following three convergences in $L^2(P)$, the convergences of the processes being with respect to the uniform norm over any compact set in R_+:

$$b_n(\omega_\bullet) \to b(\omega_\bullet);\quad \int_0^\bullet \nabla b_n(\omega_s)\, ds \to \int_0^\bullet \nabla b(\omega_s)\, ds;\quad \int_0^\bullet \partial^2 b_n(\omega_s)\, ds \to B_\bullet(\omega).$$

Proof. These are consequences of results of Hirsch-Song [13], [14]. □

Let Λ denote the family of stopping times τ such that

$$\text{P-esssup}_\omega \{\sup_n \int_0^{\tau(\omega)} \|\partial^2 b_n\|(\omega_s)\, ds + \sup_n \sup_{0\le s\le \tau(\omega)} \left| b_n(\omega_s)\right| \} < \infty.$$

Lemma 2. For any $\tau \in \Lambda$, P-esssup$_{\omega}$ sup$_n$ $N_{\tau}^n(\omega)$ is finite.

Proof. It is enough to notice that, under P,

$$\int_0^t \langle \nabla b_n(\omega_s), d\omega_s \rangle = b_n(\omega_t) - b_n(\omega_0) - \int_0^t \Delta b_n(\omega_s)\, ds. \square$$

Lemma 3. There is an increasing sequence (τ_n) in Λ such that $\lim_n \tau_n(\omega) = \infty$, P_x-a.s. ω, for dx-a.s. $x \in R^d$.

Proof. Set

$$S_t(\omega) = \sup_n \int_0^t \|\partial^2 b_n\|(\omega_s)\, ds + \sup_n \sup_{0 \le s \le t} \left| b_n(\omega_s) \right| .$$

As a consequence of Lemma 1, S_t is finite and continuous P_x-almost surely, for dx-a.s. $x \in R^d$. To prove the lemma, it is enough to set $\tau_n(\omega) = \inf\{t; S_t(\omega) \ge n\}$. $\square$

§3. C-Process.

For any $\omega \in \Omega$, let $C(\omega)$ and $C^n(\omega)$ denote respectively the following two d×d matrix-valued processes on Ω determined by the equations:

$$C_t(\omega) = I_d + 2 \int_0^t dB_s(\omega)\, C_s(\omega), \text{ P-a.s. } \omega, \text{ and}$$

$$C_t^n(\omega) = I_d + 2 \int_0^t \partial^2 b_n(\omega_s)\, C_s^n(\omega)\, ds, \text{ P-a.s. } \omega.$$

It can be shown that the solutions of these equations exist and are unique. Moreover, we have:

Lemma 1. Let A_t be the unique solution of the equation: $A_t = I_d - 2\int_0^t A_s\, dB_s$, $t \ge 0$.

Then, for any $t \ge 0$, $A_t = C_t^{-1}$ the inverse of C_t. Similar result holds also for C_t^n.

Proof. It is because $d(A_tC_t) = 0$. □

Lemma 2. $C^n(\omega) \to C(\omega)$ uniformly on each compact intervals for P-a.s. ω.

Proof. Consider the difference between C and C^n:

$$C_t(\omega) - C_t^n(\omega) = 2\int_0^t dB_s(\omega)\, C_s(\omega) - 2\int_0^t \partial^2 b_n(\omega_s)\, C_s^n(\omega)\, ds$$

$$= 2\int_0^t (dB_s(\omega) - \partial^2 b_n(\omega_s)ds)\, C_s(\omega) + 2\int_0^t \partial^2 b_n(\omega_s)\, (C_s(\omega) - C_s^n(\omega))\, ds.$$

Consider the two integrals in the last term. The second one is overestimated by

$$2 \sup_n \int_0^t \|\partial^2 b_n\|(\omega_s)\, ds \quad \sup_{0\le s\le t} \left| C_s(\omega) - C_s^n(\omega) \right|,$$

while the first one converges uniformly to zero in any compact interval for P-a.s. ω, which is the consequence of Lemma §2.1. Now, to finish the proof of the lemma, it is enough to apply Wendroff inequality (cf. Mao [20], p.24-32). □

Lemma 3. For any integer $p \ge 1$, set $\phi(p,s) = k2^{-p}$, where k is the unique integer such that $s \in [k2^{-p}, (k+1)2^{-p}[$. For a fixed integer $n \ge 1$, let $C_t(n,p)(\omega)$ be solution of the equation

$$C_t(n,p)(\omega) = I_d + 2\int_0^t \partial^2 b_n(\omega_{\phi(p,s)})\, C_s(n,p)(\omega)\, ds.$$

Then, $C^n(\omega)$ is the limit of $C_t(n,p)(\omega)$ when p tends to the infinity.

Proof. This can be proved by the usual Gronwall inequality method. □

Lemma 4. Let M be a d×d real symmetric matrix. Let $\lambda_1(M) \ge \lambda_2(M) \ge ... \ge \lambda_d(M)$ be the eigenvalues of M. Let D be the diagonal matrix corresponding to (λ_i) and let U be an orthonormal matrix such that $UMU^* = D$. Let $\exp\{tD\}$ denote the diagonal matrix corresponding to $(\exp\{t\lambda_i\})$. Let v(t) be the solution of the equation:

$$v(t) = I_d + \int_0^t M\, v(s)\, ds, \quad t \geq 0.$$

Then, $v(t) = U^* \exp\{tD\} U$.

Proof. Let $w(t) = Uv(t)U^*$. Then, $w(t)$ satisfies the equation:

$$w(t) = I_d + \int_0^t UMU^*\, w(s)\, ds = I_d + \int_0^t D\, w(s)\, ds, \quad t \geq 0.$$

Clearly, $w(t) = \exp\{tD\}$ is the unique solution of the equation. The lemma is proved. □

Corollary 5. $\|v(t)\| \leq \exp\{t\lambda_1(M)\}$.

Let $\rho = \sup_n \sup_x \lambda_1(2\partial_1^2 b_n(x))$. Remark that, by Hy5, $\rho < \infty$. We have the following lemma.

Lemma 6. We have $\|C_t\| \leq e^{\rho t}$, P-almost surely. Consequently, $\alpha(Q_\bullet) \leq \rho$.

Proof. Fix $n \geq 1$ and $p \geq 1$. Consider the process $C_t(n,p)$ introduced in Lemma 3. Set $A(t) = C_t(n,p)$. Let $M_k = \partial^2 b_n(\omega_{k2^{-p}})$, $k \geq 1$. Let $v_k(t)$ be the solution associated with M_k defined in Lemma 4. We check easily that $A(t + k2^{-p})\, A^{-1}(k2^{-p})$ and $v_k(t)$ for $t \in [0,2^{-p}[$ satisfy the same equation. By uniqueness, we conclude, if $h = \phi(p,t)2^p$ and $t_i = i2^{-p}$,

$$A(t) = A(t)A^{-1}(t_h)\, A(t_h)A^{-1}(t_{h-1}) \dots A^{-1}(t_2)A(t_1)A^{-1}(t_1)A(t_0)$$

$$= v_h(t - t_h)\, v_{h-1}(t_h - t_{h-1}) \dots v_1(t_2 - t_1)v_0(t_1).$$

Now, by Corollary 5 applied to each of v_i, we see $\|C_t(n,p)(\omega)\| \leq e^{\rho t}$, for any $t \geq 0$. Now, it is enough to apply Lemma 3 and Lemma 2 to complete the proof. □

Lemma 7. Let $T > 0$. Let ι_T denote the inversion operator : $\iota_T(\omega)_t = \omega_{T-t}$. Then, $C_{T-t} \circ \iota_T = (C_t')^{-1} C_T'$, $0 \leq t \leq T$, P-almost surely, where ' denotes the transposition of matrix.

Proof. We have P-almost surely

$$C_{T-t}\circ\iota_T = I_d + 2\int_0^{T-t} B\circ\iota_T(ds)\, C_s\circ\iota_T$$

$$= I_d + 2\int_0^{T} B\circ\iota_T(ds)\, C_s\circ\iota_T - 2\int_{T-t}^{T} B\circ\iota_T(ds)\, C_s\circ\iota_T$$

$$= C_T\circ\iota_T - 2\int_{T-t}^{T} B\circ\iota_T(ds)\, C_s\circ\iota_T = C_T\circ\iota_T - 2\int_0^{t} B(du)\, C_{T-u}\circ\iota_T.$$

From this formula we see that $C_{T-t}\circ\iota_T\,(C_T\circ\iota_T)^{-1}$ and $(C_t')^{-1}$ satisfy the same equation. By uniqueness, we conclude $C_{T-t}\circ\iota_T\,(C_T\circ\iota_T)^{-1} = (C_t')^{-1}$, $0 \le t \le T$. In particular, when $t = T$, we have $(C_T\circ\iota_T)^{-1} = (C_T')^{-1}$. This proves the lemma. □

Lemma 8. Let θ_t, $t \ge 0$, denote the translation operator on Ω: $\theta_t(\omega)_s = \omega_{t+s}$. Then, we have the relation $C_{t+s} = (C_s\circ\theta_t)C_t$, $s \ge 0$, P-almost surely.

Proof. It is enough to notice that, under the measure P, $C_{t+s}\, C_t^{-1}$ and $C_s\circ\theta_t$ satisfy both the same equation: $A_s = I_d + \int_0^s D\circ\theta_t(du)\, A_u$, $s \ge 0$. □

§4. Derivative of the process X.

Lemma 1. For any $n \ge 1$, for $x \in R^d$, let $X^n(x,\omega)$ be the unique solution (under P) of the equation:

$$dX_t^n(x) = \sqrt{2}\, d\beta_t + 2\nabla b_n(X_t^n)\, dt, \quad X_0^n = x.$$

Then, $\partial_i(X_t^n)_j(x,\omega) = (C_t^n)_{ij}(\omega)$, P-a.s. ω.

Proof. This can be proved using the results in Ikeda-Watanabe [15], using the localisation, and finally using the fact that X^n is conservative. □

Lemma 2. We have $\tilde{\nabla} Q_t^n u(x) = Q_x^n[\tilde{\nabla} u(\omega_t) C_t^n(\omega)]$, $u \in W(\mu)$, where the vector $\tilde{\nabla} u(x)$ in H is represented horizontally.

Proof. It is enough to prove it for $u \in C_c^1(R^d)$. But, by Lemma §3.6, we have

$$Q_x^n[\,\|\nabla u(\omega_t)C_t^n(\omega)\|^2\,] \le e^{2\alpha t}\, Q_x^n[\,\|\nabla u(\omega_t)\|^2\,] < \infty.$$

So, we can apply Fubini's lemma to finish the proof. □

Lemma 3. For any $\tau \in \Lambda$, for $t \ge 0$, for any $u, v \in C_c^1(R^d, H)$,

$$\lim_n Q^n[\,\langle v(\omega_0), \nabla u(\omega_t)C_t^n(\omega)\rangle\,;\, \tau(\omega) > t] = Q[\,\langle v(\omega_0), \nabla u(\omega_t)C_t(\omega)\rangle\,;\, \tau(\omega) > t].$$

Proof. Indeed,

$$\lim_n Q^n[\,\langle v(\omega_0), \nabla u(\omega_t)C_t^n(\omega)\rangle\,;\, \tau(\omega) > t]$$

$$= \lim_n P[\,e^{b_n(\omega_0)}\, N_t^n(\omega)\,\langle v(\omega_0), \nabla u(\omega_t)C_t^n(\omega)\rangle\,;\, \tau(\omega) > t]$$

$$= P[\,e^{b(\omega_0)}\, N_t(\omega)\,\langle v(\omega_0), \nabla u(\omega_t)C_t(\omega)\rangle\,;\, \tau(\omega) > t]$$

$$= Q[\,\langle v(\omega_0), \nabla u(\omega_t)C_t(\omega)\rangle\,;\, \tau(\omega) > t].$$

Lemma 4. $\mathrm{Lim}_n\, Q^n[\tau(\omega) > t] = Q[\tau(\omega) > t]$.

Proof. Indeed

$$\mathrm{Lim}_n\, Q^n[\tau(\omega) > t] = \mathrm{Lim}_n\, P[N_t^n(\omega); \tau(\omega) > t] = P[N_t(\omega); \tau > t] = Q[\tau(\omega) > t].$$ □

Corollary 5. For any $\varepsilon > 0$, for any $t > 0$, there is a $\tau \in \Lambda$ such that

$$\limsup_n Q^n[\tau(\omega) \le t] < \varepsilon.$$

Proof. This is because that, by Lemma §2.3, there is an increasing sequence $(\tau_k) \subset \Lambda$ which tends to infinity P_x-almost surely, for dx-a.s. $x \in R^d$. Since Q_x is locally absolutely continuous with respect to P_x, we have therefore $\lim_k Q[\tau_k \le t] = 0$. Now, the corollary follows from Lemma 4. □

Theorem 6. $\tilde{\nabla} Q_t u(x) = Q_x[\tilde{\nabla} u(\omega_t)C_t(\omega)]$, $u \in W(\mu)$.

Proof. It is enough to prove it for $u \in C_c^1(R^d)$. Let $1 \le i \le d$. Set $\partial_i{}^*v = \partial_i v + v\partial_i b$, $v \in C_c^1(R^d)$. Let $\varepsilon > 0$. Let $\tau \in \Lambda$ such that $\limsup_n Q^n[\tau \le t] < \varepsilon$. We have, for $v \in C_c^1(R^d)$,

$$\int \partial_i{}^*v(x)\, Q_x[u(\omega_t)]\, \mu(dx)$$

$$= \int \partial_i{}^*v(x)\, Q_x[u(\omega_t);\, \tau(\omega) > t]\, \mu(dx) + \int \partial_i{}^*v(x)\, Q_x[u(\omega_t);\, \tau(\omega) \le t]\, \mu(dx)$$

$$= \lim_n \int \partial_i{}^*v(x)\, Q_x^n[u(\omega_t);\, \tau(\omega) > t]\, \mu_n(dx) + O(\varepsilon)$$

$$= \lim_n \int \partial_i{}^*v(x)\, Q_x^n[u(\omega_t)]\, \mu_n(dx) + 2\, O(\varepsilon)$$

$$= -\lim_n \int v(x)\, (Q_x^n[\nabla u(\omega_t)\, C_t^n(\omega)])_i\, \mu_n(dx) + 2\, O(\varepsilon)$$

$$= -\lim_n \int v(x)\, (Q_x^n[\nabla u(\omega_t)\, C_t^n(\omega);\, \tau(\omega) > t])_i\, \mu_n(dx) + 3\, O(\varepsilon)$$

$$= -\int v(x)\, (Q_x[\nabla u(\omega_t)\, C_t(\omega);\, \tau(\omega) > t])_i\, \mu(dx) + 3\, O(\varepsilon)$$

$$= -\int v(x)\, (Q_x[\nabla u(\omega_t)\, C_t(\omega)])_i\, \mu(dx) + 4\, O(\varepsilon).$$

Since ε is arbitrary, we have proved

$$\int \partial_i{}^*v(x)\, Q_x[u(\omega_t)]\, \mu(dx) = -\int v(x)\, (Q_x[\nabla u(\omega_t)\, C_t(\omega)])_i\, \mu(dx).$$

This is equivalent to $\tilde{\nabla} Q_t u(x) = Q_x[\nabla u(\omega_t) C_t(\omega)]$, $u \in C_c^1(R^d)$. □

Now, we can claim that Theorem §1.1 is proved.

§5. Tengent semigroup.

Let us prove Lemma 2 in Introduction.

Lemma 1. The operators (Θ_t) is a semigroup.

Proof. By the relation $C_{t+s} = (C_s \circ \theta_t) C_t$, we have

$$\Theta_{t+s}[F](x) = Q_x[F(\omega_{t+s}) C_{t+s}] = E_x[\, F(\omega_s) \circ \theta_t\, (C_s \circ \theta_t)\, C_t\,]$$

$$= Q_x[\, Q_{X_t}(F(\omega_s) C_s)\, C_t] = \Theta_t \Theta_s [F](x).\ \square$$

Lemma 2. Θ_t is μ-symmetric.

Proof. Notice that the law Q is invariant under the operator ι_T, for any T > 0. By the relation $C_T \circ \iota_T = C_T'$, we obtain that, for any $F, G \in L^2(R^d, H, \mu)$,

$$\int \langle G, \Theta_T(F)\rangle \, d\mu = \int \langle G(x), Q_x[F(\omega_T)\, C_T]\rangle \, \mu(dx)$$

$$= Q[\, \langle G(\omega_0), F(\omega_T) C_T\rangle \,] = Q[\, \langle G(\omega_0 \circ \iota_T), F(\omega_T) \circ \iota_T \; C_T \circ \iota_T \rangle \,]$$

$$= Q[\, \langle G(\omega_T), F(\omega_0) C_T'\rangle \,] = Q[\, \langle G(\omega_T) C_T, F(\omega_0)\rangle \,] = \int \langle \Theta_T(G), F\rangle \, d\mu. \; \square$$

Lemma 3. $\langle \Theta_t F(x), \Theta_t F(x)\rangle^{1/2} \le e^{\alpha t}\, Q_t[\langle F, F\rangle^{1/2}](x)$, with $\alpha = \alpha(Q_\bullet)$.

Proof. Indeed, for any G ∈ H, we have

$$\langle G, \Theta_t F(x)\rangle = \langle G, Q_x[F(\omega_t) C_t]\rangle \le Q_x[\, \langle G, G\rangle^{1/2} \langle F(\omega_t) C_t, F(\omega_t) C_t\rangle^{1/2} \,]$$

$$\le e^{\alpha t}\, Q_x[\, \langle G, G\rangle^{1/2} \langle F(\omega_t), F(\omega_t)\rangle^{1/2} \,] = \langle G, G\rangle^{1/2} e^{\alpha t}\, Q_t[\langle F, F\rangle^{1/2}](x). \; \square$$

Let us now give a description of the generator of the tengent semigroup Θ_t.

Lemma 4. For any bounded continuous $F, G \in L^2(R^d, H, \mu)$, for any $t \ge 0$,

$$\int \langle G(x), Q_x[\int_0^t F(\omega_s)\, B(ds)\, C_s]\rangle \, \mu(dx) = \int_0^t ds \int \langle \Theta_s G(x), F(x)\, \partial^2 b(dx)\rangle.$$

Proof. It is enough to look at the limit state of the expression:

$$\int \langle G(x), Q_x^n[\int_0^t F(\omega_s)\, \partial^2 b_n(\omega_s)\, C_s^n \, ds]\rangle \, \mu_n(dx). \; \square$$

Proposition 5. For any $u \in C_c^3(R^d)$, $G \in L^2(R^d, H, \mu)$ continuous and bounded, we have

$$\frac{\partial}{\partial t} \int \langle G(x), \Theta_t[\nabla u](x)\rangle \, \mu(dx) = \frac{\partial}{\partial t} \int \langle G(x), Q_x[\nabla u(\omega_t) C_t]\rangle \, \mu(dx)$$

$$= \int \langle \Theta_t G(x), \nabla u(x) \partial b(dx)\rangle + \int \langle \Theta_t G(x), L\nabla u(x)\rangle \, \mu(dx),$$

where L denotes the infinitesimal generator of Q_t.

Proof. Apply Ito's formula and use the preceding lemma. □

Part II. Functional inequalities.

In this part we consider a well-admissible probability measure μ on a separable real riggid Banach space B. We consider the classical Dirichlet space $(\mathcal{E}_\mu, W(\mu))$ on $L^2(B,\mu)$ and its associated semigroup P_t (we use P_t instead of Q_t to denote the semigroup, the latter will be used to denote the Cauchy semigroup associated with P_t). We assume that there is a diffusion process X in B whose transition semigroup coincides with P_t.

Under the assumption:

$$P_t \text{ is a C-semigroup,}$$

we shall study three types of functional inequalities: Poincaré's inequality, logarithmical Sobolev inequality, and Stein-Meyer-Bakry inequalities.

§1. Poincaré's inequality.

Let us begin with Poincaré's inequality and logarithmical Sobolev inequality. We shall see that a C-semigroup behaves like the Ornstein-Uhlenbeck semigroups. We refer to Bakry-Emery [2], Davies [6], Rothaus [27], etc., on this subject. Recall the number $\alpha = \alpha(Q_\bullet)$ introduced in Definition 1,iv.

Theorem 1. Suppose $\alpha < 0$. Then, for any $t > 0$,

$$\int u^2(x)\,\mu(dx) - \int (P_t u)^2(x)\,\mu(dx) \le \frac{1}{-\alpha}\,\mathcal{E}_\mu(u, u),\ u \in W(\mu).$$

Proof. Let us denote the points in $C(R_+,B)$ by ω and the law of X started from μ on $C(R_+,B)$ by P. Notice that the following inequality holds:

$$\langle \tilde\nabla u(\omega_t)C_t, \tilde\nabla u(\omega_t)C_t\rangle \le e^{2\alpha t}\langle \tilde\nabla u(\omega_t), \tilde\nabla u(\omega_t)\rangle,\ \forall\, u \in W(\mu),\ \text{P-a.s.}.$$

If we denote by L the infinitesimal generator of P_t on $L^2(B,\mu)$, we have:

$$\int u^2(x)\,\mu(dx) - \int (P_t u)^2(x)\,\mu(dx) = -2\int_0^t \int P_s u(x)\, LP_s u(x)\,\mu(dx)\,ds$$

$$= 2\int_0^t ds \int \langle \tilde{\nabla} P_s u(x), \tilde{\nabla} P_s u(x)\rangle\,\mu(dx)$$

$$= 2\int_0^t ds \int \langle P_x[\tilde{\nabla} u(\omega_s) C_s], P_x[\tilde{\nabla} u(\omega_s) C_s]\rangle\,\mu(dx)$$

$$\leq 2\int_0^t ds \int P_x[\langle \tilde{\nabla} u(\omega_s) C_s, \tilde{\nabla} u(\omega_s) C_s\rangle]\,\mu(dx)$$

$$\leq 2\int_0^t ds \int P_x[e^{2\alpha s}\langle \tilde{\nabla} u(\omega_s), \tilde{\nabla} u(\omega_s)\rangle]\,\mu(dx) \leq \frac{1}{-\alpha}\,\mathcal{E}_\mu(u,u). \quad \square$$

Corollary 2. (Spectral gap) Suppose $\alpha < 0$. Let $(E_\lambda, \lambda \geq 0)$ denote the spectral family of the self-adjoint operator L. Let $f \in E_{]0,-\alpha[}[L^2(B,\mu)]$. Then, $f = 0$.

Proof. By Theorem 1, for such a function f, we have

$$\int_{0+}^{(-\alpha)-} \lambda\, d(E_\lambda f,f) = \mathcal{E}_\mu(f,f) \geq (-\alpha)\left(\int_{0+}^{(-\alpha)-} d(E_\lambda f,f) - \int_{0+}^{(-\alpha)-} e^{-2t\lambda} d(E_\lambda f,f)\right)$$

for any $t \geq 0$. Let t tend to infinity, we obtain

$$\int_{0+}^{(-\alpha)-} \lambda\, d(E_\lambda f,f) \geq (-\alpha) \int_{0+}^{(-\alpha)-} d(E_\lambda f,f).$$

But, this can hold only if $f = 0$. $\square$

§2. Logarithmical Sobolev inequality.

Theorem 1. Suppose $\alpha < 0$. Then, for any $t > 0$, for any function $f \in W(\mu)$ such that $f \geq \varepsilon$ for some constant $\varepsilon > 0$, we have

$$\int f(x) \log f(x)\mu(dx) - \int P_t f(x) \log P_t f(x)\mu(dx) \leq \frac{1}{-\alpha} \int \langle \tilde{\nabla} f(x), \tilde{\nabla} f(x)\rangle \frac{1}{f(x)} \mu(dx).$$

Proof. First of all we remark that P_t is μ-invariant. So, we have

$$\int f(x) \log f(x)\, \mu(dx) - \int P_t f(x) \log P_t f(x)\, \mu(dx)$$

$$= -\int_0^t ds \left[\int LP_s f(x) \log P_s f(x)\, \mu(dx) + \int LP_s f(x)\, \mu(dx) \right]$$

$$= -\int_0^t ds \int LP_s f(x) \log P_s f(x)\, \mu(dx) = \int_0^t ds \int \langle \tilde{\nabla} P_s f(x), \tilde{\nabla} P_s f(x)\rangle \frac{1}{P_s f(x)} \mu(dx).$$

We can overestimate:

$$\langle \tilde{\nabla} P_s f(x), \tilde{\nabla} P_s f(x)\rangle^{1/2} = \langle P_x[\tilde{\nabla} f(\omega_s) C_s], P_x[\tilde{\nabla} f(\omega_s) C_s]\rangle^{1/2}$$

$$\leq P_x[\langle \tilde{\nabla} f(\omega_s) C_s, \tilde{\nabla} f(\omega_s) C_s\rangle^{1/2}] \leq e^{\alpha s} P_x[\langle \tilde{\nabla} f(\omega_s), \tilde{\nabla} f(\omega_s)\rangle^{1/2}]$$

$$\leq e^{\alpha s} P_x[\langle \tilde{\nabla} f(\omega_s), \tilde{\nabla} f(\omega_s)\rangle \frac{1}{f(\omega_s)}]^{1/2} P_x[f(\omega_s)]^{1/2}.$$

So, we obtain

$$\int f(x) \log f(x)\, \mu(dx) - \int P_t f(x) \log P_t f(x)\, \mu(dx)$$

$$\leq \int_0^t ds \int e^{\alpha s} P_x[\langle \tilde{\nabla} f(\omega_s), \tilde{\nabla} f(\omega_s)\rangle \frac{1}{f(\omega_s)}] P_x[f(\omega_s)] \frac{1}{P_s f(x)} \mu(dx)$$

$$= \int_0^t ds \int e^{\alpha s} P_x[\langle \tilde{\nabla} f(\omega_s), \tilde{\nabla} f(\omega_s)\rangle \frac{1}{f(\omega_s)}]\, \mu(dx)$$

$$\leq \frac{1}{-\alpha} \int \langle \tilde{\nabla} f(x), \tilde{\nabla} f(x) \rangle \frac{1}{f(x)} \, \mu(dx). \quad \square$$

§3. Stein-Meyer-Bakry inequalities.

Let $\lambda \geq 0$. We define the Cauchy semigroup associated with $e^{-\lambda t} P_t$:

$$Q_t^{\lambda} = \int_0^{\infty} m(t,s)\, e^{-\lambda s} P_s \, ds, \text{ where } m(t,s) = \frac{t}{2\sqrt{\pi}} s^{-3/2} \exp\{-\frac{t^2}{4s}\}.$$

Let C^{λ} be the infinitesimal generator of Q_t^{λ}. The domain of C^{λ} is denoted by $D(C^{\lambda})$.

When $\lambda = 0$, we denote $C^0 = C$. Let $\tilde{P}_t$ be the tangent semigroup associated with P_t. The Cauchy semigroup associated with $\tilde{P}_t$ and its generator are denoted by respectively $\tilde{Q}_t^{\lambda}$ and $\tilde{C}^{\lambda}$.

The Stein-Meyer-Bakry inequalities state the mutual overestimates between the operators $\tilde{\nabla}$ and C^{λ} for a suitable $\lambda \geq 0$. There are already many studies on Stein-Meyer-Bakry inequalities. We can refer to Stein [32], Meyer [21], [22], [23], [24], Bakry [3], [4], Feyel [8], Pisier [25], Wu [37], Gundy [12], Varopoulos [35], Lohoué [18], Dellacherie-Maisonneuve-Meyer [7], Gundy-Silverstein [10], Gundy-Varopoulos [11], Banuelos [5], Strichartz [33], etc. These studies concern various type of semigroups on different spaces. The motivation for us to write again about Stein-Meyer-Bakry inequality comes from the desire of understanding the papers of Bakry [3] and [4], in the former of which, Bakry had already used implicitely the C-semigroup notion (see also Dellacherie-Maisonneuve-Meyer [7]). We have remarked that, for C-semigroups, explicit formulas exist relating the gradient $\tilde{\nabla}$ with the Cauchy operator C^{λ}. Using these formulas, we shall prove hereafter that C-semigroups satisfy Stein-Meyer-Bakry inequalities.

When $f(x,t)$ is a function of two variables (x,t), the operators $\tilde{\nabla}$, P_s, Q_s^{λ}, etc, will operate on the variable x. The resulted functions will be denoted by $\tilde{\nabla} f(x,t)$, $P_s f(x,t)$, $Q_s^{\lambda} f(x,t)$, etc. We shall use also the operator D which is the differential with respect to the variable t of the function $f(x,t)$. When $f(x)$ is a function in $L^2(B,\mu)$ (resp. in $L^2(B,H,\mu)$), we shall write $f(x,t)$ for the function $(x,t) \to Q_t^{\lambda} f(x)$ (resp. $(x,t) \to \tilde{Q}_t^{\lambda} f(x)$) defined on $B \times R_+$ (we shall not write the parameter λ), and we shall call it the <u>extension</u> of f onto $B \times R_+$. For p > 1, we denote the various L^p-norms below by $\|\bullet\|_p$ as well as by N_p.

In what follows, we shall calculate many times integrals and derivatives. We shall not each time prove that they are meaningful, because the technique to do so is routine.

For any $f \in W(\mu)$, we denote $\partial_i f = \langle \tilde{\nabla} f, k_i \rangle$, $i \geq 1$, where (k_i) is the basis of H introduced in Introduction.

As in Meyer [21], we introduce the process (X_t, B_t), where X_t is the B-valued μ-symmetric diffusion associated with P_t, and B_t is a Brownian motion started from a point $a > 0$ such that $\langle B_t, B_t \rangle = 2t$. We set $\tau = \inf\{t; B_t = 0\}$. We denote the law of $(X_t, B_t)_{t \geq 0}$ by E^a.

The corner stone of Stein-Meyer-Bakry inequalities are martingale inequalities. We shall use constantly three of them.

Lemma 1. (Doob's inequality) For any $p > 1$, for any non negative submartingale S,

$$N_p[\sup_{s \geq 0} S_s^p] \leq \frac{p}{p-1} N_p[S_\infty^p].$$

Lemma 2. (vector Burkholder-Davis-Gundy inequality) For any $p > 1$, there is constants c_p and C_p such that, for any sequence of continuous real martingales $(M_i(t); i \geq 1)$,

$$c_p N_p[(\sum_{i \geq 1} M_i^2(\infty))^{1/2}] \leq N_p[(\sum_{i \geq 1} \langle M_i \rangle(\infty))^{1/2}] \leq C_p N_p[(\sum_{i \geq 1} M_i^2(\infty))^{1/2}],$$

here obviously $\langle M_i \rangle$ denotes the increasing process associated with M_i.

Lemma 3. (Lenglart-Lépingle-Pratelli [17], Théorème 3.2) For any $p > 0$, there is a constant c_p such that, for any continuous submartingale $Z = M + A$, where M is a continuous local martingale and A is a continuous increasing process started from zero,

$$E[A_\infty^p] \leq c_p E[\sup_{s \geq 0} |Z_s|^p].$$

There are numerous forms of Stein-Meyer-Bakry inequalities. Let us begin with a general result which holds for any classical Dirichlet form.

Theorem 4. (Without C-semigroup assumption) Let $p > 1$. There are constants c_p such that, for any $f \in W(\mu)$, for any $a > 0$, we have

$$N_p[\langle \tilde{\nabla} f, \tilde{\nabla} f \rangle^{1/2}] \leq c_p N_p[\langle Q_a \tilde{\nabla} f, Q_a \tilde{\nabla} f \rangle^{1/2}] + c_p N_p[Cf],$$

$$N_p[Cf] \leq c_p N_p[Q_a Cf] + c_p N_p[\langle \tilde{\nabla} f, \tilde{\nabla} f \rangle^{1/2}].$$

Proof. Recall the following two inequalities: there are constants C such that, for any vector valued function $h = (h_i)$, with $h_i \in L^p(B,\mu)$, for any $a > 0$, we have firstly

$$N_p[(\int_0^\tau \sum_i (Dh_i(X_s,B_s))^2\, ds)^{1/2}] \leq C\, N_p[(\sum_i h_i^2)^{1/2}]$$

$$\leq C\, N_p[(\sum_i (Q_a h_i)^2)^{1/2}] + C\, N_p[(\int_0^\tau \sum_i Dh_i(X_s,B_s)^2\, ds)^{1/2}];$$

and secondly

$$N_p[(\int_0^\tau \sum_i \langle\tilde{\nabla} h_i, \tilde{\nabla} h_i\rangle(X_s,B_s)\, ds)^{1/2}] \leq C\, N_p[(\sum_i h_i^2)^{1/2}]$$

$$\leq C\, N_p[(\sum_i (Q_a h_i)^2)^{1/2}] + C\, N_p[(\int_0^\tau \sum_i \langle\tilde{\nabla} h_i, \tilde{\nabla} h_i\rangle(X_s,B_s)\, ds)^{1/2}].$$

These inequalities have been proved in Bakry [4], when h is scalar valued. But, it is easy to generalize them to the case of vector valued functions, thanks to the corresponding Burkholder-Davis-Gundy inequality.

Based on that remark, the truth of the theorem results immediately from the identity $\tilde{\nabla} Cf(x,t) = D\tilde{\nabla} f(x,t)$, valid for any $f \in D(L)$, the domain of the generator L. For example, to prove the left side inequality of the theorem, applying the first inequality to $\tilde{\nabla} f = (\partial_i f, i \geq 1)$, we write

$$N_p[\langle\tilde{\nabla} f, \tilde{\nabla} f\rangle^{1/2}]$$

$$\leq C\, N_p[\langle Q_a \tilde{\nabla} f, Q_a \tilde{\nabla} f\rangle^{1/2}] + C\, N_p[(\int_0^\tau \langle D\tilde{\nabla} f, D\tilde{\nabla} f\rangle(X_s,B_s)\, ds)^{1/2}]$$

$$\leq C\, N_p[\langle Q_a \tilde{\nabla} f, Q_a \tilde{\nabla} f\rangle^{1/2}] + C\, N_p[(\int_0^\tau \langle\tilde{\nabla} Cf, \tilde{\nabla} Cf\rangle(X_s,B_s)\, ds)^{1/2}]$$

$$\leq C\, N_p[\langle Q_a \tilde{\nabla} f, Q_a \tilde{\nabla} f\rangle^{1/2}] + C\, N_p[Cf],$$

where the last step is obtained by applying the second inequalities to the function Cf. □

Remark. The simplicity of the proof of the inequalities is due to the simple form of the "opérateur carré du champ" associated with the classical Dirichlet form.

Remark. If $\alpha < 0$, there is a gap in the spectrum of the generator L (cf. Corollary §1.2). We can hence take limits in the inequalities of Theorem 4, when a tends to infinity. Then, under some boundedness condition and ergodicity, the term $N_p[Q_a Cf]$ tends to zero, while the term $N_p[\langle Q_a \tilde{\nabla} f, Q_a \tilde{\nabla} f\rangle^{1/2}]$ tends to zero if $\mu(B) = \infty$, tends to $\langle E\nabla f, E\nabla f\rangle$ if $\mu(B) < \infty$. In particular cases such as the Ornstein-Ulenbeck semigroup, $\langle E\nabla f, E\nabla f\rangle$ can be easily controled by $\|Cf\|^2$. This provides a proof of Meyer inequality (cf. Meyer [24]).

How to cancel the terms $N_p[Q_a Cf]$ and $N_p[\langle Q_a \tilde{\nabla} f, Q_a \tilde{\nabla} f\rangle^{1/2}]$ from the right hand sides of the inequalities in Theorem 4, if α is not necessarily negative? To answer the question, the following formula, in which we recognize the intervention of the C-process, gives a good starting point.

Lemma 5. Let $\lambda > \alpha \vee 0$. Then, we have the formula:

$$\frac{1}{2}\tilde{\nabla} f(x) = E^{\infty}\Big[\int_0^{\tau} e^{\lambda s}\, \tilde{\nabla} C^{\lambda} f(X_s, B_s)\, (C_s')^{-1}\, dB_s\, C_\tau'\, e^{-\lambda\tau} \,\Big|\, X_\tau = x\Big],\ f \in D(L),$$

where $E^{\infty} = \lim_{a\uparrow\infty} E^a$.

Proof. Let g be a function in $L^2(B,\mu)$. We consider its extension g onto $B\times R_+$. Recall (cf. Bakry [3]) that, because $\lambda > 0$, the L^p-norm of $Q_t^{\lambda} g$ decreases exponentially to zero, when t tends to infinity. This will justify the convergences of various integrals which will be coming.

The formula of the lemma is the differentiable form of the following one, which is well known when $\lambda = 0$:

$$E^a\Big[\int_0^{\tau} e^{\lambda s}\, g(X_s, B_s)\, dB_s\, e^{-\lambda\tau} \,\Big|\, X_\tau = x\Big]$$

$$= E^a\Big[\int_0^{\tau} e^{\lambda s}\, g(X_s, B_s)\, 2\,\partial \log m(B_s, \tau - s)\, ds\, e^{-\lambda\tau} \,\Big|\, X_\tau = x\Big]$$

obtained by enlarging the filtration $\sigma(B)_t$ by the variable τ (see Jeulin [16]), where ∂ denotes the derivative with respect to the variable whose place is occupied by B_s;

$$= E^a[\int_0^\tau e^{-\lambda(\tau-s)} P_{\tau-s}g(x,B_s)\, 2\, \partial \log m(B_s, \tau - s)\, ds\,]$$

by the symmetry of the process X;

$$= E[\int_0^{L_a} e^{-\lambda(L_a-s)} P_{L_a-s}g(x,\sqrt{2}Z_{L_a-s})\, 2\, \partial \log m(\sqrt{2}Z_{L_a-s}, L_a - s)\, ds\,]$$

by the "retournement du temps" (cf. Revuz-Yor [26]), where Z is a 3-dimensional Bessel process started from zero and $L_a = \sup\{t;\, Z_t = a\}$;

$$= E[\int_0^{L_a} e^{-\lambda u} P_u g(x,\sqrt{2}Z_u)\, 2\, \partial \log m(\sqrt{2}Z_u, u)\, du\,]$$

$$\to \int_0^\infty du \int_0^\infty dy\; y\; 2\, m(\sqrt{2}y, u)\, e^{-\lambda u} P_u g(x,\sqrt{2}y)\, 2\, \partial \log m(\sqrt{2}y, u)$$

when a tends to the infinity; (for the potential of Z_t, see Revuz-Yor [26])

$$= 4\int_0^\infty y\, dy \int_0^\infty du\, e^{-\lambda u} P_u g(x,\sqrt{2}y)\, \partial\, m(\sqrt{2}y, u)$$

$$= 4\int_0^\infty y\, dy \int_0^\infty du\, e^{-\lambda u}\, \partial(P_u g(x,\bullet)\, m(\bullet, u)\,)_{\bullet=\sqrt{2}y}$$

$$- 4\int_0^\infty y\, dy \int_0^\infty du\, e^{-\lambda u}\, \partial(P_u g(x,\bullet)\,)_{\bullet=\sqrt{2}y}\, m(\sqrt{2}y, u)$$

$$= 4\int_0^\infty y\, dy\, \partial(Q^\lambda_{\bullet+\bullet} g(x)\,)_{\bullet=\sqrt{2}y} - 4\int_0^\infty y\, dy \int_0^\infty du\, e^{-\lambda u}\, P_u C^\lambda g(x,\sqrt{2}y)\, m(\sqrt{2}y, u)$$

$$= 4 \int_0^\infty y\, dy\, 2\, C^\lambda Q^\lambda_{2\sqrt{2y}} g(x) - 4 \int_0^\infty y\, dy\, Q^\lambda_{2\sqrt{2y}} C^\lambda g(x)$$

$$= 4 \int_0^\infty y\, dy\, Q^\lambda_{2\sqrt{2y}} C^\lambda g(x) = \frac{1}{2} (C^\lambda)^{-1} g(x).$$

Replacing g by $C^\lambda f$, we obtain:

$$\frac{1}{2} f(x) = E^\infty[\int_0^\tau e^{\lambda s}\, C^\lambda f(X_s, B_s)\, dB_s\, e^{-\lambda\tau} \,|\, X_\tau = x],\ f \in D(C^\lambda).$$

Now, to prove the lemma, it is enough to take the gradient $\tilde{\nabla}$ on both sides of this formula when $f \in D(L)$. On the left hand side, we obtain $\frac{1}{2}\tilde{\nabla} f$. To compute the right hand side, we first employ the above technique of enlargement of filtration, next, we use the C-semigroup property, then, we invert the time, finally, we obtain the formula of the lemma. □

Before studying the consequence of the formula in Lemma 5 on a general C-semigroup, let us first try it with the Ornstein-Uhlenbeck process on B (so X is now O-U). In such case, $C_s = e^{-s}$ and $\alpha = -1$. Taking the limit in the formula when λ decreases to zero. We obtain:

$$\frac{1}{2} \tilde{\nabla} f(x) = E^\infty[\int_0^\tau \tilde{\nabla} C f(X_s, B_s)\, e^s\, dB_s\, e^{-\tau} \,|\, X_\tau = x].$$

(In fact, this formula can be proved directly and very easily. Chronologically, this formula was the germ of that in Lemma 5.) Set $N_t = \int_0^t \tilde{\nabla} C f(X_s, B_s)\, dB_s$. By integration by parts, we can write:

$$\frac{1}{2} \tilde{\nabla} f(x) = E^\infty[N_\tau - \int_0^\tau N_s\, e^{s-\tau}\, ds \,|\, X_\tau = x].$$

It yields immediately:

$$\|\tilde{\nabla} f\|_p \le 2\|N_\tau\|_p + 2\|\sup_{s\le\tau}|N_s|\,\|_p.$$

To finish the estimate, we first apply the Burkholder-Davis-Gundy inequality to the vector martingale N, then apply the inequalities mentioned in the beginning of the proof of Theorem 4. We conclude $\|\tilde{\nabla} f\|_p \le c_p \|Cf\|_p$. By duality, we conclude also the inverse inequality: $\|Cf\|_p \le c_p \|\tilde{\nabla} f\|_p$. The formula given by Lemma 5 provides thus a second proof of Meyer's inequality.

Give up the Ornstein-Uhlenbeck semigroup and consider again our C-semigroup P_t. We notice that the above technique remains applicable, if the C-process has a bounded log-derivative and $\alpha < 0$. We can therefore claim our second form of Stein-Meyer-Bakry inequalities:

Theorem 6. Assume that the C-process has a bounded log-derivative γ and $\alpha < 0$. Then, the following inequality holds:

$$\|\tilde{\nabla} f\|_p \le \Big(2\,c_p + 2\,\frac{1}{-\alpha}\,c_p\,\|\gamma\|_\infty\Big)\|Cf\|_p,\quad \forall f \in W(\mu),$$

where c_p is a martingale inequality constant. An inverse inequality also holds by duality.

Nevertheless, Theorem 6 is not the optimal form of the inequalities, while all power of martingale inequalities has not been exhausted yet. Let us start up off again with the following representation of the norm $\|\tilde{\nabla} f\|_p$:

$$\|\tilde{\nabla} f\|_p = 2\,\sup_h E^\infty\Big[\Big\langle \int_0^\tau e^{\lambda s}\,\tilde{\nabla} C^\lambda f(X_s,B_s)\,(C_s')^{-1}\,dB_s\,C_\tau'\,e^{-\lambda\tau},\ h(X_\tau)\Big\rangle\Big],$$

where the supremum is taken over the family of h in $L^q(R^d,H,\mu)$ such that $\|h\|_q = 1$, where q is the conjugate number of p. Let us set $h(x,y) = \tilde{Q}^\lambda_y h(x)$ for such a function h. The function h is related with the martingale: for $t < \tau$,

$$M_t = E[e^{-\lambda\tau}\,h(X_\tau)C_\tau \,|\, F_t] = e^{-\lambda t}\,\tilde{Q}^\lambda_{B_t} h(X_t)\,C_t = e^{-\lambda t}\,h(X_t,B_t)\,C_t.$$

The martingale M_t, for $t < \tau$, has another expression:

$$M_t = \int_0^t e^{-\lambda s}\,\tilde{C}^\lambda\,h(X_s,B_s)\,C_s\,dB_s + \int_0^t e^{-\lambda s}\sum_i \partial_i h(X_s,B_s)\,C_s\,d\sqrt{2}\beta^i_t,$$

where β^i are independent brownian motions. Using the function h(x,t), we have the following estimation:

$$E^\infty[\langle \int_0^\tau e^{\lambda s}\, \tilde\nabla C^\lambda f(X_s,B_s)\,(C_s')^{-1}\, dB_s\, C_\tau'\, e^{-\lambda\tau}, h(X_\tau)\rangle]$$

$$= E^\infty[\langle \int_0^\tau e^{\lambda s}\, \tilde\nabla C^\lambda f(X_s,B_s)\,(C_s')^{-1}\, dB_s, e^{-\lambda\tau}\, h(X_\tau) C_\tau\rangle]$$

$$= E^\infty[\int_0^\tau \langle e^{\lambda s}\, \tilde\nabla C^\lambda f(X_s,B_s)\,(C_s')^{-1}, e^{-\lambda s}\, \tilde C^\lambda\, h(X_s,B_s)\, C_s\rangle\, 2\, ds]$$

$$= 2\, E^\infty[\int_0^\tau \langle \tilde\nabla C^\lambda f(X_s,B_s), \tilde C^\lambda h(X_s,B_s)\rangle\, ds]$$

$$\le 2\, \|(\int_0^\tau \langle \tilde\nabla C^\lambda f, \tilde\nabla C^\lambda f\rangle(X_s,B_s)\, ds)^{1/2}\|_p\; \|(\int_0^\tau \langle \tilde C^\lambda h, \tilde C^\lambda h\rangle(X_s,B_s)\, ds)^{1/2}\|_q .$$

Let us estimate separately the above two norms. We need the following notation. Let λ be a real number. We shall write $\lambda dt \ge - C_t\, dC_t^{-1}$, if, for any H-valued process g_t, we have

$$\int_0^\infty (\lambda \langle g_t, g_t\rangle\, dt + \langle g_t, g_t C_t\, dC_t^{-1}\rangle) \ge 0.$$

Lemma 7. If $\lambda dt \ge - C_t\, dC_t^{-1}$, we have $\|C_t\|^2 \le e^{2\lambda t}$.

Proof. Let $\varepsilon > 0$. Set $f(t) = (\langle gC_t^{-1}, gC_t^{-1}\rangle + \varepsilon)^{-1}$. Set $g_t = f(t)\, g\, C_t^{-1} 1_{[0,a]}(t)$ in the above definition, where $a > 0$, $g \in H$. We have

$$-2\int_0^a \lambda\, f(t)\, \langle gC_t^{-1}, gC_t^{-1}\rangle\, dt \le 2\int_0^a f(t)\, \langle gC_t^{-1}, g dC_t^{-1}\rangle$$

$= \log[\langle gC_a^{-1}, gC_a^{-1}\rangle + \varepsilon] - \log[\langle g,g\rangle + \varepsilon]$.

Taking the limit in the above inequality when ε tends to zero, and replacing g by hC_a, we obatin: $-2\lambda a \le \log\langle h,h\rangle - \log\langle hC_a, hC_a\rangle$. This proves the lemma. □

Remark. This lemma provides another proof of Lemma §3.6, Part I.

Lemma 8. Set $h_t = h(X_t,B_t)$. There is constant c_q such that, for any $\lambda > 0$ such that $\lambda dt \ge - C_t\, dC_t^{-1}$, we have:

$$\|(\int_0^\tau \langle \tilde{C}^\lambda h_s, \tilde{C}^\lambda h_s\rangle\, ds)^{1/2}\|_q \le c_q\, \|\langle h(X_\tau),h(X_\tau)\rangle^{1/2}\|_q \le c_q\, \|h\|_q.$$

Proof. Indeed, by Itô's formula, we have:

$$\begin{aligned} d\langle h_t,h_t\rangle = {} & 2\,\langle h_t, \tilde{C}^\lambda h_t\rangle\, dB_t + 2\sum_i \langle h_t, \partial_i h_t\rangle\, d\sqrt{2}\beta_t^i \\ & + 2\,\lambda\,\langle h_t, h_t\rangle\, dt + 2\,\langle h_t, h_t C_t\, dC_t^{-1}\rangle \\ & + 2\sum_j \sum_i (\partial_i h_t^j)^2\, dt + 2\,\langle Dh_t, Dh_t\rangle dt. \end{aligned}$$

Remark $Dh_t = \tilde{C}^\lambda h(X_t,B_t)$. Since $\lambda dt \ge - C_t\, dC_t^{-1}$, this formula implies that $\langle h_t,h_t\rangle$ is a submartingale. Applying Lemma 3 to the couple $\langle h_t,h_t\rangle$ and $\int \langle Dh_t, Dh_t\rangle dt$, we obtain:

$$\|(\int_0^\tau \langle \tilde{C}^\lambda h_s, \tilde{C}^\lambda h_s\rangle\, ds)^{1/2}\|_q \le C\,\|\sup_{s\le\tau}\langle h_s,h_s\rangle^{1/2}\|_q.$$

In order to replace $\sup_{s\le\tau}\langle h_s,h_s\rangle^{1/2}$ by $\langle h_\tau,h_\tau\rangle^{1/2}$, we shall prove that $\langle h_t,h_t\rangle^{1/2}$ also is a submartingale. Look at the bounded variation part of the semi-martingale $\langle h_t,h_t\rangle^{1/2}$. It is given by:

$$\begin{aligned} & \tfrac{1}{2}\langle h_t,h_t\rangle^{-1/2}\,(2\,\lambda\,\langle h_t, h_t\rangle\, dt + 2\,\langle h_t, h_t C_t\, dC_t^{-1}\rangle) \\ & \quad + \tfrac{1}{2}\langle h_t,h_t\rangle^{-1/2}\, 2\sum_j \sum_i (\partial_i h_t^j)^2\, dt + \tfrac{1}{2}\langle h_t,h_t\rangle^{-1/2}\, 2\,\langle Dh_t, Dh_t\rangle dt \\ & \quad\quad - \tfrac{1}{8}\langle h_t,h_t\rangle^{-3/2}\, 8\,[\langle h_t, \tilde{C}^\lambda h_t\rangle^2\, dt + \sum_i \langle h_t, \partial_i h_t\rangle^2\, dt]. \end{aligned}$$

The Schwarz inequality yields that this is bigger than

$$\langle h_t,h_t\rangle^{-1/2} (\lambda \langle h_t, h_t\rangle dt + 2 \langle h_t, h_t C_t\, dC_t^{-1}\rangle)$$

which is non negative. So, $\langle h_t,h_t\rangle^{1/2}$ is a submartingale. Now, by Doob's inequality, we can write:

$$\|\sup_{s\le\tau}\langle h_s,h_s\rangle^{1/2}\|_q \le C\|\langle h_\tau,h_\tau\rangle^{1/2}\|_q,$$

which is what we wanted. □

Lemma 9. There is a constant c_p such that:

$$\|(\int_0^\tau \langle\tilde\nabla C^\lambda f,\tilde\nabla C^\lambda f\rangle(X_s,B_s)\, ds)^{1/2}\|_p \le c_p \|C^\lambda f\|_p, \quad f\in D(L).$$

Proof. Notice that $C^\lambda f(X_t,B_t) = e^{\lambda t} E^a[C^\lambda f(X_\tau) e^{-\lambda\tau} | X_t,B_t]$, $t<\tau$. We can now make the same arguments as that used in the proof of Lemma 6, to the submartingale $|C^\lambda f(X_t,B_t)|$. □

Corollary 10. Set $\alpha' = \inf\{\lambda > 0;\ \lambda dt \ge -C_t\, dC_t^{-1}\}$. Then, there is constant c_p such that, for any $f\in W(\mu)$, $\|\tilde\nabla f\|_p \le c_p \|C^{\alpha'} f\|_p$.

Proof. We need only to consider $f\in D(L)$. Let $\lambda > 0$ such that $\lambda dt \ge -C_t\, dC_t^{-1}$. According to Lemma 7, the formula in Lemma 5 is valid. We can therefore write:

$$\|\tilde\nabla f\|_p = 2\sup_h E^\infty[(\int_0^\tau e^{\lambda s}\, \tilde\nabla C^\lambda f(X_s,B_s)\, (C_s')^{-1}\, dB_s\, C_\tau' e^{-\lambda\tau}, h(X_\tau))]$$

$$\le 4\sup_h \|(\int_0^\tau \langle\tilde\nabla C^\lambda f,\tilde\nabla C^\lambda f\rangle(X_s,B_s)\, ds)^{1/2}\|_p\, \|(\int_0^\tau \langle\tilde C^\lambda h_s, \tilde C^\lambda h_s\rangle ds)^{1/2}\|_q$$

$$\le C\|C^\lambda f\|_p.$$

But the constant in the above inequality is the same for all $\lambda > \alpha'$. The lemma is proved by taking the limit when λ decreases to α'. □

Theorem 11. There is constants $c_p > 0$ such that, for any $f \in W(\mu)$, we have

$$\frac{1}{c_p} \|\tilde{\nabla} f\|_p \le \|C^{\alpha'} f\|_p \le (c_p \|\tilde{\nabla} f\|_p + \alpha' \|f\|_p).$$

Proof. The left hand side inequality is proved in the preceding corollary. The right hand side inequality can be proved by duality. □

Remark. We can substitute $\|Cf\|_p + \|f\|_p$ for $\|C^{\alpha'} f\|_p$ (see Bakry [3]).

References.

[1] Albeverio (S.), Röckner (M.): Classical Dirichlet forms on topological vector spaces – closability and a Caméron-Martin formula, J.F.A. 88, p.395-436, 1990.

[2] Bakry (D.), Emery (M.): Diffusions hypercontractives, Sém. Prob. XIX, LN 1123, p.179-206, Springer, 1985.

[3] Bakry (D.): Etude probabiliste des transformation de Riesz dans les variétés à courbure de Ricci minorée, Sém. Prob. XXI, LN 1247, p.137-171, Springer, 1987.

[4] Bakry (D.): Transformations de Riesz pour les semi-groupes symétriques, Sém. Prob. XIX, LN 1123, p.131-174, Springer, 1985.

[5] Banuelos (R.): Martingale transforms and related singular integrals, Trans. Amer. Math. Soc. Vol.293, N°2, p.547-563, 1986.

[6] Davies (E.): Heat kernels and spectral theory, Cambridge University Press, 1989.

[7] Dellacherie (C.), Maisonneuve (B.), Meyer (P.A.): Probabilités et Potentiel, Ch.XVII à XXIV, Hermann, 1992.

[8] Feyel (D.), de La Pradelle (A.): Capacités gaussiennes, Ann. Inst. Fourier, Grenoble 41, 1, p.49-76, 1991.

[9] Fukushima (M.): Dirichlet forms and Markov processes, North-Holland/Kodansha, 1980.

[10] Gundy (R.F.), Silverstein (M.L.): On a probabilistic interpretation for the Riesz transforms, Functional Analysis in Markov Processes, LN 923, p.199-203, 1982.

[11] Gundy (R.F.), Varopoulos (N.T.): Les transformations de Riesz et les intégrales stochastiques, CRAS 289, p.13-16, 1979.

[12] Gundy (R.F.): Sur les transformation de Riesz pour le semi-groupe d'Ornstein-Uhlenbeck, CRAS 303, 1986.

[13] Hirsch (F.), Song (S.): Inequalities for Bochner's subordinates of two-parameter symmetric Markov processes, To appear in Ann. I.H.P..

[14] Hirsch (F.), Song (S.): Markov properties of multiparameter processes and capacities, To appear in Probab. Th. Rel. Fields.

[15] Ikeda (N.), Watanabe (S.): Stochastic Differential Equations and diffusion processes, North-Holland/Kodansha, 1981.

[16] Jeulin (T.): Semimartingales et grossissements de filtrations, LN 833, Springer, 1980.

[17] Lenglart (E.), Lépingle (D.), Pratelli (M.): Présentation unifiée de certaines inégalités de la théorie des martingales, Sém. Prob. XIV, LN 784, p.26-48, Springer, 1980.

[18] Lohoué (N.): Comparaison des champs de vecteurs et des puissances du Laplacien sur une variété riemannienne à courbure non positive, J.F.A. 61, p.164-201, 1985.

[19] Ma (Z.), Röckner (M.): Introduction to the theory of (non symmetric) Dirichlet forms, Berlin - Heidelberg, Springer - Verlag, 1992.

[20] Mao (X.): Exponential stability of stochastic differential equations, Pure and Applied Mathematics, Marcel Dekker, 1994.

[21] Meyer (P.A.): Démonstration probabiliste de certaines inégalités de Littlewood-Paley, Sém. Prob. X, LN 511, p.125-183, Springer 1976.

[22] Meyer (P.A.): Note sur les processes d'Ornstein-Uhlenbeck, Sém. Prob. XVI, LN 920, p.95-132, Springer, 1982.

[23] Meyer (P.A.): Retour sur la théorie de Littlewood-Paley, Sém. Prob. XV, LN 850, p.151-166, Springer, 1981.

[24] Meyer (P.A.): Transformations de Riesz pour les lois gaussiennes, Sém. Prob. XVIII, LN 1059, p.179-193, 1983.

[25] Pisier (G.): Riesz transforms: a simpler analytic proof of P.A. Meyer's inequality, Sém. Prob. XXII, LN 1321, p.485-501, Springer, 1988.

[26] Revuz (D.), Yor (M.): Continuous martingales and Brownian motion, Springer-Verlag, 1991.

[27] Rothaus (O.S.): Diffusion on compact Riemannian manifolds and logarithmic Sobolev inequalities, J.F.A. 42, p.102-120, 1981.

[28] Song (S.): A study on Markovian maximality, change of probability and regularity, Potential Analysis, 3(4): 391-422, 1994.

[29] Song (S.): Admissible vectors and their associated Dirichlet forms, Potential Analysis 1, p.319-336, 1992.

[30] Song (S.): C-semigroups and Markovian uniqueness, preprint, 1994.

[31] Song (S.): Construction d'un processus à deux paramètres à partir d'un semi-group à un paramètre, in Classical and Modern Potential Theory and Applications, Gowrisankaran and al. eds., Kluwer Acad. Publ., p.419-451, 1994.

[32] Stein (E.M.): Topics in harmonic analysis related to the Littlewood-Paley theory, Princeton University Press, 1970.

[33] Strichartz (R.): Analysis of the Laplacian on the complete Riemannian manifold, J.F.A. 52, p.48-79, 1983.

[34] Takeda (M.): The maximum Markovian self-adjoint extensions of generalized Schrödinger operators, J. Math. Soc. Japan, 44, p.113-130, 1992.

[35] Varopoulos (N.T.): Aspects of probabilistic Littlewood-Paley theory, J.F.A. 38, p.25-60, 1980.

[36] Wu (L.M.): Construction de l'opérateur de Malliavin sur l'espace de Poisson, Sém. Prob. XXI, LN 1247, p.100-113, Springer, 1987.

[37] Wu (L.M.): Inégalité de Sobolev sur l'espace de Poisson, Sém. Prob. XXI, LN 1247, p.114-136, Springer, 1987.

Lecture Notes in Mathematics

For information about Vols. 1–1431
please contact your bookseller or Springer-Verlag

Vol. 1432: K. Ambos-Spies, G.H. Müller, G.E. Sacks (Eds.), Recursion Theory Week. Proceedings, 1989. VI, 393 pages. 1990.

Vol. 1433: S. Lang, W. Cherry, Topics in Nevanlinna Theory. II, 174 pages.1990.

Vol. 1434: K. Nagasaka, E. Fouvry (Eds.), Analytic Number Theory. Proceedings, 1988. VI, 218 pages. 1990.

Vol. 1435: St. Ruscheweyh, E.B. Saff, L.C. Salinas, R.S. Varga (Eds.), Computational Methods and Function Theory. Proceedings, 1989. VI, 211 pages. 1990.

Vol. 1436: S. Xambó-Descamps (Ed.), Enumerative Geometry. Proceedings, 1987. V, 303 pages. 1990.

Vol. 1437: H. Inassaridze (Ed.), K-theory and Homological Algebra. Seminar, 1987–88. V, 313 pages. 1990.

Vol. 1438: P.G. Lemarié (Ed.) Les Ondelettes en 1989. Seminar. IV, 212 pages. 1990.

Vol. 1439: E. Bujalance, J.J. Etayo, J.M. Gamboa, G. Gromadzki. Automorphism Groups of Compact Bordered Klein Surfaces: A Combinatorial Approach. XIII, 201 pages. 1990.

Vol. 1440: P. Latiolais (Ed.), Topology and Combinatorial Groups Theory. Seminar, 1985–1988. VI, 207 pages. 1990.

Vol. 1441: M. Coornaert, T. Delzant, A. Papadopoulos. Géométrie et théorie des groupes. X, 165 pages. 1990.

Vol. 1442: L. Accardi, M. von Waldenfels (Eds.), Quantum Probability and Applications V. Proceedings, 1988. VI, 413 pages. 1990.

Vol. 1443: K.H. Dovermann, R. Schultz, Equivariant Surgery Theories and Their Periodicity Properties. VI, 227 pages. 1990.

Vol. 1444: H. Korezlioglu, A.S. Ustunel (Eds.), Stochastic Analysis and Related Topics VI. Proceedings, 1988. V, 268 pages. 1990.

Vol. 1445: F. Schulz, Regularity Theory for Quasilinear Elliptic Systems and – Monge Ampère Equations in Two Dimensions. XV, 123 pages. 1990.

Vol. 1446: Methods of Nonconvex Analysis. Seminar, 1989. Editor: A. Cellina. V, 206 pages. 1990.

Vol. 1447: J.-G. Labesse, J. Schwermer (Eds), Cohomology of Arithmetic Groups and Automorphic Forms. Proceedings, 1989. V, 358 pages. 1990.

Vol. 1448: S.K. Jain, S.R. López-Permouth (Eds.), Non-Commutative Ring Theory. Proceedings, 1989. V, 166 pages. 1990.

Vol. 1449: W. Odyniec, G. Lewicki, Minimal Projections in Banach Spaces. VIII, 168 pages. 1990.

Vol. 1450: H. Fujita, T. Ikebe, S.T. Kuroda (Eds.), Functional-Analytic Methods for Partial Differential Equations. Proceedings, 1989. VII, 252 pages. 1990.

Vol. 1451: L. Alvarez-Gaumé, E. Arbarello, C. De Concini, N.J. Hitchin, Global Geometry and Mathematical Physics. Montecatini Terme 1988. Seminar. Editors: M. Francaviglia, F. Gherardelli. IX, 197 pages. 1990.

Vol. 1452: E. Hlawka, R.F. Tichy (Eds.), Number-Theoretic Analysis. Seminar, 1988–89. V, 220 pages. 1990.

Vol. 1453: Yu.G. Borisovich, Yu.E. Gliklikh (Eds.), Global Analysis – Studies and Applications IV. V, 320 pages. 1990.

Vol. 1454: F. Baldassari, S. Bosch, B. Dwork (Eds.), p-adic Analysis. Proceedings, 1989. V, 382 pages. 1990.

Vol. 1455: J.-P. Françoise, R. Roussarie (Eds.), Bifurcations of Planar Vector Fields. Proceedings, 1989. VI, 396 pages. 1990.

Vol. 1456: L.G. Kovács (Ed.), Groups – Canberra 1989. Proceedings. XII, 198 pages. 1990.

Vol. 1457: O. Axelsson, L.Yu. Kolotilina (Eds.), Preconditioned Conjugate Gradient Methods. Proceedings, 1989. V, 196 pages. 1990.

Vol. 1458: R. Schaaf, Global Solution Branches of Two Point Boundary Value Problems. XIX, 141 pages. 1990.

Vol. 1459: D. Tiba, Optimal Control of Nonsmooth Distributed Parameter Systems. VII, 159 pages. 1990.

Vol. 1460: G. Toscani, V. Boffi, S. Rionero (Eds.), Mathematical Aspects of Fluid Plasma Dynamics. Proceedings, 1988. V, 221 pages. 1991.

Vol. 1461: R. Gorenflo, S. Vessella, Abel Integral Equations. VII, 215 pages. 1991.

Vol. 1462: D. Mond, J. Montaldi (Eds.), Singularity Theory and its Applications. Warwick 1989, Part I. VIII, 405 pages. 1991.

Vol. 1463: R. Roberts, I. Stewart (Eds.), Singularity Theory and its Applications. Warwick 1989, Part II. VIII, 322 pages. 1991.

Vol. 1464: D. L. Burkholder, E. Pardoux, A. Sznitman, Ecole d'Eté de Probabilités de Saint- Flour XIX-1989. Editor: P. L. Hennequin. VI, 256 pages. 1991.

Vol. 1465: G. David, Wavelets and Singular Integrals on Curves and Surfaces. X, 107 pages. 1991.

Vol. 1466: W. Banaszczyk, Additive Subgroups of Topological Vector Spaces. VII, 178 pages. 1991.

Vol. 1467: W. M. Schmidt, Diophantine Approximations and Diophantine Equations. VIII, 217 pages. 1991.

Vol. 1468: J. Noguchi, T. Ohsawa (Eds.), Prospects in Complex Geometry. Proceedings, 1989. VII, 421 pages. 1991.

Vol. 1469: J. Lindenstrauss, V. D. Milman (Eds.), Geometric Aspects of Functional Analysis. Seminar 1989-90. XI, 191 pages. 1991.

Vol. 1470: E. Odell, H. Rosenthal (Eds.), Functional Analysis. Proceedings, 1987-89. VII, 199 pages. 1991.

Vol. 1471: A. A. Panchishkin, Non-Archimedean L-Functions of Siegel and Hilbert Modular Forms. VII, 157 pages. 1991.

Vol. 1472: T. T. Nielsen, Bose Algebras: The Complex and Real Wave Representations. V, 132 pages. 1991.

Vol. 1473: Y. Hino, S. Murakami, T. Naito, Functional Differential Equations with Infinite Delay. X, 317 pages. 1991.

Vol. 1474: S. Jackowski, B. Oliver, K. Pawałowski (Eds.), Algebraic Topology, Poznań 1989. Proceedings. VIII, 397 pages. 1991.

Vol. 1475: S. Busenberg, M. Martelli (Eds.), Delay Differential Equations and Dynamical Systems. Proceedings, 1990. VIII, 249 pages. 1991.

Vol. 1476: M. Bekkali, Topics in Set Theory. VII, 120 pages. 1991.

Vol. 1477: R. Jajte, Strong Limit Theorems in Noncommutative L_2-Spaces. X, 113 pages. 1991.

Vol. 1478: M.-P. Malliavin (Ed.), Topics in Invariant Theory. Seminar 1989-1990. VI, 272 pages. 1991.

Vol. 1479: S. Bloch, I. Dolgachev, W. Fulton (Eds.), Algebraic Geometry. Proceedings, 1989. VII, 300 pages. 1991.

Vol. 1480: F. Dumortier, R. Roussarie, J. Sotomayor, H. Żoładek, Bifurcations of Planar Vector Fields: Nilpotent Singularities and Abelian Integrals. VIII, 226 pages. 1991.

Vol. 1481: D. Ferus, U. Pinkall, U. Simon, B. Wegner (Eds.), Global Differential Geometry and Global Analysis. Proceedings, 1991. VIII, 283 pages. 1991.

Vol. 1482: J. Chabrowski, The Dirichlet Problem with L^2-Boundary Data for Elliptic Linear Equations. VI, 173 pages. 1991.

Vol. 1483: E. Reithmeier, Periodic Solutions of Nonlinear Dynamical Systems. VI, 171 pages. 1991.

Vol. 1484: H. Delfs, Homology of Locally Semialgebraic Spaces. IX, 136 pages. 1991.

Vol. 1485: J. Azéma, P. A. Meyer, M. Yor (Eds.), Séminaire de Probabilités XXV. VIII, 440 pages. 1991.

Vol. 1486: L. Arnold, H. Crauel, J.-P. Eckmann (Eds.), Lyapunov Exponents. Proceedings, 1990. VIII, 365 pages. 1991.

Vol. 1487: E. Freitag, Singular Modular Forms and Theta Relations. VI, 172 pages. 1991.

Vol. 1488: A. Carboni, M. C. Pedicchio, G. Rosolini (Eds.), Category Theory. Proceedings, 1990. VII, 494 pages. 1991.

Vol. 1489: A. Mielke, Hamiltonian and Lagrangian Flows on Center Manifolds. X, 140 pages. 1991.

Vol. 1490: K. Metsch, Linear Spaces with Few Lines. XIII, 196 pages. 1991.

Vol. 1491: E. Lluis-Puebla, J.-L. Loday, H. Gillet, C. Soulé, V. Snaith, Higher Algebraic K-Theory: an overview. IX, 164 pages. 1992.

Vol. 1492: K. R. Wicks, Fractals and Hyperspaces. VIII, 168 pages. 1991.

Vol. 1493: E. Benoît (Ed.), Dynamic Bifurcations. Proceedings, Luminy 1990. VII, 219 pages. 1991.

Vol. 1494: M.-T. Cheng, X.-W. Zhou, D.-G. Deng (Eds.), Harmonic Analysis. Proceedings, 1988. IX, 226 pages. 1991.

Vol. 1495: J. M. Bony, G. Grubb, L. Hörmander, H. Komatsu, J. Sjöstrand, Microlocal Analysis and Applications. Montecatini Terme, 1989. Editors: L. Cattabriga, L. Rodino. VII, 349 pages. 1991.

Vol. 1496: C. Foias, B. Francis, J. W. Helton, H. Kwakernaak, J. B. Pearson, H_∞-Control Theory. Como, 1990. Editors: E. Mosca, L. Pandolfi. VII, 336 pages. 1991.

Vol. 1497: G. T. Herman, A. K. Louis, F. Natterer (Eds.), Mathematical Methods in Tomography. Proceedings 1990. X, 268 pages. 1991.

Vol. 1498: R. Lang, Spectral Theory of Random Schrödinger Operators. X, 125 pages. 1991.

Vol. 1499: K. Taira, Boundary Value Problems and Markov Processes. IX, 132 pages. 1991.

Vol. 1500: J.-P. Serre, Lie Algebras and Lie Groups. VII, 168 pages. 1992.

Vol. 1501: A. De Masi, E. Presutti, Mathematical Methods for Hydrodynamic Limits. IX, 196 pages. 1991.

Vol. 1502: C. Simpson, Asymptotic Behavior of Monodromy. V, 139 pages. 1991.

Vol. 1503: S. Shokranian, The Selberg-Arthur Trace Formula (Lectures by J. Arthur). VII, 97 pages. 1991.

Vol. 1504: J. Cheeger, M. Gromov, C. Okonek, P. Pansu, Geometric Topology: Recent Developments. Editors: P. de Bartolomeis, F. Tricerri. VII, 197 pages. 1991.

Vol. 1505: K. Kajitani, T. Nishitani, The Hyperbolic Cauchy Problem. VII, 168 pages. 1991.

Vol. 1506: A. Buium, Differential Algebraic Groups of Finite Dimension. XV, 145 pages. 1992.

Vol. 1507: K. Hulek, T. Peternell, M. Schneider, F.-O. Schreyer (Eds.), Complex Algebraic Varieties. Proceedings, 1990. VII, 179 pages. 1992.

Vol. 1508: M. Vuorinen (Ed.), Quasiconformal Space Mappings. A Collection of Surveys 1960-1990. IX, 148 pages. 1992.

Vol. 1509: J. Aguadé, M. Castellet, F. R. Cohen (Eds.), Algebraic Topology - Homotopy and Group Cohomology. Proceedings, 1990. X, 330 pages. 1992.

Vol. 1510: P. P. Kulish (Ed.), Quantum Groups. Proceedings, 1990. XII, 398 pages. 1992.

Vol. 1511: B. S. Yadav, D. Singh (Eds.), Functional Analysis and Operator Theory. Proceedings, 1990. VIII, 223 pages. 1992.

Vol. 1512: L. M. Adleman, M.-D. A. Huang, Primality Testing and Abelian Varieties Over Finite Fields. VII, 142 pages. 1992.

Vol. 1513: L. S. Block, W. A. Coppel, Dynamics in One Dimension. VIII, 249 pages. 1992.

Vol. 1514: U. Krengel, K. Richter, V. Warstat (Eds.), Ergodic Theory and Related Topics III, Proceedings, 1990. VIII, 236 pages. 1992.

Vol. 1515: E. Ballico, F. Catanese, C. Ciliberto (Eds.), Classification of Irregular Varieties. Proceedings, 1990. VII, 149 pages. 1992.

Vol. 1516: R. A. Lorentz, Multivariate Birkhoff Interpolation. IX, 192 pages. 1992.

Vol. 1517: K. Keimel, W. Roth, Ordered Cones and Approximation. VI, 134 pages. 1992.

Vol. 1518: H. Stichtenoth, M. A. Tsfasman (Eds.), Coding Theory and Algebraic Geometry. Proceedings, 1991. VIII, 223 pages. 1992.

Vol. 1519: M. W. Short, The Primitive Soluble Permutation Groups of Degree less than 256. IX, 145 pages. 1992.

Vol. 1520: Yu. G. Borisovich, Yu. E. Gliklikh (Eds.), Global Analysis – Studies and Applications V. VII, 284 pages. 1992.

Vol. 1521: S. Busenberg, B. Forte, H. K. Kuiken, Mathematical Modelling of Industrial Process. Bari, 1990. Editors: V. Capasso, A. Fasano. VII, 162 pages. 1992.

Vol. 1522: J.-M. Delort, F. B. I. Transformation. VII, 101 pages. 1992.

Vol. 1523: W. Xue, Rings with Morita Duality. X, 168 pages. 1992.

Vol. 1524: M. Coste, L. Mahé, M.-F. Roy (Eds.), Real Algebraic Geometry. Proceedings, 1991. VIII, 418 pages. 1992.

Vol. 1525: C. Casacuberta, M. Castellet (Eds.), Mathematical Research Today and Tomorrow. VII, 112 pages. 1992.

Vol. 1526: J. Azéma, P. A. Meyer, M. Yor (Eds.), Séminaire de Probabilités XXVI. X, 633 pages. 1992.

Vol. 1527: M. I. Freidlin, J.-F. Le Gall, Ecole d'Eté de Probabilités de Saint-Flour XX – 1990. Editor: P. L. Hennequin. VIII, 244 pages. 1992.

Vol. 1528: G. Isac, Complementarity Problems. VI, 297 pages. 1992.

Vol. 1529: J. van Neerven, The Adjoint of a Semigroup of Linear Operators. X, 195 pages. 1992.

Vol. 1530: J. G. Heywood, K. Masuda, R. Rautmann, S. A. Solonnikov (Eds.), The Navier-Stokes Equations II – Theory and Numerical Methods. IX, 322 pages. 1992.

Vol. 1531: M. Stoer, Design of Survivable Networks. IV, 206 pages. 1992.

Vol. 1532: J. F. Colombeau, Multiplication of Distributions. X, 184 pages. 1992.

Vol. 1533: P. Jipsen, H. Rose, Varieties of Lattices. X, 162 pages. 1992.

Vol. 1534: C. Greither, Cyclic Galois Extensions of Commutative Rings. X, 145 pages. 1992.

Vol. 1535: A. B. Evans, Orthomorphism Graphs of Groups. VIII, 114 pages. 1992.

Vol. 1536: M. K. Kwong, A. Zettl, Norm Inequalities for Derivatives and Differences. VII, 150 pages. 1992.

Vol. 1537: P. Fitzpatrick, M. Martelli, J. Mawhin, R. Nussbaum, Topological Methods for Ordinary Differential Equations. Montecatini Terme, 1991. Editors: M. Furi, P. Zecca. VII, 218 pages. 1993.

Vol. 1538: P.-A. Meyer, Quantum Probability for Probabilists. X, 287 pages. 1993.

Vol. 1539: M. Coornaert, A. Papadopoulos, Symbolic Dynamics and Hyperbolic Groups. VIII, 138 pages. 1993.

Vol. 1540: H. Komatsu (Ed.), Functional Analysis and Related Topics, 1991. Proceedings. XXI, 413 pages. 1993.

Vol. 1541: D. A. Dawson, B. Maisonneuve, J. Spencer, Ecole d' Eté de Probabilités de Saint-Flour XXI - 1991. Editor: P. L. Hennequin. VIII, 356 pages. 1993.

Vol. 1542: J.Fröhlich, Th.Kerler, Quantum Groups, Quantum Categories and Quantum Field Theory. VII, 431 pages. 1993.

Vol. 1543: A. L. Dontchev, T. Zolezzi, Well-Posed Optimization Problems. XII, 421 pages. 1993.

Vol. 1544: M.Schürmann, White Noise on Bialgebras. VII, 146 pages. 1993.

Vol. 1545: J. Morgan, K. O'Grady, Differential Topology of Complex Surfaces. VIII, 224 pages. 1993.

Vol. 1546: V. V. Kalashnikov, V. M. Zolotarev (Eds.), Stability Problems for Stochastic Models. Proceedings, 1991. VIII, 229 pages. 1993.

Vol. 1547: P. Harmand, D. Werner, W. Werner, M-ideals in Banach Spaces and Banach Algebras. VIII, 387 pages. 1993.

Vol. 1548: T. Urabe, Dynkin Graphs and Quadrilateral Singularities. VI, 233 pages. 1993.

Vol. 1549: G. Vainikko, Multidimensional Weakly Singular Integral Equations. XI, 159 pages. 1993.

Vol. 1550: A. A. Gonchar, E. B. Saff (Eds.), Methods of Approximation Theory in Complex Analysis and Mathematical Physics IV, 222 pages, 1993.

Vol. 1551: L. Arkeryd, P. L. Lions, P.A. Markowich, S.R. S. Varadhan. Nonequilibrium Problems in Many-Particle Systems. Montecatini, 1992. Editors: C. Cercignani, M. Pulvirenti. VII, 158 pages 1993.

Vol. 1552: J. Hilgert, K.-H. Neeb, Lie Semigroups and their Applications. XII, 315 pages. 1993.

Vol. 1553: J.-L- Colliot-Thélène, J. Kato, P. Vojta. Arithmetic Algebraic Geometry. Trento, 1991. Editor: E. Ballico. VII, 223 pages. 1993.

Vol. 1554: A. K. Lenstra, H. W. Lenstra, Jr. (Eds.), The Development of the Number Field Sieve. VIII, 131 pages. 1993.

Vol. 1555: O. Liess, Conical Refraction and Higher Microlocalization. X, 389 pages. 1993.

Vol. 1556: S. B. Kuksin, Nearly Integrable Infinite-Dimensional Hamiltonian Systems. XXVII, 101 pages. 1993.

Vol. 1557: J. Azéma, P. A. Meyer, M. Yor (Eds.), Séminaire de Probabilités XXVII. VI, 327 pages. 1993.

Vol. 1558: T. J. Bridges, J. E. Furter, Singularity Theory and Equivariant Symplectic Maps. VI, 226 pages. 1993.

Vol. 1559: V. G. Sprindžuk, Classical Diophantine Equations. XII, 228 pages. 1993.

Vol. 1560: T. Bartsch, Topological Methods for Variational Problems with Symmetries. X, 152 pages. 1993.

Vol. 1561: I. S. Molchanov, Limit Theorems for Unions of Random Closed Sets. X, 157 pages. 1993.

Vol. 1562: G. Harder, Eisensteinkohomologie und die Konstruktion gemischter Motive. XX, 184 pages. 1993.

Vol. 1563: E. Fabes, M. Fukushima, L. Gross, C. Kenig, M. Röckner, D. W. Stroock, Dirichlet Forms. Varenna, 1992. Editors: G. Dell'Antonio, U. Mosco. VII, 245 pages. 1993.

Vol. 1564: J. Jorgenson, S. Lang, Basic Analysis of Regularized Series and Products. IX, 122 pages. 1993.

Vol. 1565: L. Boutet de Monvel, C. De Concini, C. Procesi, P. Schapira, M. Vergne. D-modules, Representation Theory, and Quantum Groups. Venezia, 1992. Editors: G. Zampieri, A. D'Agnolo. VII, 217 pages. 1993.

Vol. 1566: B. Edixhoven, J.-H. Evertse (Eds.), Diophantine Approximation and Abelian Varieties. XIII, 127 pages. 1993.

Vol. 1567: R. L. Dobrushin, S. Kusuoka, Statistical Mechanics and Fractals. VII, 98 pages. 1993.

Vol. 1568: F. Weisz, Martingale Hardy Spaces and their Application in Fourier Analysis. VIII, 217 pages. 1994.

Vol. 1569: V. Totik, Weighted Approximation with Varying Weight. VI, 117 pages. 1994.

Vol. 1570: R. deLaubenfels, Existence Families, Functional Calculi and Evolution Equations. XV, 234 pages. 1994.

Vol. 1571: S. Yu. Pilyugin, The Space of Dynamical Systems with the C^0-Topology. X, 188 pages. 1994.

Vol. 1572: L. Göttsche, Hilbert Schemes of Zero-Dimensional Subschemes of Smooth Varieties. IX, 196 pages. 1994.

Vol. 1573: V. P. Havin, N. K. Nikolski (Eds.), Linear and Complex Analysis – Problem Book 3 – Part I. XXII, 489 pages. 1994.

Vol. 1574: V. P. Havin, N. K. Nikolski (Eds.), Linear and Complex Analysis – Problem Book 3 – Part II. XXII, 507 pages. 1994.

Vol. 1575: M. Mitrea, Clifford Wavelets, Singular Integrals, and Hardy Spaces. XI, 116 pages. 1994.

Vol. 1576: K. Kitahara, Spaces of Approximating Functions with Haar-Like Conditions. X, 110 pages. 1994.

Vol. 1577: N. Obata, White Noise Calculus and Fock Space. X, 183 pages. 1994.

Vol. 1578: J. Bernstein, V. Lunts, Equivariant Sheaves and Functors. V, 139 pages. 1994.

Vol. 1579: N. Kazamaki, Continuous Exponential Martingales and *BMO*. VII, 91 pages. 1994.

Vol. 1580: M. Milman, Extrapolation and Optimal Decompositions with Applications to Analysis. XI, 161 pages. 1994.

Vol. 1581: D. Bakry, R. D. Gill, S. A. Molchanov, Lectures on Probability Theory. Editor: P. Bernard. VIII, 420 pages. 1994.

Vol. 1582: W. Balser, From Divergent Power Series to Analytic Functions. X, 108 pages. 1994.

Vol. 1583: J. Azéma, P. A. Meyer, M. Yor (Eds.), Séminaire de Probabilités XXVIII. VI, 334 pages. 1994.

Vol. 1584: M. Brokate, N. Kenmochi, I. Müller, J. F. Rodriguez, C. Verdi, Phase Transitions and Hysteresis. Montecatini Terme, 1993. Editor: A. Visintin. VII. 291 pages. 1994.

Vol. 1585: G. Frey (Ed.), On Artin's Conjecture for Odd 2-dimensional Representations. VIII, 148 pages. 1994.

Vol. 1586: R. Nillsen, Difference Spaces and Invariant Linear Forms. XII, 186 pages. 1994.

Vol. 1587: N. Xi, Representations of Affine Hecke Algebras. VIII, 137 pages. 1994.

Vol. 1588: C. Scheiderer, Real and Étale Cohomology. XXIV, 273 pages. 1994.

Vol. 1589: J. Bellissard, M. Degli Esposti, G. Forni, S. Graffi, S. Isola, J. N. Mather, Transition to Chaos in Classical and Quantum Mechanics. Montecatini Terme, 1991. Editor: S. Graffi. VII, 192 pages. 1994.

Vol. 1590: P. M. Soardi, Potential Theory on Infinite Networks. VIII, 187 pages. 1994.

Vol. 1591: M. Abate, G. Patrizio, Finsler Metrics – A Global Approach. IX, 180 pages. 1994.

Vol. 1592: K. W. Breitung, Asymptotic Approximations for Probability Integrals. IX, 146 pages. 1994.

Vol. 1593: J. Jorgenson & S. Lang, D. Goldfeld, Explicit Formulas for Regularized Products and Series. VIII, 154 pages. 1994.

Vol. 1594: M. Green, J. Murre, C. Voisin, Algebraic Cycles and Hodge Theory. Torino, 1993. Editors: A. Albano, F. Bardelli. VII, 275 pages. 1994.

Vol. 1595: R.D.M. Accola, Topics in the Theory of Riemann Surfaces. IX, 105 pages. 1994.

Vol. 1596: L. Heindorf, L. B. Shapiro, Nearly Projective Boolean Algebras. X, 202 pages. 1994.

Vol. 1597: B. Herzog, Kodaira-Spencer Maps in Local Algebra. XVII, 176 pages. 1994.

Vol. 1598: J. Berndt, F. Tricerri, L. Vanhecke, Generalized Heisenberg Groups and Damek-Ricci Harmonic Spaces. VIII, 125 pages. 1995.

Vol. 1599: K. Johannson, Topology and Combinatorics of 3-Manifolds. XVIII, 446 pages. 1995.

Vol. 1600: W. Narkiewicz, Polynomial Mappings. VII, 130 pages. 1995.

Vol. 1601: A. Pott, Finite Geometry and Character Theory. VII, 181 pages. 1995.

Vol. 1602: J. Winkelmann, The Classification of Three-dimensional Homogeneous Complex Manifolds. XI, 230 pages. 1995.

Vol. 1603: V. Ene, Real Functions – Current Topics. XIII, 310 pages. 1995.

Vol. 1604: A. Huber, Mixed Motives and their Realization in Derived Categories. XV, 207 pages. 1995.

Vol. 1605: L. B. Wahlbin, Superconvergence in Galerkin Finite Element Methods. XI, 166 pages. 1995.

Vol. 1606: P.-D. Liu, M. Qian, Smooth Ergodic Theory of Random Dynamical Systems. XI, 221 pages. 1995.

Vol. 1607: G. Schwarz, Hodge Decomposition – A Method for Solving Boundary Value Problems. VII, 155 pages. 1995.

Vol. 1608: P. Biane, R. Durrett, Lectures on Probability Theory. VII, 210 pages. 1995.

Vol. 1609: L. Arnold, C. Jones, K. Mischaikow, G. Raugel, Dynamical Systems. Montecatini Terme, 1994. Editor: R. Johnson. VIII, 329 pages. 1995.

Vol. 1610: A. S. Üstünel, An Introduction to Analysis on Wiener Space. X, 95 pages. 1995.

Vol. 1611: N. Knarr, Translation Planes. VI, 112 pages. 1995.

Vol. 1612: W. Kühnel, Tight Polyhedral Submanifolds and Tight Triangulations. VII, 122 pages. 1995.

Vol. 1613: J. Azéma, M. Emery, P. A. Meyer, M. Yor (Eds.), Séminaire de Probabilités XXIX. VI, 326 pages. 1995.